Principles and Methods of Plant Breeding

PRINCIPLES AND METHODS OF PLANT BREEDING

With 58 Figures and 84 Tables

Slavko Borojević

Akademie-Verlag Berlin

Acad. Slavko Borojević
College of Agriculture
University of Novi Sad (Yugoslavia)

ISBN 3-05-500821-9

Lizenzausgabe für den Akademie-Verlag Berlin

Printed in the German Democratic Republic
Gesamtherstellung: VEB Druckhaus Köthen 4370
LSV 4205
Bestellnummer: 764 128 6 (9297)

CONTENTS

PREFACE

A course I have developed on the Theory of Plant Breeding for postgraduate students in the Plant Genetics and Breeding group at the University of Novi Sad forms the basis of this book.

However, it can also be said to be the result of 30 years of research on the development of high-yielding wheat varieties, which have been grown commercially in Yugoslavia and other European countries for more than two decades, and which are used in many breeding centres around the world as germ plasm.

The first edition, entitled Principles and Methods of Plant Breeding, was published in 1981 by Cirpanov Publishing House, Novi Sad, in Serbo-Croat. This work has become a textbook for postgraduate students in Yugoslavia and is also an indispensable manual for plant breeders. As Serbo-Croat and Russian are linguistically similar, the book aroused great interest in the Soviet Union so that Kolos Publishing House in Moscow issued a Russian edition in 1984.

This present edition has been very considerably modified. As a result, its volume has increased by almost 40% because of the important developments in the field of plant genetics and breeding that have taken place in recent years.

However, the book does not pretend to offer an integral theory of plant breeding, especially with regard to the breeding of horticultural and forest plant species. But it is specifically designed to help researchers and plant breeders who wish to make new varieties that are superior in genetic yield potential and quality to the currently cultivated varieties of field crops and various other plant species. The ultimate goal of these studies is to increase food production for the expanding world population whilst taking account of the need to secure crops minimally contaminated with substances detrimental to human and animal health.

Plant breeding has reached notable heights in a number of countries, so it is not a simple task to create a new and superior variety. The development of such a variety that will become popular not only in its country of origin but expand outside the national borders and contribute to the welfare of society is a source of immense pleasure to a plant breeder. A failed breeding programme causes frustration much more intense than that felt, for example, upon receiving a lukewarm or bad review of a research paper.

I sincerely hope that this book will help postgraduate students and researchers all over the world to achieve new successes based on genetic laws rather than chance and empiricism, so that newly developed genetic materials will spring from genetic rather than non-genetic variability. In a nutshell, if successful, it should reduce wondering and spur success.

Novi Sad, 1989 — Slavko Borojević

1. IMPORTANCE OF PLANT BREEDING

The expression "plant breeding" implies the conscious human effort aimed at improving old and developing new varieties of crops in order to satisfy the demands for human food and animal feed.

The beginnings of plant breeding, which are intimately related to the domestication of wild plants and the beginnings of agriculture, date back some 10,000 years. Initially, it was only through selecting the best fruits and seeds, which were then cultivated, that more food could be provided than by merely gathering it from the wild. The processes of selection and the improvement of conditions for plant growth which led to the development of cultivated plants was quite slow. It has been estimated that cultivated plants grew to exceed 50% of the total human diet only in the second century B.C. (Harlan, 1976).

Plant breeding in the sense of deliberately creating something new, something that could not be found in natural populations, is much more recent. The first study in Europe on sexual reproduction in plants was conducted by R. J. Camerarius and was published in 1694. The fact that plants reproduce sexually was accepted only after 1760, when K. Linneaus, a Swedish botanist, published his works in which he described the sexual organs of a number of plant species and some hybrids that he had made by crossing different varieties. J. G. Kölreuter (1760) is considered to be the pioneer of hybridization (sexual crossing between genetically dissimilar parents). He made and described a large number of crosses within and between various plant species. The work of Charles Darwin, published in his famous book "On the Origin of Species by Means of Natural Selection" (1859), gave a strong impetus to plant breeding. In 1865, Gregor Mendel made important discoveries by crossing pea varieties possessing different traits. However, the contemporary community did not realise the importance of his discoveries even for genetics, let alone plant breeding. Only in 1900 did E. von Tschermak, C. Correns, and H. de Vries confirm the results of G. Mendel and point out the importance of them for the development of genetics as an independent scientific discipline as well as for the improvement of organisms on the basis of genetic principles.

In this century, plant breeding based on methods of hybridization of different varieties within the same species together with special methods of selection in subsequent segregating generations has achieved extraordinary results which have considerably promoted agricultural production.

In the previous century, sugar beet varieties contained about 9% sugar; today, there are varieties containing over 20%. Between the two world wars, sunflower varieties contained about 30% oil whereas the currently grown hybrids contain 50% and there are individual genotypes containing over 60% oil (Pustovoit, 1975). The record yields of maize (*Zea mays*) of 5,0 t/ha achieved in the first half of this century have nowadays become average yields in the United States of America and Europe; the current record yields exceed 20.0 t/ha (Table 1.1). New wheat varieties ensure average yields between 6.0 and 8.0 t/ha and the record yields exceed 10.0 t/ha. Similar improvements have been achieved with many other crops (Table 1.1).

Table 1.1 Record yield of some field crops

Species	Tons/ hectare	Ratio record vs. average yield	Literature
Maize	23.9	4.3	Wittwer, 1975
Maize	21.2	3.9	Chou et al., 1977
Sorghum	21.5	6.5	Chou et al., 1977
Rice	14.4	5.8	Chou et al., 1977
Rice	9.9		Murata & Matzushima, 1975
Wheat	14.1	7.0	Anonymous, 1966
Wheat	11.7	2.3	Jones, 1978
Wheat	10.9	2.9	Borojević, 1978
Wheat	9.6	2.8	Borojević, 1978
Barley	11.4	4.8	Chou et al., 1977
Oats	10.6	6.2	Chou et al., 1977
Soyabean	7.4	3.9	Chou et al., 1977
Potato	94.1	3.5	Chou et al., 1977
Sugar beet	120.0	2.4	Chou et al., 1977
Sugar cane	150.0	3.0	Chou et al., 1977

The contribution of the new varieties to the advancement of agricultural production is illustrated by the research results of Ishizuka (1969) who found that the new rice varieties were 50 to 60% higher yielding than the old varieties, i.e., that rice breeding brought about a one percent annual yield increase over a period of 60 years (Table 1.2). A study conducted in Yugoslavia showed that the new wheat varieties, given the same agricultural conditions, outyield the old varieties by 50 to 60% (Borojević, 1978).

Table 1.2 Yield of old and new rice varieties in a comparative trial conducted in Japan

Variety	Period of commercial production	Yield	%
Akage	1890–1930	3.4	100
Bohzu 5	1910–1935	4.1	120
Fukoku	1935–1940	4.7	138
Eikoh	1940–1959	5.1	150
Tayokihari	1950	5.0	147
Mimasari	1959	5.5	162

Source: Ishizuka, 1969

Furthermore, a large number of high yielding and high quality varieties of apple, pear, peach, and other fruits have been developed. Also, in addition to numerous wine varieties, many table varieties of grape have been developed. These varieties are high yielding, seedless, and suitable for transportation and storage. Still more spectacular developments have been achieved in the production of vegetables and ornamental plants where a change in colour or form has frequently brought about sizable economic profits. Let us mention the excellence of new varieties of tomato, pepper and lettuce, and the beauty of new varieties of gladiolus, rose and carnation.

After World War II, new methods of plant breeding started to be used alongside the conventional hybridization methods. The new techniques include: induced polyploidy, induced mutations, chromosome engineering (chromosome addition and substitution), male sterility, etc.; and most recent methods of protoplast fusion and gene manipulation (recombinant DNA techniques) will allow for the development of new varieties or synthetic species that have not existed before.

Table 1.3 Changes in cereal mean yields per crop season in different countries from 1948/52 to 1981/85

Crop	Country	Mean yields in t/ha	in % of yield for 1948/52				
		1948/52*	1961/65*	1966/70	1971/75	1976/80	1981/85
Rice	India	1.11	133	140	156	166	187
	Indonezia	1.61	109	124	158	183	239
	Japan	4.25	118	130	134	137	139
	Philippines	1.18	107	125	136	174	202
	Thailand	1.31	134	138	142	146	155
	USA	2.56	170	194	198	198	215
Wheat	Bulgaria	1.24	146	218	274	306	290
	France	1.83	160	186	235	257	300
	Hungary	1.38	135	174	233	253	326
	India	0.66	127	166	197	227	273
	Romania	1.02	143	166	225	265	265
	USA	1.12	152	169	187	187	223
	USSR	0.84	114	155	166	190	178
	Yugoslavia	1.14	158	210	254	289	298
Maize	Bulgaria	0.98	258	319	406	416	553
	France	1.36	222	350	295	370	415
	Hungary	1.77	147	183	235	273	346
	Italy	1.84	178	224	239	356	379
	Romania	0.81	218	275	331	416	505
	USA	2.49	167	195	218	242	271
	Yugoslavia	1.34	169	224	260	304	342

* Data from Sehgal (1977), In: J. Sneep and A. J. T. Hendricksen (Editors), 1979. Plant Breeding Perspectives. Pudoc, Wageningen, p. 432. Other data are calculated on the basis of mean yields supplied from the FAO yearbooks.

Plant breeding is thus a creative work essential to solving the global problem of food provision. Its importance is further emphasized by the fact that there is less than 0.4 hectare of arable land per capita while the potentially arable land cannot do more than merely double the presently available cultivated acreages (Jackson et al., 1975). Therefore, a permanent increase of yield per unit area is a solution to the problem of providing food to the rapidly expanding world population. Significant breakthroughs have already been achieved. In many countries, the yields of the main crops have been doubled or tripled in the last thirty years (Table 1.3). Further yield increases will be achieved if new varieties with high yield potentials are developed and if cultural practices are so improved that the maximum utilization of these new genetic potentials can be obtained.

The development of varieties with high yield potential is the main way to increase the production of food. However, such varieties require intensive production conditions which cannot always be provided. There exist sizable areas under sandy, saline, acid or swampy soils which are used only for primitive agriculture or not at all. These soils should be ameliorated or varieties should be developed which could be profitably grown on them. Another challenge is the development of varieties adapted to growing in regions prone to stress, i.e., aridity, long periods of frost, strong winds, etc. (Mussel et al., 1979).

On the other hand, numerous fruit crops, medicinal plants, herbs, forages and similar crops which keep expanding their acreages in fertile plains should be relocated to mountain regions and poor soils where they may grow successfully, unlike more demanding crops such as sugar beet or maize.

It is also necessary to further diversify the crop germplasm by breeding varieties which satisfy specific requirements of human and animal diets and of the food industries.

In the following chapters, the theoretical basis of plant breeding will be reviewed and methods of developing new plant varieties will be discussed.

References

Anonymous, 1966. New record wheat yields – 216 bushels per acre. Farm J. 90: 42.

Borojević, S., 1978. Genetics, environment and modeling in breeding of organisms Savremena poljoprivreda (Novi Sad), 11–12: 5–28.

Chou, M., Harmon, D. P. Jr., Kalm, H., and Wittwer, S. H., 1977. World Food Prospects and Agricultural Potential. Praeger Publishers, New York.

Correns, C., 1900. Mendels Regel über das Verhalten der Nachkommenschaft der Rasenbastarde; Berichte d. deutsch. bot. Ges., 18: 158–168.

Darwin, C., 1895. On the Origin of Species by Means of Natural Selection. London. Translation to Serbocroatian, Prosveta, Belgrade, 1948.

De Vries, H., 1900. Sur la loi de disjunctions des hybrides. Comp. Rend. Aca. Sci., Paris.

Harlan, J. R., 1976. The plants and animals that nourish man. Scientific American, 235: 88–97.

Ishizuka, Y., 1969. Engineering for higher yields. In: J. S. Eastin et al. (Editors). Physiological Aspects of Grain Yield, p. 15–26.

Jackson, W. A., Knezek, B. D., and van Schilfgaarde, J., 1975. Water, soil and mineral input. In: Crop Productivity ... Research Imperatives, East Lansing, Mich., p. 201–274.

Jones, M. J., 1978. Pointers to profitable wheat. Imperial Chemical Ind. Ltd., England, p. 1–17.
Mendel, G., 1865. Versuche über Pflanzenhibriden. Verhandlungen naturforschender Verein. Brünn. Serbocroatian translation, Belgrade, 1934.
Murata, Y., and Matsushima, S., 1975. Rice. In: L. T. Evans (Editor), Crop Physiology. Cambridge Univ. Press, Cambridge, p. 73–99.
Mussel, H., and Staples, R. C. (Editors), 1979. Stress Physiology in Crop Plants. Wiley, New York.
Pustovoit, Galina, 1975. Podsolnechnik, Izd. Kolos, Moskva.
Sehgal, S. M., 1977. Private sector international agricultural research: the genetic supply industry. In: T. M. Arndt, D. G. Dalrymple and W. V. Rutton (Editors), Resource Allocation and Productivity in National and International Agricultural Research. Univ. Minnesota Press, Minneapolis, p. 404–415.
Tschermak, E. von, 1900. Über künstliche Kreuzung bei Pisum sativum. Berichte d. deutsch. bot. Ges., 18: 232–239.
Wittwer, S. H., 1975. Food production: technology and resources base. Science, 4188: 579–584.
Wittwer, S. H., 1979. Future technological advances in agriculture and their impact on the regulatory environment. Bio. Science, 29: 603–610.

2. PHYTOGEOGRAPHICAL DIFFERENTIATION AS A BASIS OF PLANT BREEDING

The importance of plant breeding for the development of human society is aptly illustrated by the statement that "plant breeding is evolution in human hands" said by a great Russian geneticist and breeder N. I. Vavilov. Having gained knowledge of the laws of inheritance and using various methods of hybridization and selection, man has developed innumerable varieties in a matter of several decades and centuries. These new varieties would have needed thousands or even millions of years to be developed in nature or they even might not have developed at all.

Since the evolution of organisms depends on variability, as discovered by Darwin, success in plant breeding is based both upon the variability within and between the species existing in nature and the variability created by the man.

Centres of origin of cultivated plants

The studies of N. I. Vavilov and his associates clearly demonstrated the genetic diversity of cultivated plants and the importance of that diversity for further breeding. In the period from 1923 to 1931, Vavilov organized many expeditions to over sixty countries and collected over 300,000 samples of cultivated plants and their wild relatives. The collected materials were studied for several years at the All-Union Institute of Plant Production in Leningrad and a large number of regional institutes and out-stations in the Soviet Union. The project rendered valuable data on the distribution of various cultivated plants in the world, the number of varieties within individual species, the variability in specific traits, the frequency and distribution of genes for specific traits, the occurrence of traits in response to the environment, etc. On the basis of this project Vavilov (1935) concluded that there exist in the world eight independent centres of origin of cultivated plants (Fig. 2.1).

The main characteristics of these centres will be outlined in this chapter; for a more detailed reading refer to the capital publication of N. I. Vavilov et al., "Teoreticheskie osnovi selekcii rastenii", Volumes 1 and 2, Moscow – Leningrad, 1935, or the English translation, "The Origin, Variation, Immunity and Breeding of Cultivated Plants" translated from the Russian by K. Starr Chester, The Ronald Press Co., New York, 1951.

1. The Chinese Centre

The mountainous regions of central and western China, together with the adjacent lowlands are the earliest and largest independent centre of the world's agriculture and of the origin of cultivated plants.

Fig. 2.1 World centers of origin of cultivated plants.
Source: Vavilov, 1935

The most important endemic plants of the Chinese Centre are millet, buckwheat, soyabean and a large number of legumes. The centre also includes many root and tuber species. China takes the first place in the world according to the variability of pears, apples, stoned fruits (peach, etc.) and citruses. In most cases, these species have a large number of botanical varieties and genotypes.

2. The Indian Centre

The Indian Centre covers the territories of India (without Punjab and northwestern India), Assam and Burma. It is the site of origin for wild and cultivated species of rice possessing multitudes of dominant genes, as well as jute, kenaf, sugar cane, numerous grain legumes, and citruses. The Indian Centre includes the Indo-Malayan Sub-Centre.

2a The Indo-Malayan Sub-Centre

The Indo-Malayan Sub-Centre includes Indo-China and the entire Malay Archipelago. Banana and several citrus species originate from this centre.

3. The Central Asiatic Centre

The Central Asiatic Centre covers the territories of Punjab, Kashmir, Afghanistan, Tadzhikistan, western Uzbekistan and western Tian-Shan.

This centre gave rise to bread wheat (*Triticum aestivum* subsp. *vulgare*), other hexaploid wheats and chief grain legumes (e.g., pea, lentil, mung bean, chick pea, etc.). The centre is not rich in the number of species but there is a high variability within each species found.

4. The Near-Eastern Centre

The Near-Eastern Centre includes the Near-East, Asia Minor, Transcaucasia, Iran and the highlands of Turkestan.

A large number of *Triticum* species, with both 14 and 28 chromosomes, are found in this centre. Asia Minor is the homeland of cultivated rye and several wild ryes, the Near-East of grape vine, lucerne (*alfalfa*) and other legumes, melons and water-melons. This centre is exceedingly rich in fruit species, e.g., pears, cherries, walnuts, almonds, etc.

5. The Mediterranean Centre

The Mediterranean Centre is the homeland of the olive, numerous legumes, forage plants and vegetables. The species originating in this centre are notable for their large seeds.

6. The Abyssinian Centre

Ethiopia, together with Eritrea, is a distinct centre of origin of a number of plant species. There has been found a variety of barleys, 28-chromosome wheats and several legumes. The centre does not include many species but all of them are highly variable.

7. The South Mexican and Central American Centre

This centre is the homeland of maize, cotton, cacao, beans, pepper, pumpkins, etc.

8. The South American (Peruvian-Ecuadorean-Bolivian) Centre

This centre includes a large number of species of potato (*Solanum* subspecies with 24, 36, and 60 chromosomes). It is famous for the tomato and tobacco.

The South American Centre includes two sub-centres, the Chiloe Sub-Centre and the Brazilian-Paraguayan Sub-Centre.

8a The Chiloe Sub-Centre

It is the homeland of the potato (*Solanum tuberosum*, 48 chromosomes) and some other plant species.

8b The Brazilian-Paraguayan Sub-Centre

This sub-centre is the homeland of the peanut, manioc, pineapple, etc.

Of about the 640 most important cultivated plant species, more than 500 originate from the Old World, about 400 of them from Asia alone.

Many species have been found to originate from several centres; thus there are primary and secondary centres of origin.

Later studies have shown that it is more accurate to refer to "centres of diversity" rather than to "centres of origin" (Harlan, 1951; Harlan and Zohary, 1966). Namely, centres of diversity include the cultivated species as well as their putative wild predecessors and present day wild relatives. Vavilov also found that many crops belong to centres quite far from the centres in which their corresponding wild relatives were found. For example, barley, wheat, pea, flax and other species exhibited high variability in characters in the Abyssinian Centre but their wild relatives were not found there. It is an indication that these species had not been domesticated there but elsewhere. It has been proved that the species in question had been domesticated in the Near East in the course of the neolithic agricultural revolution and subsequently brought to Ethiopia, probably by the Hamitic settlers. This is an example of a secondary centre of diversity.

Harlan (1951, 1970) points out that present day cultivated species have undergone large changes since the time of their domestication; they were closer to their wild relatives in the beginning. In the process of domestication, many species spread out from their centre of origin, came in contact with related wild species and thus included new germplasm. It is known that hexaploid wheat developed as a consequence of the inclusion of germplasm from *Aegilops squarrosa* into already domesticated tetraploid wheats by the process of introgressive hybridization. The inclusion of germplasm from *Tripsacum* into maize (*Zea mays*) has followed the same pattern.

The importance of introgressive hybridization for evolution and the development of cultivated plants may be illustrated by an almost contemporary example. One to two hundred years ago, at the time of the European conquest of North America, the annual sunflower *Helianthus annuus* began to spread rapidly, crossing with several other *Helianthus* species as it did so. The genes distributed in various regions made *H. annuus* adaptable to a wide range of ecological conditions and allowed it to spread over an enormous area (Zohary, 1970).

Therefore, Harlan (1970) considers the origin of cultivated plants to be diffuse in time and space. He believes that the problems associated with the centres of origin will never be completely solved and that it is thus better to talk of centres of original domestication and of centres of diversity as two separate phenomena.

Vavilov too referred to centres of genetic diversity but he adhered to the classification according to centres of origin.

Vavilov's explorations of gene centres gained knowledge of the accumulation and distribution of genes (for colour, form, resistance to pathogens, etc.) in specific regions which is important for the current search for gene sources to be used in contemporary breeding programmes.

The law of homologous series in the variation of traits

Studying the evolution of different plant species, Darwin noticed a parallelism in the variation of traits within a single species as well as among species. Vavilov concluded that the variation of traits both within and among species follows a certain homology which recurs in simple series. For example, the group of 42-chromosome wheats (*Triticum*) includes several species which originate from the Central Asian Centre; the group of 28-chromosome wheats comprises materials from Ethiopia, Transcaucasia and the Mediterranean and the group of 14-chromosome wheats comes from Asia Minor. Although these species are distinct in their geographical origin and many other traits, there exists a number of traits in common. Each species contains awned and awnless types, brown and white spikes, winter and spring habits, early and late forms, resistant and susceptible genotypes, etc.

Comparing two species from different genera, e.g., wheat and rye, it can be seen that both species have winter and spring forms, early and late varieties, large and small grains, etc. Apples and peaches, which belong to different genera, possess many common traits. Homologous variations may be found among different families but, understandably, these are less numerous than those within or among genera (Table 2.1).

In summarizing the regularities about homologous series, Vavilov came to the following conclusions: "Species and genera that are genetically closely related are characterized by similar series of heritable variations with such regularity that knowing the series of forms within the limits of one species, we can predict the occurrence of parallel forms in other species and genera. The more closely related the species and Linneons in the general system, the more resemblance will there be in the series of variations.

Whole families of plants in general are characterized by definite cycles of variability occurring through all genera and species making up the family.

The regularities in polymorphism of closely related species and genera make it possible to predict the possibility of finding corresponding forms in nature or obtaining them artificially by means of mutations, inbreeding, or hybridization" (English edition, 1951, p. 75).

To demonstrate this we can take the following example. The species *Triticum turgidum* includes forms with a branched spike. Such forms have not been found in *T. aestivum* subsp. *vulgare* but they were obtained by crossing it with *T. turgidum* and *T. vavilovii*. Likewise, forms of leaves without the ligula have not been found

Table 2.1 Homologous series. General scheme of variability in species of the Papilionaceae

Heritable variable character			*Pisum sativum* L. – Pea	*Vicia sativa* L. – Vetch	*Vicia faba* L. – Broadbean	*Lens esculenta* Moench. – Lentil	*Lathyrus sativus* L. – Grass pea	*Cicer arietinum* L. – Chick pea	*Glycine hipida* Max. – Soybean	*Phaseolus vulgaris* L. – Bean	*Canavalia gladiata* DC – Canavalia	*Stizolobium hassjoo* Piper – Velvet bean	*Cajanus indicus* Spreng. – Peigeon pea	*Medicago sativa* L. – Blue alfalfa	*Trifolium pratense* L. – Red clover	*Lotus corniculatus* L. – Bird's foot trefoil
Blossom characters	Color of blossom	White	+	+	+	+	+	+	+	+	+	+	..	+	+	+
		Rose	+	+	+	+	+	+	+	+	+	..	..	+	+	..
		Red	+	..	+	..	+	+	..	..	..	..	..	+	+	..
		Violet-blue	+	+	+	+	+	+	+	+	+	+	..	+	+	..
		Yellow	+	..	+	..	..	+	..	..	..	..	+	+	..	+
		Variegated:														
		Color of standards quite different from color of wings	+	+	+		+		+	+						
		Wings and standard spotted or striped	..	+	..	..	+	..	+	..	..	..	+	+	..	+
	Size of blossom	Large	+	+	+	+	+	+	+	+	+	+	+	+	+	+
		Small	+	+	+	+	+	+	+	+	..	..	+	+	+	+
Fruit characters	Structure of pood wall	With parchment layer	+		+					+	+	+				
		Without parchment layer	+	+	+	+	+	+	..	+	..	..	+	..	..	..
	Form of pod	Linear	+	+	+	..	+	..	..	+	+	+	+	..	..	+
		Rhombic	..	+	+	+	+	+	..	..	..	..	..	+	..	..
		Crescent shaped	+	+	+	..	..	..	..	+	..	+	..	..	..	..
		Sword shaped	+	..	+	..	..	..	..	+	+	+	..	..	..	..
		Bead-like	+	+	+	..	..	..	..	+	..	+	+	..	..	..
		Hairy	..	+	+	..	+	+	+	+	..	+	+	..	..	..
		Smooth	+	+	+	+	+	..	..	+	+	..	+	..	..	+

Source: Vavilov, 1951

in barley but they were found in wheat and rye. However, barley mutants without the ligula have been induced by exposing barley seeds to X-rays.

Therefore, the knowledge of the existence of the homologous series in the variation of traits within species, among species and genera is important for the development of plant varieties possessing new traits.

New agricultural plants

Possibilities offered by nature have not been fully exploited as yet. According to a study conducted by the National Academy of Sciences of the United States (1975), out of more than 3,000 potentially usable plant species only 20 to 30 of them form the staple crops for the mankind (Fig. 2.2). Jain (1983) and Hinman (1986) describe several plant species which have been underexploited, e.g., winged bean (*Phosphocarpus* species), buffalo gourd (*Cucurbita foetidissima*), amaranths, jojoba (*Simmondsia chinensis*), crambe (*Crambe abyssinica*), veronia, guayule (*Parthenium argentatum*) gopher plant, etc. The winged bean (*Psophocarpus*) contains 37% protein and 18% oil, almost as much as soyabean, which is not adapted to the humid tropics where winged bean may be grown successfully. *Amaranthus edulis* contains 6% lysine and it is an excellent source of vitamins A and C. *Crambe abyssinica* gives a valuable seed oil with 55% to 60% of erucic acid. Guayule (family *Compositae*) produces natural rubber equivalent to *Hevea* rubber. According to Calvin (1979) the gopher plant which produces a petroleum-like fuel from its leaves is of particular interest today.

However, the majority of these crop plants possess some negative characteristics which have precluded their earlier introduction to agriculture. Thanks to new scientific methods or techniques, some of those disadvantages may be eliminated or changed. Jain (1983) gives examples of lupins, safflower, and rapeseed which have become important agricultural crop plants after the genes responsible for deleterious alkaloids or other undesirable compounds had been changed by induced mutation or other methods.

A danger of restricted genetic variability

The knowledge of the centres of genetic diversity offers a possibility of using genes for particular traits in breeding programmes. This possibility gains importance because numerous local populations have either disappeared or have been reduced to a limited number of varieties in the course of the modernization of plant production and agriculture. Striving to improve the uptake of mineral nutrients and the use of machinery, in order to increase the production per unit area, crop producers resorted to monoculture as a means of intensifying the agricultural production. This practice ultimately made the varieties grown quite homogeneous; with many crops the

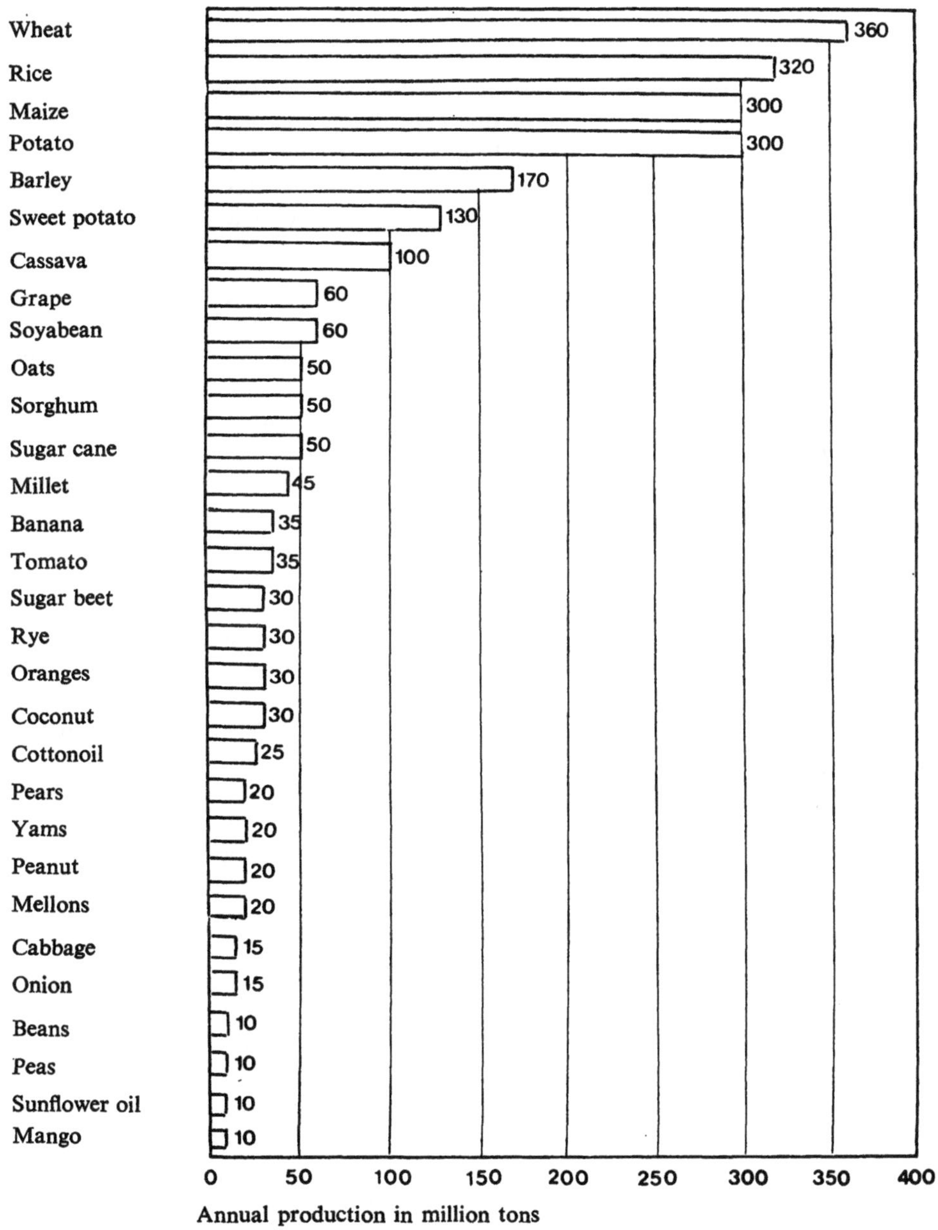

Fig. 2.2 Staple plants wich nourish the world

Source: Harlan 1976

variety (cultivar) represents a pure line. A good example for this is the introduction of Mexican high-yielding dwarf wheat varieties to India, Pakistan, and some other Asian and African countries in which these varieties have spread over millions of hectares. Although they did not cause a slump in wheat production but, on the contrary, brought significant increases in yield and overall production, some ecologists and geneticists have become concerned. They have pointed out that the new varieties are genetically uniform and that genetic variability is constantly being lowered and this may lead to epidemics of plant diseases, reductions in production, and even hunger (Frankel and Bennett, 1970; Frankel and Hawkes, 1975).

There are certainly good reasons for alarm since reductions in genetic variability have allowed outbreaks of diseases resulting in drastic reductions in production. Chang (1979) states:

- "The Irish famine of the 1840s due to the potato late blight epidemic (*Phytophthora infestans*).
- The wheatless days of 1917 in the USA, due to stem rust epidemics (*Puccinia graminis*).
- The great Bengal famine of India in 1943 associated with the brown spot disease of rice (*Cochliobolus miyabeanus*).
- The complete elimination in the mid-1940s of all oats derived from the variety Victoria in the USA, due to Victoria blight disease (*Cochliobolus victoriae*).
- The southern corn leaf blight epidemic (*Cochliobolus heterostrophus*) og 1970–1971 on all US maize hybrids carrying T-type cytoplasmic male sterility.
- The rapid shift in the rice brown plant-hopper (*Nilaparvata lugens*) from biotype 1 to biotype 2 during 1974–1976, when large areas in the Philippines and in Indonesia were planted to a few semidwarf rices". (p. 91).

Although it is justifiable to emphasize the risks of decreased genetic variability, brought about by the development of high yielding varieties, these risks are made even greater by the system of production, market demand and the heedless search for profit, i.e., the high yielding varieties are not in themselves to blame.

It should be mentioned that every human expansion to new areas has caused reductions or even the complete extinction of certain plant species (grasses, weeds, trees, etc.), animals, birds, insects and so on. On the other hand, usable species and varieties have been expanded over territories that they naturally would have never conquered. Just remember the expansion of the maize, potato, sunflower, tomato and tobacco crops from the New World into the Old World and the expansion of the crops of wheat, rice and fruit trees from the Old World into the New World in the post-Columbian times.

High-yielding varieties and hybrids that have recently been developed in different breeding centres around the world are not genetically less different amongst themselves than were the previously grown extensive and local varieties. As regards the resistance to pathogenic organisms, local varieties again do not offer any advantage over the new high-yielding varieties and hybrids. A screening test which included local and other wheat varieties grown in Yugoslavia (Borojević, 1955) showed that resistance to rusts existed in only two varieties, Thatcher and Regent, which had been imported from the United States of America. An equilibrium has been reached between local populations of crops and pathogenic organisms at the level of medium resistance or tolerance, as confirmed, by yield stability. However, that yield is below the level of acceptability and new varieties must thus be developed.

Therefore, the risk of epidemics is not due to the lack of diverse varieties but rather due to the choices made by farmers who, insisting on growing the best variety, concentrate on one or two varieties in a region. As much as it is desirable to develop a large number of divergent varieties of a crop, capable of counteracting the main limiting factors of the environment, it is equally important, for the sake of yield stability, to grow several genetically different varieties in a region or at a large farm.

Germplasm maintenance and exploitation

The maintenance of available plant germplasm is an important task for a number of reasons. Social and scientific developments impose new requirements, offer new possibilities, and put into reality things that have not been thought of in the past. Some examples are now given to show this important aspect of germplasm.

Since wheat is not only the main food item but also the main source of protein for a large portion of mankind, an idea was engendered to increase the protein content in wheat by breeding. The increase of the nutritional value of wheat, however, implies an increase in the content of lysine, an aminoacid contained in the presently grown varieties at a level below 3% of the percentage of crude proteins. Out of a collection of over 15,000 entries analysed in the United States of America only a few entries, coming from Argentina and India, contained about 4% lysine (Johnson et al., 1968). Of course, these entries should be preserved as sources of genes for high lysine for use in breeding programmes.

Before it was discovered that male sterility may be used for the production of hybrids in plants which cannot be emasculated economically (sorghum, wheat, etc.), male sterile plants encountered in breeding work were discarded and nobody tried to preserve them. Before the need for fertility restoration genes was developed, associated with the exploitation of the phenomenon of heterosis, plants with these genes were neither looked for nor was it known that genotypes differ in this trait.

Here is a recent example illustrating how valuable genetic material may be lost. A few years ago we exchanged several new high-yielding wheat lines with the Agricultural Research Institute in New Delhi, India, where they underwent a series of tests including the test for resistance to plant viruses. Of the entire collection tested, only the line NS-879 was found to be resistant to yellow barley dwarf virus. When Indian colleagues asked for more seed, we found that the line had been eliminated from our collection. Fortunately, the line was found at the Agricultural Research Station in Tamworth, Australia, with which we also exchanged seed materials at that time.

Such examples may be found for all crops and it is quite certain that valuable materials have been and will be lost for ever. Plant breeders became aware of that risk a long time ago. Already at the beginning of this century and especially after

World War I, seed collections of staple crops were established in many countries. As the intensification of agriculture following World War II caused a rapid loss of local populations, primitive and selected varieties, as well as new breeding materials, organized efforts have been undertaken, especially through the Food and Agriculture Organization of the United Nations (FAO), to preserve the genetic variability in cultivated plants and their wild relatives by establishing so-called "gene banks".

The main sources of genetic variability that should be preserved are:

- the previously and currently grown commercial varieties;
- early breeding materials which have not gone through official testing and registration but which are known to possess desirable genes and are used in hybridization programmes;
- breeding materials featuring specific genes (disease resistance, high protein content, etc.);
- local populations (land races), especially those whose genetic variability has not been introduced into commercially cultivated varieties;
- special genetic material (mutants, genetic testers, cytoplasmic sources, etc.);
- polyploids and aneuploids;
- synthetic species hybrids;
- wild and weedy species.

Well-organized institutions are needed for the maintenance of plant materials. They should be well-equipped in order to maintain seeds for scores of years without frequent regeneration. Furthermore, efficient data systems should be established to ensure an easy access to necessary information and plant materials.

Such research institutions, which maintain collections of seeds and vegetative parts, have already been established in several countries, to name those keeping a large number of entries:

Vavilov All-Union Institute for Plant Production, Leningrad, USSR

Agricultural Research Service, United States Department of Agriculture, Beltsville, Maryland, USA

International Rice Research Institute (IRRI), Los Banos, Philippines

International Crops Research Institute for Semi-Arid Tropics (ICRISAT), Hyderabad, India

Centro Internacional de Mejoramento de Maiz y Trigo (CIMMYT), El Batan, Ciudad Mexico, Mexico

Dutch/German Potato Gene Bank, Braunschweig, FRG

International Potato Center (CIP), Lima, Peru

International Institute of Tropical Agriculture (IITA), Ibadan, Nigeria

Nordic Gene Bank, Lund, Sweden

Asian Vegetable Research and Development Centre (AVRDC), Shenghwa, Taiwan

Aegean Regional Agricultural Research Institute (ARARI), Izmir, Turkey.

In addition, there are several national centres which keep rich collections of plant genetic resources, such as:

Royal Botanic Garden, Kew, England

Plant Breeding Institute, Cambridge, England

Institute National de Recherche Agronomique (INRA), Versailles, France
Zentralinstitut für Genetik und Kulturpflanzenforschung, Gatersleben, GDR
Foundation for Agricultural Breeding, Wageningen, the Netherlands
National Agricultural Research Centre Yatabe, Tsukuba, Japan.

Thanks to these institutions and several others not mentioned here, the world gene bank accessions contain thousands of samples of major crops as may be seen in Table 2.2.

Table 2.2 Estimated number of gene-bank accessions worldwide

	Number of			Coverage Percentages	
Crop	Accessions	Distinct Samples	Collections of 200 + Accessions	Landraces	Wild Spp
CEREALS					
Wheat	410,000	125,000	37	95	60
Barley	280,000	55,000	51	85	20
Rice	215,000	90,000	29	75	10
Maize	100,000	50,000	34	95	15
Sorghum	95,000	30,000	28	80	10
Oats	37,000	15,000	22	90	50
Pearl millet	31,500	15,500	10	80	10
Finger millet	9,000	3,000	8	60	10
Other millets	16,500	5,000	8	45	2
Rye	18,000	8,000	17	80	30
PULSES					
Phaseolus	105,500	40,000	22	50	10
Soybean	100,000	18,000	28	60	30
Groundnut	34,000	11,000	7	70	50
Chickpea	25,000	13,500	15	80	10
Pigeonpea	22,000	11,000	10	85	10
Pea	20,500	6,500	11	70	10
Cowpea	20,000	12,000	12	75	1
Mungbean	16,000	7,500	10	60	5
Lentil	13,500	5,500	11	70	10
Faba bean	10,000	5,000	10	75	15
Lupin	3,500	2,000	8	50	5
ROOT CROPS					
Potato	42,000	30,000	28	95	40
Cassava	14,000	6,000	14	35	5
Yams	10,000	5,000	12	40	5
Sweet potato	8,000	5,000	27	50	1
VEGETABLES					
Tomato	32,000	10,000	28	90	70
Cucurbits	30,000	15,000	23	50	30
Cruciferae	30,000	15,000	32	60	25
Capsicum	23,000	10,000	20	80	40

Table 2.2 (Continued)

	Number of			Coverage Percentages	
Crop	Accessions	Distinct Samples	Collections of 200 + Accessions	Landraces	Wild Spp
Allium	10,500	5,000	14	70	20
Amaranths	5,000	3,000	8	95	10
Okra	3,600	2,000	4	60	10
Eggplant	3,500	2,000	10	50	30
INDUSTRIAL CROPS					
Cotton	30,000	8,000	12	75	20
Sugar cane	23,000	8,000	12	70	5
Cacao	5,000	1,500	12	*	*
Beet	5,000	3,000	8	50	10
FORAGES					
Legumes	130,000	n.a.	47	n.a.	n.a.
Grasses	85,000	n.a.	44	n.a.	n.a.

Notes: The coverage percentages are estimates derived from a consensus of scientific opinion (Lyman, 1984). No precise figures can be given until all accessions have been described. The coverage of wild species related to those largely in the primary gene pool – that is, those species that were progenitors of crops, have co-evolved with cultivated species by continuously exchanging genes, or are otherwise closely related. Such wild species are more easily used in breeding. In addition, hundreds of more distantly related species, with great potential for crop improvement, exist in secondary or tertiary gene pools, but they have not been adequately considered. Estimates of these are not provided.

* Coverage difficult to estimate because many selections are from the wild.

n.a. = data not available.

Source: Plucknett et al., 1987

References

Borojević, S., 1955. Plan proizvodnje sortnog semena i selekcija pšenice u Jugoslaviji. Informator, (Zagreb), 155: 1–3.

Calvin, M., 1979. Petroleum plantations for fuel and materials. BioScience, 29: 533–538.

Chang, T. T., Brown, W. L., Boonman, J. G., Sneep, J., and Lamberts, H., 1979. Crop genetic resources. In: J. Sneep and J. T. Hendriksen (Editors), Plant Breeding Perspectives. Pudoc, Wageningen, p. 83–104.

Frankel, O. H., and Bennett, E., 1970. Genetic Resources in Plants – their Exploration and Conservation. Blackwell, Oxford. 554 pp.

Frankel, O. H., and Hawkes, J. G. (Editors), 1975. Crop Genetic Resources for Today and Tomorrow. Cambridge Univ. Press, Cambridge, 492 pp.

Harlan, J. R., 1951. Anatomy of gene centers. American Nat., 85: 97–103.

Harlan, J. R., 1970. Evolution of cultivated plants. In: O. H. Frankel and E. Bennet (Editors), Genetic Resources in Plants. Blackwell, Oxford, p. 19–32.

Harlan, J. R., 1976. The plants and animals that nourish man. Scientific American, 235: 88–97.

Harland, J. R., and Zohary, D., 1966. Distribution of wild wheat and barley. Science 153: 1074–1080.

Hinman, C. W., 1986. Potential new crops. Scientific American, 255: 25–29.

Jain, S. K., 1983. Domestication and breeding of new crop plants. In: D. R. Wood (Editor), Crop Breeding. American Society Agronomy, Madison, USA, p. 177–197.
Johnson, V. A., Whited, D. A., Mattern, P. J., and Schmidt, J. W., 1968. Nutritional improvement of wheat by breeding. Proc. 3rd. Int. Wheat Genet. Symp., Canberra, p. 457–461.
Lyman, J. M., 1984. Progress and planning for germplasm conservation of major food crops. Plant Genetic Resources Newsletter (Food and Agriculture Organisation/International Board for Plant Genetic Resources), 60: 3–21.
National Academy of Science, 1975. Underexploited Tropical Plant with Promising Economic Value. NAS, Washington, D.C., 189 pp.
Plucknett, D. L., Smith, N. J. H., Williams, J. T., and Amishetty, N. M., 1987. Gene Banks and the World's Food. Princeton Univ. Press, Princeton, 247 pp.
Vavilov, N. T. (Editor), 1935. Teorethicheskie osnovi selekcii rastenii. Tom 1, 2 and 3, Seljskohoz. Moscow-Leningrad, 2615 pp.
Vavilov, N. I., 1951. The Origin, Variation, Immunity and Breeding of Cultivated Plants. Selected writings translated from the Russian by K. Starr Chester, The Ronald Press Company, New York, 364 pp.
Zeven, A. C., and Zhukovsky, P. M., 1975. Dictionary of Cultivated Plants and their Centers of Diversity. Pudoc, Wageningen, 219 pp.
Zeven, A. C., and de Wet, I. M. I., 1982. Dictionary of cultivated plants and their regions of diversity. Pudoc. Wageningen, 263 pp.
Zohary, D., 1970. Centers of diversity and centers of origin. In: O. H. Frankel and E. Bennett (Editors), Genetic Resource in Plants. Blackwell, Oxford. p. 33–42.

3. SOURCES OF GENETIC VARIABILITY

Success in plant breeding is directly dependent on the genetic variability in the source populations. A major breakthrough in plant breeding was made when man realized that the main sources of genetic variability were gene mutations and recombinations and that these can be manipulated without waiting for them to occur naturally. This, however, only took place at the beginning of this century, when genetics became a separate scientific discipline.

Today, other sources of genetic variability are at our disposal, i.e., direct gene manipulation or gene transfer by the use of recombinant DNA techniques and somaclonal variations ensuing from the application of tissue and cell culture techniques.

Gene mutation

Spontaneous mutations occur in nature at random or due to environmental effects. An accidental mutation may take place after a tautomeric shift of the hydrogen atom which prevents the pairing of complementary bases in desoxyribonucleic acid (DNA). For example, adenine (A) does not pair with thymine (T) but with guanine (G). In the next replication, A pairs with T and G pairs with C (cytosine) which brings about a change in nucleotides (Watson, 1970).

The changes in the nucleotides of the DNA molecule cause changes in the code which in its turn brings a change in the codon of the messenger ribonucleic acid (mRNA). Consequently, different amino acids are bound resulting in the synthesis of different proteins. Examples of it are the sickle-cell anaemia in humans, colour mutations in plants, etc. According to Watson (1970), spontaneous mutations resulting only from tautomeric shifts occur at the frequency of 10^{-9} at the nucleotide level and at the frequency of 10^{-6} at the gene level (about 1,000 nucleotides) per generation. Mutations are also caused by chromosomal aberrations (deletion, duplication, inversion, translocation).

In microorganisms, which have an enormous reproduction rate, the number of mutations occurring in a single reproduction cycle is exceedingly large. For example, the number of leaf rust (*Puccinia recondita* f.sp. *tritici*) uredospores on wheat, under the assumption that it infects only 1% of the leaf area per 1 m^2 of land area, is about 300 spores per 1 mm^2/day, i.e., 10^{11} spores/day/hectare. If the estimated frequency of spontaneous mutations per locus is 10^{-8}, there will be 1,000 mutants/locus/day/hectare (Parlevliet and Zadoks, 1977). When it is taken into account that each organism contains several thousand genes, it may be stated that mutations occur continuously. As pointed out by Wright (1931), spontaneous mutations are always in excess of the evolutionary needs of the species. Table 3.1 shows the frequencies of spontaneous mutations in some genes of three types of organisms.

When Muller (1927) succeeded in inducing mutations in a fruit fly *Drosophila melanogaster* by *X*-rays, it marked the beginning of a new era in genetics and induced mutations became the focus of genetic studies. Since then a number of other mutagens have been discovered: irradiation, chemical agents, physical treatments, etc. Pioneer studies on induced mutations in plants were undertaken by Stadler (1928).

The frequency of induced mutations is much higher than the frequency of spontaneous mutations, ranging from 10^{-3} to 10^{-5} per gene per generation. Some genes mutate more readily than others, as in the case of spontaneous mutations (Table 3.1). The spectra of spontaneous and induced mutations are similar and man-induced mutations have either been found or may be found in nature.

Table 3.1 Frequency of natural mutations in some organisms

Organism	Trait	No. of mutations per 1,000,000 gametes
Zea mays (after Stadler, 1942)	Red (*R*) to colourless aleurone (*r*)	492
	Colour inhibitor (*I*) to non-inhibitor (*i*)	106
	Purple (*Pr*) to colourless aleurone (*pr*)	11
	Normal (*Su*) to sugar endosperm (*su*)	2.4
	Yellow (*Y*) to white endosperm (*y*)	2.2
	Normal (*Sh*) to shrunken endosperm (*sh*)	1.2
	Normal (*Wx*) to waxy endosperm (*wx*)	0
Drosophyla melanogaster after Glass and Ritterhoff, 1956)	Wild type (+) to yellow colour (*y*)	120
	Wild type (+) to browncolour (*bw*)	30
	Wild type (+) to ebony (*e*)	20
	Normal eyes (+) to eyeless (*ey*)	60
Escherichia coli (*K*12) (various sources)	Penicillin resistant	0.1
	Ti FAG resistant	0.03
	Triptophane independent	0.06
	Arginine independent	0.004
	Leucine independent	0.0007
	Streptomycin resistant	0.0004

According to their usefulness, mutations are divided into useful, detrimental, and neutral. As the mutation process is random and uncontrolable in the cases of spontaneous and induced mutations alike, most of the mutations produced are neutral, i.e., they are neither advantageous nor disadvantageous for the individuals carrying them. Such mutations may be maintained in a population for a long time, while their genes are in the heterozygous state. When the genes come to the homozygous state, these mutations may be expressed and may even bring an advantage if environmental conditions are changed. Since mutations result not only from a change of genes in relation to the wild type but also from deletion of one or more bases in DNA and from chromosome aberrations, many mutations are detrimental and disadvantageous to their carriers. Such mutations are eliminated in the processes of natural and artificial selection.

Natural selection favours mutations that are advantageous to their carriers. These mutations are potential sources of something new and they play an important role in the evolution of species. The possible uses of natural and induced mutations in breeding will be discussed afterwards.

Gene recombination

Unlike mutations which are associated with changes in genes themselves and which lead to new genetic variability, gene recombinations do not imply changes in genes but in combinations of genes which come from different parents. The result is again new genetic variability.

If two parents differ in two pairs of allelic genes which are non-linked and one of them is fully dominant over the other, then as a result of random chromosome segregation during meiosis (reduction division) and random mating of the female and male gametes in the F_1 generation, two new homozygous genotypes containing new recombinations of parental genes will develop in the F_2 generation:

P_1 *AAbb* × *aaBB* P_2

F_1 *AaBb*

F_2 *AABB* – a new genotype
AAbb – parental genotype
aaBB – parental genotype
aabb – a new genotype

When crossing parents which differ in three pairs of allelic genes, six new genotypes differing from the parents will develop in the F_2 generation:

P_1 *AAbbCC* × *aaBBcc* P_2

F_1 *AaBbCc*

F_2 *AABBCC* – a new genotype
AABBcc – a new genotype
AAbbCC – parental genotype
AAbbcc – a new genotype
aaBBCC – a new genotype
aaBBcc – parental genotype
aabbCC – a new genotype
aabbcc – a new genotype.

If plants selected for crossing differ in 4, 5, 10 or more pairs of allelic genes (n), which is usually the case, a large number of different gametes (2^n) develops because of heterozygosity in the F_1 generation which, in conditions of random mating, may result in the development of a large number (3^n) of genotypes and 2^n of different phenotypes in the F_2 generation (see Table 11.1). Thus, a tremendous new genetic variability is created which is usable for selection of new combinations of traits, i.e., for development of new lines and cultivars.

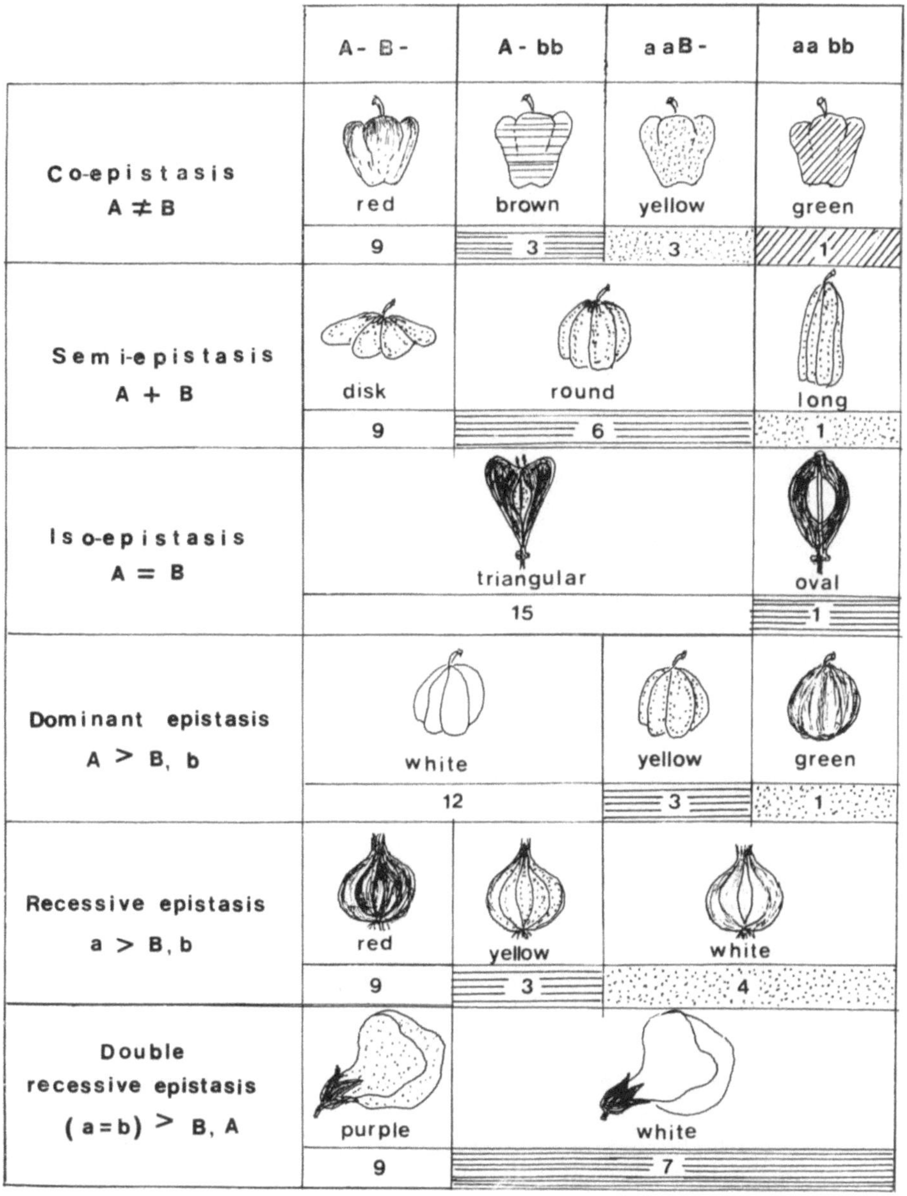

Fig. 3.1 Examples of interallelic gene interactions.
Source: Fasoulas, 1976, adapted

Although the recombinations of genes which form a new genotype from two (or more) parents create a large genetic variability which may be used for developing varieties with new traits, all combinations include only those genes which were present in the parents. In the case of non-linked genes, new gene recombinations are foreseeable in frequency and time (generations). In the case of linked genes, gene recombinations are foreseeable as the result of crossing-over but their frequencies cannot be estimated unless distances between gene loci on the chromosomes have been found previously.

As a result of intra-allelic and inter-allelic interactions, it is possible to obtain genotypes with traits which were not possessed by either parent. This happens when two or more different gene loci having different effects control the expression of a certain trait, e.g., co-epistatic, semi-epistatic, iso-epistatic, hypostatic or inhibitory and complementary gene effects. These are sources of new genetic variability and some possibilities are presented in Fig. 3.1. These phenomena have been exploited to obtain coloured flowers from colourless ones, different fruit forms, different characteristics of grain, etc.

Gene transgression deserves to be mentioned as another source of genetic variability which leads to the segregation of traits in one or both directions with regard to the parental traits. It happens frequently that some F_2 genotypes are earlier than the earlier parent, shorter than the shorter parent, superior in vitamin content than the better parent, etc. Assuming that one quantitative trait is controlled by five genes with additive effect, the following combinations, diverging from the parents in both directions, may be obtained in the F_2 generation:

P_1	*aa BB cc dd ee* 60 cm	×	*AA bb CC DD ee* 80 cm	P_2
F_1		*Aa Bb Cc Dd ee*		
F_2	*aa bb cc dd ee* 50 cm shorter than the shorter parent		*AA BB CC DD ee* 90 cm taller than the taller parent	

This was the case when the Italian wheat cultivar S-13 (plant height 63 cm) was crossed with the Japanese cultivar Aobakomughi (83 cm) (Fig. 3.2).

Plant height as a quantitative trait is mainly controlled by minor genes with additive effect, but a few major genes may also be present in some genotypes. Gale et al. (1981) found that the wheat cultivar Aobakomughi possesses the major gene for stem height reduction, Rht2, coming from Norin 10, while the cultivar S-13 probably possesses the gene Rht1S, derived from the Japanese variety Saitama 27 (Worland and Petrović, 1988) although most Italian cultivars possess the Akakomughi-derived gene Rht8 (Gale and Youssefian, 1985). Therefore, progeny no. 20 from the F_3 generation (which is shorter than S-13) probably contains the genes Rht1S and Rht2 and progeny no. 171 (which is taller than Aobakomughi) the gene Rht0, while the other progenies with intermediate height the gene Rht1S or Rht2 (Fig. 3.2).

Examples of gene recombinations and interactions leading to new recombinations of traits at the diploid level have been given in the preceding paragraphs. However, genetic variability may be further enlarged if interactions take place at different levels of ploidy, e.g.:

Superdominance at the level of

diploids: $AA < Aa > aa$

tetraploids: $AA \cdot AA < AA \cdot aa > aa \cdot aa$

hexaploids: $AA \cdot AA \cdot AA < AA \cdot AA \cdot aa = AA \cdot aa \cdot aa > aa \cdot aa \cdot aa$

There are certainly many other examples of gene actions and recombinations as sources of genetic variability but they exceed the scope of this book. The reader may refer to genetics textbooks and other special publications dealing with the subject matter.

In addition to variability found in natural plant populations, planned hybridization of genetically diverse individuals is the richest source of new genetic variability needed for plant breeding.

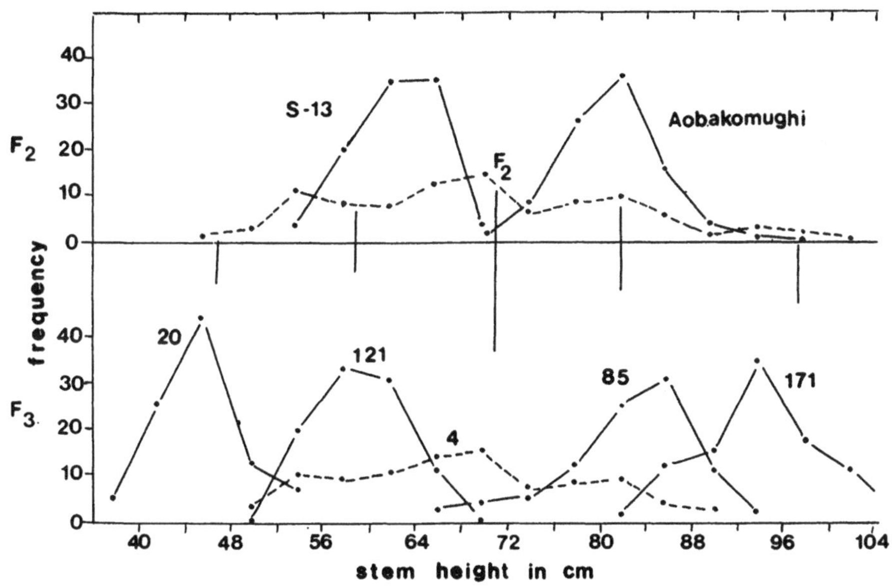

Fig. 3.2 Transgressive segregation in both directions for the height of stem in the F_2 and some progenies in the F_3 from the wheat cross, S-13 × Aobakomughi.
Source: Borojević and Mikić, 1965

Gene transfer

In recent decades, molecular genetics has developed to the point that it has become possible to isolate individual genes from an organism, clone them, and incorporate them into another organism. Such achievements have been preceeded by pioneer studies conducted by many researchers. Some of them will be mentioned here, according to a review made by Watson et al. (1983).

The investigation of O. T. Avery, C. M. Mac Leod, and M. MacCarty (1943–1944) and A. D. Hershey and Martha Chase (1952) supplied convincing proofs of DNA being the primary genetic material. J. D. Watson and F. H. C. Crick (1953) established a double helix structure of the DNA molecule. M. Meselson and F. W.

Stahl (1958) explained DNA replication. Thanks to investigations of M. W. Nirenberg and J. H. Mathaei on the genetic code, and the investigations of F. H. C. Crick on codon-anticodon pairing, it was already realized in the sixties that the DNA structure, via RNA, controls the synthesis of proteins. F. Jacob and J. Monod (1961) offered the first explanation on genetic regulatory mechanisms in the synthesis of proteins. The investigations of F. Sanger and a group of associates in the seventies provided information on the sequencing of DNA. The enzymology of DNA replication (A. Kornberg, 1959) allowed for the discovery of restriction enzymes which cut the DNA molecule into fragments at certain points. Individual genes were isolated in that way, and cloned to produce the so-called recombinant DNA molecule (J. E. Mertz and R. W. Davis, 1972; H. O. Smith and D. Nathans, 1978). H. G. Khorana (1979) were the first to succeed in synthetizing a gene chemically.

These investigations gave an additional impetus to gene manipulation studies. Efforts have been made to try to transfer recombinant DNA from one organism to another by suitable methods in order to provoke an increased production of proteins usable in the chemical industry, pharmacy, etc. The procedure itself has not been a simple one and it was necessary to solve series of biochemical and purely technical problems. In principle, the technique of recombinant DNA requires the isolation of a gene from an organism, its cloning and transfer to a vector to be incorporated into the recipient (Fig. 3.3).

At the beginning, viruses used to serve as vectors of desirable genes. Nowadays, plasmids are mostly used for that purpose but recent studies indicate that in some cases it is possible to transfer genes by a direct injection into the cell (protoplast) or into organelles.

A large number of experiments on gene transfer or genetic engineering has been conducted on the level of prokaryotes. On the level of eucaryotes, experiments are much more complex and transferred gene frequently does not come to expression.

So far, the application of the technique of recombinant DNA has been most successful in the production of human insulin, which has been available on the market since 1982 under the name of humulin. A diagram of the production of humulin presented in Fig. 3.3 was supplied by Watson et al. (1983).

Methods of genetic engineering are also successfully used in the production of human growth hormone which may become an efficient tool in the therapy of human dwarfs who are deficient in growth hormone. Steady improvements are registered in the production of interferone by the technique of recombinant DNA, which is used

---

Fig. 3.3 Production of insulin in bacteria by using synthetic insulin genes. Oligonucleotides coding for the *A* and *B* chains of insulin were synthesized chemically, and the trinucleotide ATG was added to the 5′ ends. These oligonucleotides were cloned downstream from the β-galactosidase promoter and part of the β-gal protein-coding sequence in pBR322. When the clones were introduced into Escherichia coli, the bacteria produced hybrid proteins consisting of the *N*-terminal portion of the β-gal protein, followed by a methionine residue (from the ATG codon) fused to either the *A* or the *B* chain of insulin. The β-gal protein portion was removed by treatment with CnBr, which cleaves polypeptides at methionine, liberating free *A* or *B* insulin chains. The two chains were then combined to produce biologically active insulin.
Source: Watson et al., 1983

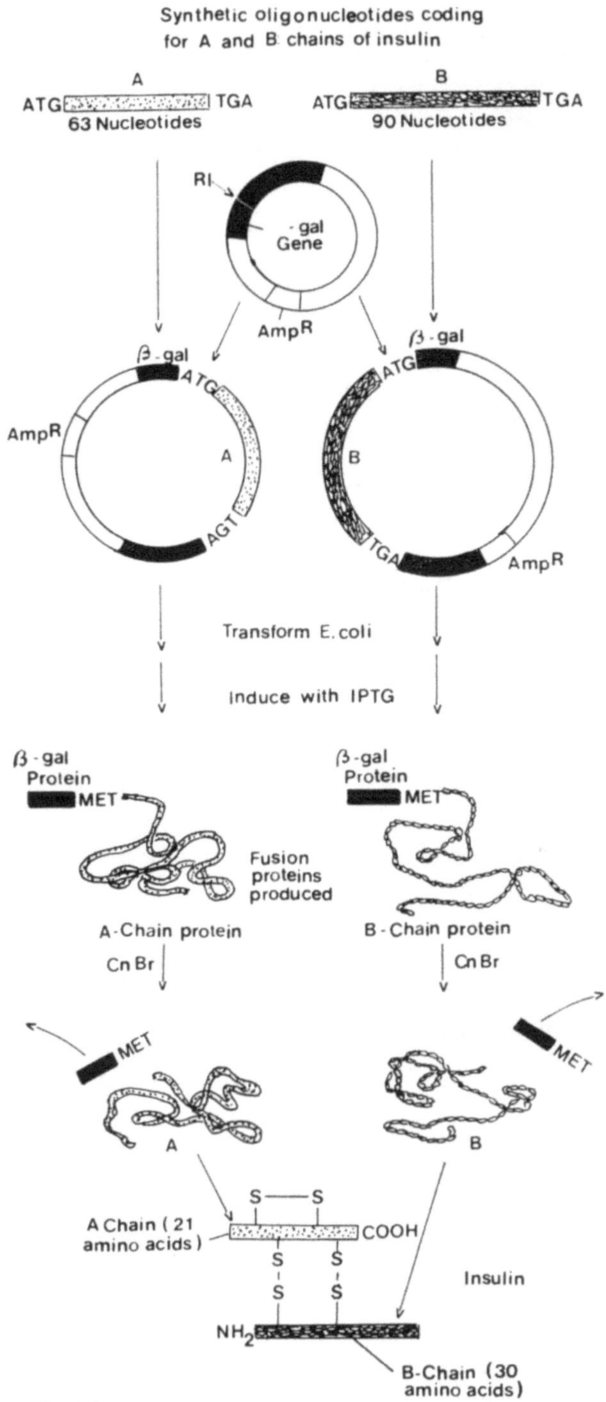

Synthetic oligonucleotides coding
for A and B chains of insulin
A
ATG
TGA
63 Nucleotides
B
ATG
TGA
90 Nucleotides
RI
-gal
Gene
AmpR
β-gal
ATG
A
AGT
AmpR
β-gal
ATG
B
TGA
AmpR
Transform E.coli
Induce with IPTG
β-gal
Protein
MET
Fusion
proteins
produced
A-Chain protein
CnBr
β-gal
Protein
MET
B-Chain protein
CnBr
MET
A
MET
B
S
S
A Chain (21
amino acids)
COOH
S
S
S
S
Insulin
NH2
B-Chain (30
amino acids)

Fig. 3.3

in the therapy against viral infections and some types of cancer. The application of recombinant DNA for the production of vaccines, e.g., for human hepatitis B and other diseases, keeps expanding.

A considerable number of plant-derived genes has been selected and synthetized. The most frequent vector for the transfer of desirable genes to another plant organism is the Ti-plasmid (Tumor inducing plasmid) isolated from *Agrobacterium tumefaciens*. Other examples of potential importance in plant breeding will be discussed in Chapter 19.

The technique of recombinant DNA allows man to induce desirable changes in an organism. Hybridization does not offer this possibility because desirable genes take part in recombinations along with undesirable ones. Induced mutations too are less satisfactory because of their random and unpredictable occurrence. Gene transfer is definitely a prospective source of genetic variability which has already been used on the level of prokaryotes. Its use on the level od eukaryotes will be intensified when more information is gathered and new methodologies are developed.

Somaclonal variation

In principle, plants derived from *in vitro* culture give plants phenotypically and genetically identical to the material from which tissues or cells have been taken. However, depending on the culture medium, the age of cells or tissue in culture, genotype, effects of temperature, light, etc., a rather large variability is observed in single-gene traits (colour, form, etc.) as well as in multi-gene traits (height, branching, etc.). Australian scientists Larkin and Scowcroft (1981) named that kind of variability as "somaclonal" variation and considered it to be of use in plant breeding as a new source of genetic variability.

Larkin and Scowcroft (1981, 1983), Scowcroft and Larkin (1982), Lörz and Brown (1986), reviewing extensive literature on somaclonal variation and cited the following events as main causes of somaclonal variation: (1) gross karyotypic changes, (2) cryptic chromosomal rearrangements, (3) somatic crossing-over and sister chromatid exchange, (4) transposable elements and (5) gene amplification and diminution. Depending on the nature of changes, somaclones are more or less stable, fertile or sterile.

There are opinions (Barbier and Dulien, 1980; Kasperbauer et al., 1981) that much of somaclonal variations are in fact preexisting mutations which become expressed in *in vitro* culture. Oono (1981), Hofmann et al. (1982), De Paepe et al. (1981) gave examples which show that new mutations occur continuously in cell culture. According to Lörz and Brown (1986), the cell culture itself has to be considered as a kind of "mutagenic treatment". The longer the tissue culture cycle, the more variation is found, which is more true for undifferentiated calli, cell suspensions, or protoplasts than for meristem and shoot tip cultures. The same authors consider gross karyological changes which occur in regenerants as a consequence of tissue culture stress and not a direct cause of somaclonal variation.

The origin of the majority of somaclonal variants is unknown but there are several examples which demonstrate that they may be considered a new source of variability. Possible uses of somaclonal variation in plant breeding will be discussed in Chapter 19.

References

Barbier, M., and Dulien, H. L., 1980. Effects génétiques observés sur des plantes de Tabac régénérées à partir de cotyledons par culture in vitro. Ann. Amelior. Plantes 30: 321–344.

Borojević, S. i Mikić, D., 1965. Pojava i osobine patuljastih linija pšenice dobivenih transgresivnim cepanjem. Arhiv za polj. nauke (Beograd), XVIII, sv. 61: 3–15.

De Paepe, R., Bleton, E., and Gnangbe, F., 1981. Basis and extent of genetic variability among doubled haploid plants obtained by pollen culture in *Nicotiana syvestris*. Theor. Appl. Genet. 59: 177–184.

Fasoulas, A., 1976. Principles and Methods of Plant Breeding. Aristotelian Univ. of Thessaloniki, Publ. No. 6, 55 pp.

Gale, M. D., Marshall, G. A., and Rao, M. V., 1981. A classification of the Norin 10 and Tom Thumb dwarfing genes in British, Mexican, Indian and other hexaploid bread wheat varieties. Euphytica 30: 355–361.

Gale, M. D., and Youssefian, S., 1985. Dwarfing genes in wheat. In: G. E. Russel (Editor), Progress in Plant Breeding-1. Butterworth, London, p. 1–35.

Glass, B., and Ritterhof, R. K., 1956. Spontaneous mutation rates at specific loci in *Drosophila* males and females. Science, 124: 314–315.

Hoffmann, F., Thomas, E., and Wenzel, G., 1982. Anther culture as a breeding tool in rape. II. Progeny analysis of androgenetic lines and induced mutations from haploid cultures. Theor. Appl. Genet. 61: 232.

Kasperbauer, M. J., Sutton, T. J., Anderson, P. A., and Gubton, C. L., 1981. Tissue culture of plants from a chimeral mutation of tobacco. Crop Sci. 21: 213–217.

Larkin, P. J., and Scowcroft, N. R., 1981. Somaclonal variation – A novel source of variability from cell cultures for plant improvement. Theor. Appl. Genet. 60: 197–214.

Larkin, P. J., and Scowcroft, W. R., 1983. Somaclonal variation and eyespot toxin tolerance in sugar cane. Plant Cell Tissue Organ Culture, 2: 111–121.

Lörz, H., and Brown, P. T. M., 1986. Variability in tissue culture derived plants – possible origins; advantages and drawbacks. In: W. Horn, C. J. Jensen, W. Odenbach and O. Schrieder (Editors), Genetic Manipulation in Plant Breeding. Walter de Gruyter, Berlin-New York, p. 217–229.

Müller, H. J., 1927. Agricultural transmutation of genes. Science, 66.

Oono, K., 1981. In vitro methods applied to rice. In: T. A. Thorpe (Editor), Plant Tissue Culturie Academic Press, New York, p. 273–298.

Parlevliet, J. E., and Zadoks, J. C., 1977. The integrated concept of disease resistance; a new view including horizontal and vertical resistance in plants. Euphytica, 26: 5–21.

Scowcroft, W. R., and Larkin, P. J., 1982. Somaclonal variation: a new option for plants. In: Indra K. Vasil, W. R. Scowcroft and K. J. Frey (Editors), Plant Improvement and Somatic Cell Genetics. Academic Press, New York, p. 159–178.

Stadler, L. J., 1928. Genetic effects of *X*-rays on maize. Proc. Nat. Acad. Sci., USA, 14: 69–75.

Stadler, L. J., 1942. Some observations on gene variability and spontaneous mutation. Michigan State College, The Spragg Memorial Lectures (Third series).

Watson, J. D., 1970. Molecular Biology of Gene. Benjamin. Menlo Park, California, 2nd. ed., 662 pp.

Watson, J. D., Tooze, J., and Kurtz, D. T., 1983. Recombinant DNA. A short course. W. H. Freeman, New York, 260 pp.

Worland, A. J., and Petrović, S., 1988. The gibberellic acid insensitive dwarfing gene from the wheat variety Saitama 27. Euphytica, 38: 55–63.

Wright, S., 1931. Evolution in Mendelian populations. Genetics, 17: 97–159.

4. GENETIC BASIS OF BREEDING SELF-POLLINATED PLANTS

The effects of selection and the success of breeding plants depends considerably on the mating system, i.e., the mode of reproduction of plants. There are two main mating systems – self pollination and open pollination.

Self-pollination or autogamy is a system in which pollination takes place between male and female sexual cells of the same individual, i.e., plant. The system implies the existence of bisexual flowers which contain a style with a stigma (the female sexual organ) and anthers with pollen grains (the male sexual organ). Self pollination may also take place if a plant has separate sexual organs.

A large number of plant species reproduce by self pollination. Some species are completely self pollinated, some have a small percentage of open pollination, and then there are some which may be counted among both self- and open pollinated plant species. The percentage of open pollination in self-pollinated species depends on genotype, environmental factors and their interactions. The percentage tends to increase in conditions favouring a protracted period of flowering and opening of flowers.

Self-pollinated plant species are those in which self pollination occurs as a rule and open pollination is an exception ranging below 4%.

Self-pollinating plant species among staple agricultural crops:

wheat	common bean	tomato
barley	broadbean	eggplant
oats	Lima bean	lettuce
rice	mung bean	endive
millet	soyabean	pepper
sorghum *	chick pea	
	vetch	peach
	peanut	apricot
flax	sweet clover	citrus
cotton *	hybrid clover	grapevine (some species)
tobacco	subterranean clover	
	annual fescue	
	mountain broomgrass	
	slender wheatgrass	

* Sorghum and cotton frequently have over 10% of open pollination.

The theory of pure lines

Self-pollination is the highest degree of inbreeding achievable by an organism. It leads to homozygosity in all gene loci and traits.

A Danish scientist Johannsen (1909, 1926) who studied the garden bean and other species, concluded that there exists a variability even among self-pollinated plants of the same variety. To establish the kind of variability observed and the genetic constitution of a self-pollinated variety, he applied the method of individual selection to the bean variety Princess. The progenies of the selected plants differed in height, earliness, seed size and other traits. Johannsen distinguished 19 types which were fairly different and called them lines (Table 4.1). The progenies of these lines, however, were quite uniform and stable from generation to generation. Such lines, genetically uniform and capable of transmitting this uniformity to their progenies, he called *pure lines.*

Table 4.1 Distribution of weight of bean seeds originating from seeds possessing different masses

Seed weight in female	seed 10	weight of progenies 20	30	40	50	60	70	80	90	Average
20	–	1	15	90	63	11	–	–	–	43,8
30	–	15	95	322	310	91	2	–	–	44,5
40	5	17	175	776	959	282	24	3	–	46,2
50	–	4	57	305	521	196	51	4	–	48,9
60	–	1	23	130	230	168	46	11	–	51,2
70	–	–	5	53	175	180	64	15	2	56,0
Total:	5	38	370	1,676	2,255	928	187	33	2	47,9

Source: Johannsen, 1909

Johannsen was further interested in estimating the effect of selection within a pure line. He divided the seeds of a pure line derived from the variety Princess into three categories – large, medium and small seeds. Each category was sown separately and the resulting plants were analysed for seed size. The plants derived from large seeds produced not only large seeds but also medium and small seeds. The same happened with the plants derived from medium and small seeds. Regarding the yield of seed per unit area, the three groups of progenies produced fairly equal yields (Table 4.2).

The results showed that after six years of individual selection within the single pure line there occurred no significant differences in the average mass of seeds originating from small and large seeds. Thus Johannsen concluded that any attempt at selection within a pure line is ineffective.

His conclusion is a fact resulting from the genetic constitution of pure lines. A pure line consists of the progenies of a single homozygous plant and all individuals

Table 4.2 The effects of selection for six generations in line one of the Princess bean. (After Johannsen, 1926)

	Mean Weight of Parent Seeds			Mean Weight of Progeny Seeds		
Harvest Year	Light Line	Heavy Line	Difference	Light Line	Heavy Line	Difference
1902	60	70	10	63.15	64.85	+1.70
1903	55	80	25	75.19	70.88	−4.31
1904	50	87	37	54.59	56.68	+2.09
1905	43	73	40	63.55	63.64	−0.09
1906	46	84	38	74.38	73.00	−1.38
1907	56	81	25	69.07	67.66	−1.41

In each generation the light line was selected for light seeds and the heavy line for heavy seeds.
Source: Allard, 1960

are homozygous for gene loci and they have the same genotype. Their homozygosity is the result of self pollination, i.e., the union of male and female gametes with identical genic alleles.

Since all individuals in a pure line are homozygous and there is no genetic variability amongst them, then selection within a pure line must inevitably be ineffective. The variability that may be observed in a pure line is due to the effect of environmental factors and the selection may have effect only if the selected material possesses heritable variability, as pointed out by Darwin.

Understandably, a pure line is not absolutely uniform and permanently homozygous for all traits. Heritable changes in a pure line occur due to mutation and open pollination. Spontaneous mutations occur regularly, a certain percentage of open pollination is inevitable, and these are sources of genetic variability in a pure line. Another source of variability in a pure line may be heterozygosity left over after hybridization. If the environment, i.e., natural selection, favours heterozygotes, they will be more productive and they will produce more progenies which will be reflected on the longer maintenance of heterozygosity than it might theoretically be expected. The maintenance of genetic variation within an inbred line will depend on the level of heterozygous advantage and on competitive advantage, very probably a frequency-dependent selection (Mayo, 1980). Nevertheless, all these sources of genetic variability in a pure line occur at very low frequencies so that such a line features a high degree of uniformity and whole fields appear to be very uniform.

It may thus be concluded that the selection from a single variety of a self-pollinated plant species (wheat, barley, soyabean, peach, etc.) will be effective only if the variety is not a pure line but if it comprises several lines genetically differing in certain traits. Once pure lines are achieved, any selection within a pure line will be ineffective and it is more profitable to try to produce something new by crossing or by other methods than to count on changes that might eventually take place in a pure line.

Genetics of hybridization of self-pollinated plants

Crossing or hybridization is the fusion of genetically different gametes resulting in hybrid organisms heterozygous for one or more allelic genes. In the system of self-pollination, homozygosity is achieved relatively quickly after hybridization, even in the case of full heterozygosity at the beginning. Let us take an example of parental varieties differing in one pair of allelic genes:

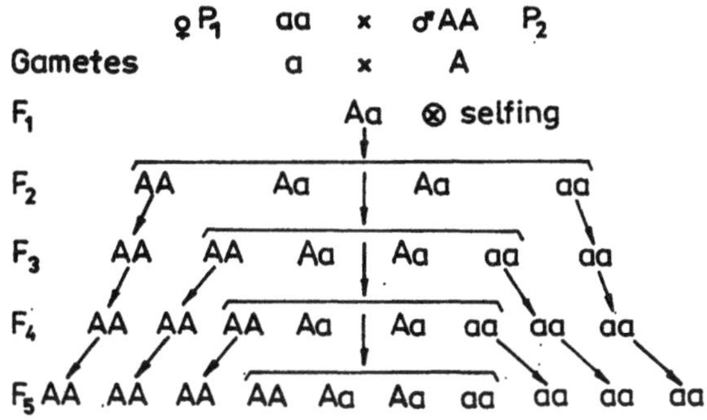

Thus, all individuals in F_1 generation are heterozygous and their genetic constitution is *Aa*. The F_1 plants form two kinds of gametes, *A* and *a*, both on the side of ovular cells and on the side of the pollen. As we are dealing with a self-pollinated plant, hybridization of F_1 plants is not necessary because pollination takes place in each flower which offers equal chances to each kind of gametes of pollinating each other kind of gametes. The result of pollination is the F_2 generation consisting of genotypes in the following ratio: 25% *AA* : 50% *Aa* : 25% *aa*. The ratio of homzygous genotypes (*AA* and *aa*) to heterozygous genotypes (*Aa*) is 50% : 50. The level of homozygosity reaches 50% already in the first generation of selfing of heterozygotes, i.e., in the second generation counting from the original parents.

Table 4.3 Effect of selfing on the source population consisting entirely of heterozygous genotypes *Aa*

Generation	Genotypes *AA*	*Aa*	*aa*	Number of heterozygotes	Percentage of homozygotes
0	–	1	–	1	0
1	1/4	2/4	1/4	1/2	50,0
2	3/8	2/8	3/8	1/4	75,0
3	7/16	2/16	7/16	1/8	87,5
4	15/32	2/32	15/32	1/16	93,8
5	31/64	2/64	31/64	1/32	96,9
10	1023/2048	2/2048	1023/2048	1/1024	99,9
:					
n	$(2^n - 1/)2^{n+1}$	$2/2^{n+1} = 1/2^n$	$(2^n - 1)/2^{n+1}$	$1/2^n$*	

* If number of homozygotes remains in generation n then $1 - 1/2^n = (2^n - 1)\ 2^n$.

In subsequent generation of selfing, the homozygous individuals will produce equally homozygous progenies (*AA AA* : *aa aa*) while the heterozygous individuals (*Aa*) will produce homozygous (*AA* and *aa*) and heterozygous progenies (*Aa*). The number of homozygous individuals increases in each generation of selfing by one half of the difference between the preceeding two generations and the number of heterozygotes decreases by half. After five generations of selfing there will be 96.9% of homozygous individuals and after ten generations 99.9% of homozygous individuals in the entire population (Table 4.3).

If the parents differ in a large number of gene pairs, the homozygosity will be achieved later. Nevertheless, even in the case of differences in 100 independent gene pairs, the population will have 50% of homozygous individuals for all these genes already after seven generations of selfing (Fig. 4.1). The approach to homozygosity will also be slowed down by linkage.

Therefore, selfing leads quickly to the state of homozygosity, regardless of the number of heterozygous pairs of genes present at the beginning. It should be pointed out that in such a population each plant is homozygous but the population is not homogeneous, i.e., it does not comprise a single genotypes but different homozygous lines or families. The actual number of these lines depends on the number of hetero-

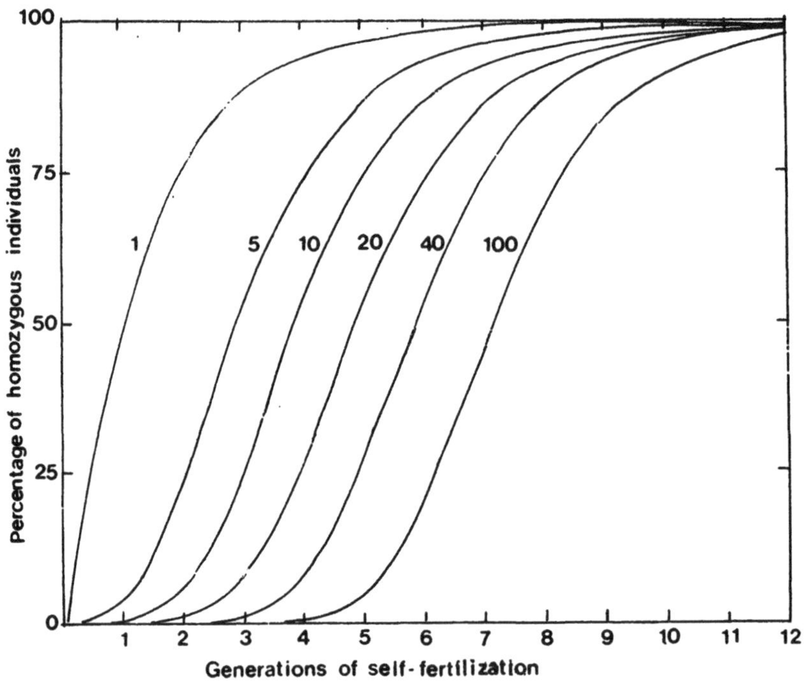

Fig. 4.1 Percentage of homozygous individuals after various generations of self-fertilization, when the number of independently inherited gene pairs is 1, 5, 10, 20, 40, or 100. The percentage of homozygosity in any selfed generation is given by the curve for one gene pair.
Source: Allard, 1960

zygous gene pairs (2*n*). If there was one gene pair, the population will consist of two inbred lines, if there were five gene pairs, the population will consist of 32 inbred lines, etc.

The percentage of homozygous individuals (*ho*) per generation after crossing may be calculated by the formula:

$$ho = (1 - he)^n$$

where *he* is the percentage of heterozygous individuals and *n* is the number of loci in which the parents differed. As the number of heterozygotes decreases by half in each generation of selfing, it ensues:

$$he = (1/2)^g$$

where *g* is the number of generations of selfing.
Thus:

$$ho = \left[1 - \left(\frac{1}{2}\right)^g\right]^n$$

For example, if the parents differed in two gene loci (*n*), the percentage of homozygous individuals after three generations of selfing (*g*) will be:

$$ho = \left[1 - \left(\frac{1}{2}\right)^3\right]^2 = 0.767 = 76.7\%$$

It means that 76.7% of lines will be completely homozygous for the two gene loci already by the F_3 generation. Even if differences existed in 40 gene loci, over 50% of lines will be homozygous for all the loci in the F_3 generation; in the F_{10} generation, differences in the number of loci are practically irrelevant (Fig. 4.1).

This formula is applicable only if all genotypes have the same rate of survival and if the genes are non-linked. Otherwise, gene linkage increases the proportion of homozygous individuals in each generation but it does not affect the percentage of homozygosity.

Since the success of breeding self-pollinated plants depends on the time necessary to achieve homozygosity for the desired traits, which in turn is conditioned by the number and effect of gene loci for these traits, the principles of breeding for qualitative traits and for quantitative traits will be discussed separately.

Inheritance of qualitative traits

Qualitative or alternative traits are those where their development is conditioned by the action of genes with a strong effect, the so-called *major genes*. Colour and form of flowers, fruits, leaves and other plant parts and in many cases the resistance to pathogenic organisms are counted among qualitative traits.

Since there are no ideal varieties, i.e., varieties which would possess the combination of genes and traits which could satisfy us for a long period, there exists a need to continuously create new gene recombinations and thus new varieties possessing new sets of traits.

To produce new recombinations of genes, parental varieties should differ in at least two pairs of gene loci, i.e., in two traits. Let us consider the target of developing a two-row barley variety possessing rust resistance having at our disposal a two-row variety susceptible to rust and a six-row variety resistant to rust. It has already been established that the gene locus for spike structure is located on one chromosome, that the gene for two rows is dominant (*D*) over the gene for six rows (*d*), that the genic locus for rust resistance is located on another chromosome and that the gene for rust resistance is dominant (*R*) over the gene for rust susceptibility (*r*). If the susceptible two-row variety is selected for the female, its anthers must be emasculated in time to prevent selfing. There are three anthers in each barley flower and they should be removed by forceps while they are still green, i.e., when the spike is still in the sheath of the flag leaf (before heading). Two to three days later, mature pollen gathered from the male parents, the resistant six-row variety, is dabbed onto the stigmata of the emasculated female parents. The anthers on the female parent could also be made non-functional by a gametocide treatment before self-pollination takes place. So, the plan of hybridization and the situation in subsequent generations look like this:

P_1 *DDrr* × *ddRR* P_2

gametes *Dr* × *dR*

F_1 *DdRr*

In the F_1 generation, plants with two-row spikes resistant to rust develop. But since the parents differed in two pairs of allelic genes the F_1 generation is heterozygous and the plants cannot be considered as a variety because of segregation in the F_2 generation. Thus the F_2 plants should be grown which is a simple task with barley because neither emasculation nor any other intervention is needed since each F_1 plant will be selfed:

F_1 *Dd Rr* ⊗

gamete

♀ \ ♂	*DR*	*Dr*	*dR*	*dr*
DR	*DDRR*	*DDRr*	*DdRR*	*DdRr*
Dr	*DDRr*	*DDrr*	*DdRr*	*Ddrr*
dR	*DdRR*	*DdRr*	*ddRR*	*ddRr*
dr	*DdRr*	*Ddrr*	*ddRr*	*ddrr*

F_2

Out of 16 possible combinations, 9/16 will possess the desired gene recombinations, i.e., they will be resistant two-row plants. However, the phenotypes of these 9/16 do not have the same genotype. Only 1/16 is homozygous for both dominant gene pairs (*DDRR*), 2/16 are homozygous for the gene for two rows and heterozygous for the gene for resistance (*DDRr*), 4/16 are heterozygous for both genes (*DdRr*), and 2/16 are heterozygous for the gene for two rows and homozygous for the gene for resistance (*DdRR*).

As the gene for two rows and the gene for resistance are fully dominant over their recessive alleles, then in either case the homozygous state cannot be distinguished phenotypically from the heterozygous state. Therefore, in the F_2 generation it is necessary to select all resistant two-row plants (*D–R–*) and only in the F_3 generation will it be possible to select genotypes homozygous for both desired traits which will not segregate in the progeny for these traits (*DDRR*). If there is a sufficient number of such plants, there is no need to select resistant two-row plants from other segregating plots since they may be heterozygous.

If the F_4 generation does not segregate for either trait, it may be considered that a line of two-row barley resistant to rust (*DDRR*) has been developed. The line remains to be multiplied in subsequent generations and, if its yield and quality are satisfactory, a new variety has been developed.

As only two major genes were considered and the desired recombination contained both dominant gene pairs, the frequency of such phenotypes in the F_2 generation was much higher than those of the other phenotypes, i.e., 9/16 out of 16/16. Therefore, it is not necessary to grow a large population of F_2 plants to find the desired gene recombination. A population of several hundred F_2 plants was quite sufficient. For example, the population of 300 plants would contain $9/16 \times 300 = 168$ resistant two-row plants (*D–R–*) and 18 of those would be *DDRR* ($1/16 \times 300$). It is quite enough to select the necessary number of lines with the desired gene recombination in subsequent generations.

For the sake of illustration, the simplest case of inheritance of traits and recombination of genes was considered. In practical work, the target is frequently to make recombinations in three or more traits. For example, if it is planned to develop a two-row barley (*DD*), resistant to rust (*RR*), with erect spike (*nn*), the cross will be as follows:

P_1 $\quad DDrrnn \times ddRRNN \quad P_2$

gametes $\quad Drn \quad \times dRN$

$\quad F_1 \quad DdRrNn$

In the F_1 generation, plants are two-rowed, resistant, with the nutans-type of the spike. To produce the desired recombinations, the F_2 generation should be grown.

The F_1 plants with the constitution *DdRrNn* will produce eight kinds of gametes: *DRN*, *DRn*, *DrN*, *Drn*, *dRN*, *dRn*, *drN*, and *drn*. These gametes in free pollination will produce 64 combinations in the F_2 generation. Of that number, 9/64 have the desired recombinations (*D–R–nn*). For further work, resistant two-row plants with erect spikes must be selected. Since the last trait is recessive, there will be no segregation for that trait in a single progeny; however, some progenies will segregate for the two-row spike and resistance. Further selection is thus needed for these two characters until homozygostity is achieved.

Theoretically speaking, when dealing with three traits and the recombinations of genes at the frequency of 9/64, it is necessary to have a larger number of plants in the F_2 generation than when dealing with two traits. With a population of 300 plants in F_2 generation, the frequency of expected recombinations would be $9/64 \times 300 = 42$ plants. This number is too low because it may happen that we fail to find some of the required type which lowers the chance of success. It is thus necessary to grow about 1000 plants in the F_2 generation to ensure that there are about 140 plants with the desired gene recombinations ($9/64 \times 1000$).

Effect of gene interaction

Independent gene action was considered in the previous example, when two or more genes are located on different chromosomes and each of them affect a trait independently.

However, series of gene loci mutually interact and they have different effects, bringing about various ratios of segregation in the F_2 generation, e.g., 9 : 7, 9 : 3 : 4, 9 : 6 : 1, 3 : 13, 15 : 1, etc. These ratios have already been shown in Chapter 3 (Fig 3.1.), discussing about the development of new genetic variability after hybridization. In such cases, the forecast of the desired gene recombination is more difficult or uncertain and published data may be helpful. Quite new traits, interesting for plant breeding, may be obtained by these interactions.

Effect of gene linkage

Considering a relatively limited number of chromosomes in an organism, there is a comparatively large number of linked genes. Parental combinations with linked agronomically good traits are desirable. If, however, a variety possesses a positive trait linked with negative one, it is necessary to make gene recombinations with another parent so as to remove the undesirable trait by progeny selection. In that case, due to linkage, the progenies will have a great number of parental gene combinations while recombinations will occur only as the result of the crossing-over of the gametes. The number of recombinants depends on the distance between the gene loci in question. In the case of two linked but fairly distant gene loci, the number of expected recombinations may be obtained in a relatively small F_2 population. If there are three or more linked gene loci, and the distances among them are small, a large population is needed and the expected recombinations are difficult to find (Table 4.4).

The effect of gene linkage on expected gene recombinations and the difference between the coupling phase and the repulsion phase, compared against independent gene loci, may be illustrated by the example of the inheritance of wheat resistance to smut given by Briggs and Knowles (1967).

It had been established that resistance to smut is controlled by two gene loci (*T* and *R*) which are linked by approximately 15% of crossing-over units. If both genes for resistance are present in a single variety (Turkey – 10016), i.e., the coupling phase *RT*/*RT*, and the genes for susceptibility are in another variety (Rio – *rt*/*rt*), their F_2 generation will contain 81.94% resistant plants possessing one of the two resistance genes or both of them (Table 4.5). Of the total number of resistant plants, 83.05% will possess both resistance genes (*R* and *T*) and 21.74% of them will be homozygous for both resistance genes.

If these two genes had been independent, i.e., if each of them was on a different chromosome, the F_2 generation would have had 93.75% resistant plants but only 56.25% would have included both resistance genes and only 6.25% would have been homozygous for both genes (Table 4.5).

Table 4.4 Effect of gene linkage on the proportion of *AB/AB* genotypes expected in F_2 from double heterozygotes *AB/ab* or *Ab/aB*

Recombination or crossing-over values	Percentage of *AB/AB* or *ab/ab* individuals in F_2 if F_1 was:	
	AB/ab (coupling phase)	*Ab/aB* (repulsion phase)
0.50	6,25	6,25
0,25	14,06	1,56
0,10	20,25	0,25
0,02	24,01	0,01
0,01	24,50	0,0025
p	$1/4(1-p)^2$	$1/4p^2$

Source: Briggs and Knowles, 1967

In the case when one gene for resistance is in one variety (*Rt/Rt*) and the other one in another variety (*rT/rT*), i.e., in the repulsion phase, then the number of desirable recombinations is quite different. Almost 99.5% F_2 plants are resistant, 50.5% contain both resistance genes, but only 0.56% are homozygous for both genes of resistance to wheat smut (Table 4.5).

Table 4.5 Proportion of the various wheat genotypes in F_2 where two genes for resistance are considered: (1) assuming linkage with 15 per cent of crossing-over when these enter the cross in both coupling and repulsion phase, and (2) assuming no linkage

Genotype	Phenotype	Coupling *RT/rt*, %	Repulsion *Rt/rT*, %	No Linkage *RrTt*, %
TTRR	Resistant	18.06	0.56	6.25
TtRR	Resistant	6.37	6.37	12.50
TTRr	Resistant	6.37	6.37	12.50
TtRr	Resistant	37.25	37.25	25.00
TTrr	Resistant	0.56	18.06	6.25
Ttrr	Resistant	6.37	6.37	12.50
ttRR	Resistant	0.56	18.06	6.25
ttRr	Resistant	6.37	6.37	12.50
ttrr	Susceptible	18.06	0.56	6.25

Source: Briggs and Knowles, 1967

This example shows how the linkage of gene loci affects the recombination of desirable traits as well as the fact that a larger F_2 population is needed with the repulsion phase than with the coupling phase in order to obtain these desired recombinations and select them.

Inheritance of quantitative traits

The majority of desirable traits, such as plant height, number of grains, protein content, percentage of oil, yield of dry matter, yield of grain, etc., fall in the group of quantitative or metric traits. These traits are as a rule controlled by several genes, each of them contributing to the formation and expression of the trait in question and usually acting additively. Such genes, because of less pronounced individual effects, are called minor genes. Since there are several of them acting together, they are also known as polygenes.

The development and the expression of a quantitative trait depends largely on environmental factors. These factors are capable of modifying the effect of polygenes. This is why differences exist within an inbred line in height, number of seeds, content of proteins, and especially the total yield. Since an inbred line is practically homozygous and it lacks genetic variability, the variability it shows is due to environmental factors, and it is expressed in the form of a normal frequency distribution curve.

Such a manifestation of quantitative traits complicates selection after hybridization because genetic and environmental variability are intertwinted in the progenies while the success of selection depends on gene effects and heritability of the trait selected. An example of the inheritance of plant height in wheat will illustrate gene effect and the occurrence of variability.

F_1 progenies of the wheat variety M-910, with a mean value of stem height of 131 cm, and the variety Accaio, with a stem height of 57 cm, had the average height

Table 4.6 Inheritance of stem height and frequency distribution in F_1 and F_2 in the cross of wheat varieties M-910 × Acciaio

	Classes for stem height in cm															
	46	50	54	58	62	66	70	74	78	82	86	90	94	98	102	106
1959																
M-910																
F_1												1	4	9	22	4
Acciaio	3	7	16	22	1	1										
1960																
M-910																
F_2		2	8	7	28	42	87	112	133	145	131	110	80	45	27	6
Acciaio	1	5	18	23	8	1										

of 102.7 cm which might be considered as partial dominance of the taller parent (Table 4.6). It should be mentioned that not all F_1 plants were 103 cm tall but rather varied in stem height like their parents (Fig. 4.2).

In the F_2 generation, the mean value for stem height was 84 cm, i.e., intermediate in relation to the parents. However, there were no clearly separated classes according to height; the variability in the F_2 generation was continuous and much higher than for the parents and the F_1 generation. The variability in the F_2 generation was primarily the result of the action of gene segregation and the effect of environmental factors went to the same extent as it had been done for the parents and the F_1 generation.

The action of genes for quantitative traits is usually additive, i.e., when effects of each of them are added, a phenotype is produced which represents the sum of positive and negative effects of individual genes. The additive action of genes and the effect of environmental factors on them render a continuous distribution of frequencies with the mean value equal to the parents' average.

Certain genes for quantitative characters may exercise a dominant effect. In that case, the mean value of the F_1 generation is closer to the mean value of either parent, and the distribution of frequencies in the F_2 generation does not show symmetrical but assymetrical continuous variation leaning towards the parent possessing the dominant genes. In the case of full dominance, the phenotypic value of F_1 equals the value of one of the parents.

The key of success in selection is genetic variability in the F_2 generation. In the example given above of stem height, the variability in the F_2 generation was high. As the parents were genetically quite distant in that trait, the prevalent part of the observed variability was due to genetic variability. The distribution of frequencies offered a possibility of selecting genotypes with the stem height quite different from that of Acciaio and all the way up to the height of M-910 (Table 4.6 and Fig. 4.2). In that particular case, we were interested in genotypes for stem height between 70 and 90 cm because they are sufficiently resistant to lodging and they are not too short,

110	114	118	121	126	130	134	138	142	n	$\bar{x}$	$m_{\bar{x}}$	σ	v
	3	6	2	9	15	17	5		57	130.9	0.84	6.38	4.9
1									41	102.7	0.57	3.67	3.8
									50	57.2	0.52	3.68	6.4
2	3	23	24	14	6	2			74	123.8	0.56	4.83	3.9
3	2								968	83.8	0.33	10.45	12.5
									56	58.6	0.43	3.27	3.0

Source: Borojević, 1965

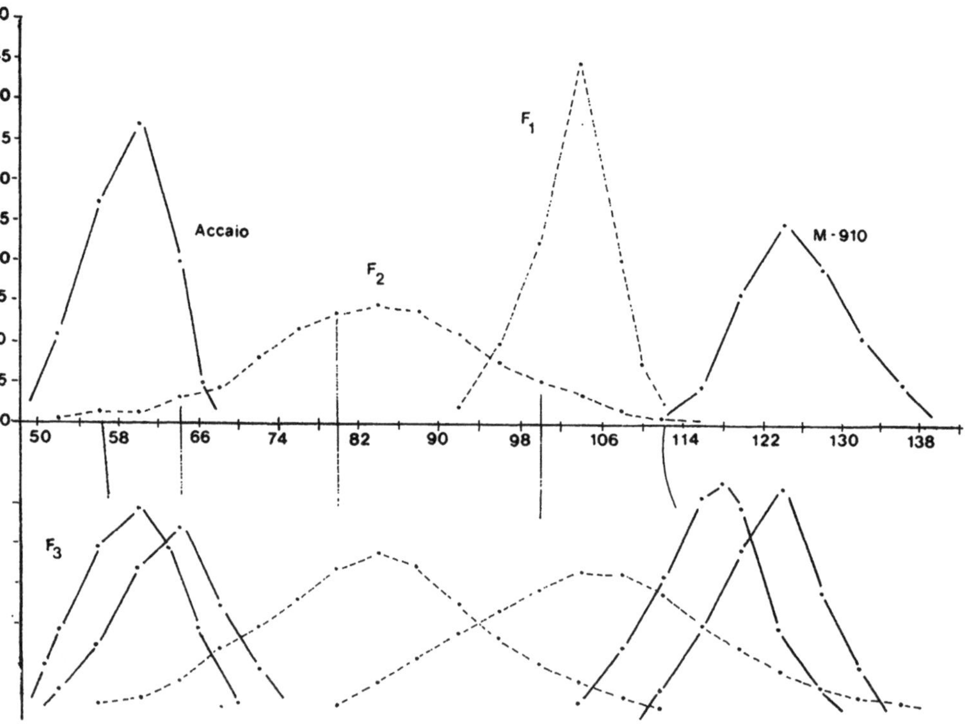

Fig. 4.2 Graphic presentation of results given in Table 4.6. The inheritance of the height of plants is partially dominant in the F_1 and intermediary in the F_2. Due to the effect of minor genes and environmental factors, the variation is continuous. In the F_3 the majority of plants are segregating like in the F_2, but also there are plants which do not segregate giving uniform progeny or line. Source: Borojević, S., and K., 1971

as is Acciaio, which encounters the problem of shading in the canopy which in turn affects negatively the photosynthetic activity and, ultimately, the yield of grain.

Owing to discoveries of cytogenetics and molecular genetics, it becomes evident that not all quantitative traits are controlled exclusively by minor genes; some of them appear to be controlled by major genes as well.

Ten major genes have already been determined to control stem height in wheat which is a typical quantitative trait. There is already a classical example of major genes for the reduction of stem height, so-called Rht genes, found in the Japanese drawf variety Norin 10 after crossing it with the tall Brevor variety (Allan, 1970). The tests by a series of monosomics showed that the genes Rht 1 and Rht 3 are located on chromosome 4A, Rht 2 and Rht 10 on 4D, and Rht 8 on 2D (Gale and Youssefian, 1985).

Major genes have also been identified for the stage of vernalization – Vrn, for the photoperiodic stage – Ppd (Pugsley, 1983), winter hardiness (Maystrenko, 1978). The position of leaves, horizontal versus erect leaves, is also controlled by at least two major genes, the horizontal position being dominant ever the erect (Borojević and Kraljević-Balalić, 1983).

The success in selection depends not only on the relationship between genetic and environmental variability but also on components of genotypic variability. The importance of individual components of variability and how they are analysed will be discussed in Chapter 6.

References

Allard, R. W., 1960. Principles of Plant Breeding. John Wiley, New York, 485 pp.

Allan, R. E., 1970. Differentiating between two Norin 10/Brevor 14 semi dwarf genes in a common genetic background. Seiken Ziho, 22: 83–90.

Borojević, S., 1965. (Mode of inheritance and heritability of quantitative characters in crosses of different wheat varieties). Contemporary Agriculture, (Novi Sad), XIII: 587–606.

Borojević, S., i Borojević, Katarina, 1971. Genetika. Kulturni centar, Novi Sad, 455 pp.

Borojević, S., and Kraljević-Balalić, Marija, *1983*. Inheritance of leaf architecture. Proc. 6th Inter. Wheat Genet. Symp. Kyoto, pp. 287–290.

Briggs, F. N., and Knowles, P. F., 1967. Introduction to Plant Breeding. Reinhold, New York, 426 pp.

Gale, M. D., and Youssefian, S., 1985. Dwarfing genes in wheat. In: G. E. Russel (Editor). Progress in Plant Breeding. Butterworth, London, p. 1–35.

Johannsen, W. L.: Elemente der exacten Erblichkeitslehre. Jena 1909 and 1926.

Mayo, O., 1980. The Theory of Plant Breeding. Clarendon Press, Oxford, 293 pp.

Maystrenko, O. I., 1978. Cytogenetic study of the growth habit and spike emergence time in wheat *Triticum aestivum L.* XIV Inter. Congress Genetics, Moscow, 1978.

Pugsley, A. T., 1983. Identification and management of major genes monitoring yield and adaptation. Proc. 6th. Inter. Genet. Symp. Kyoto, Japan, p. 971–974.

5. GENETIC BASIS OF BREEDING OPEN POLLINATED PLANTS

A large number of plant species reproduce by open pollination. Open pollination or allogamy is a system in which male sexual cells of one plant fertilize female sexual cells of another plant. Sexual cells are carried by wind, insects, or, in some instances, by water. Some plant species have completely separate sexes, as in animals, some have sexual organs separated on the same plant and some have bisexual flowers but which also have genetic mechanisms which prevent self pollination. So there are plant species which are decidedly open pollinating and on the other hand there are some which may remain open pollinated even in the system of self pollination. Here is the list of important open pollinated crops:

Pronounced open pollinating plants with bisexual flowers:

rye	red clover	lucerne (alfalfa)
buckwheat	alsike clover	
sunflower	white clover	
	hybrid clover	

Open pollinated plants with bisexual flowers and some degree of self-pollination:

sugar beet	chicory	lupin	apple
fodder beet	poppy		pear
rapeseed	potato	vetch	plum
		(some species)	
mustard	tobacco	orchard grass	
cabbage		brome grass	
		ryegrass	

Monoecious plants (male and female sexual organs separated on a single plant):

maize	water melon	walnut
	melons	chestnut
castorbean	pumpkin	

Dioecious plants (male and female sexual organs on separate plants):

hop	spinach	papaya
hemp	asparagus	date

Open pollinated plants which are mostly vegetatively propagated:

grapevine	potato
strawberry	hop.

This list shows that in the course of evolution, in order to ensure open pollination, many plant species evolved separated male and female flowers on a single plant (monoecious plants) or even went as far as to develop separate male and female plants (dioecious plants). Still, the majority of open pollinated plant species have bisexual flowers, but open pollination has been ensured or favoured in relation to self pollination. Special genetical, morphological, or physiological mechanisms have developed which cause incompatibility, (i.e., abortiveness in the case of the union of a male and a female gamete from the same flower), leading to autosterility.

It should be emphasized that progress in genetic studies led to the development of breeding methods by which it is possible to self pollinate typical open pollinated plants. Inbred lines have even been produced of sunflower, rye, etc.

Incompatibility

To ensure open pollination, especially in bisexual flowers, different barriers developed in the course of evolution to protect open pollinated plants from self-pollination. There are several causes for autosterility or incompatibility, to mention some of them:

protandry – a condition in which stamens mature before the stigma in the same flower preventing pollen grains from pollinating the stigma;

protogyny – a condition in which the stigma matures before stamens with the same result as above;

hercogamy – physical arrangement of male and female organs on the same plant preventing self pollination in the absence of an insect visit.

In addition to those morphological and mechanical mechanisms which ensure open pollination, there exist genetical and physiological mechanisms which ensure incompatibility; e.g., in a large number of varieties of apple, pear and almond, many varieties of plum, cherries, black cherry and olive, in rye, sugar beet, lucerne, some varieties of grasses, tobacco, etc.

Lewis (1954) divides incompatibility in plants into two groups: heteromorphic and monomorphic.

Heteromorphic incompatibility

This system of incompatibility is based on the difference in length between the stamens and the style, this is known as *heterostyly*. Lewis named the type of flower with a long style and short anthers – a pin, and the other type of flower with long anthers and a short style – a thrum. The genetical constitution of the pin type is *ss* and of the thrum type *Ss*, *S* being fully dominant over *s* (Fig. 5.1).

The pollination is possible only between the following genotypes:

(1) pin (*ss*) × thrum (*Ss*)
1 pin (*ss*) : 1 thrum (*Ss*)

(2) thrum (*Ss*) × pin (*ss*)
1 pin (*ss*) : 1 thrum (*Ss*)

The combination pin × pin and thrum × thrum are incompatible which means that genotypes SS do not occur (Fig. 5.1).

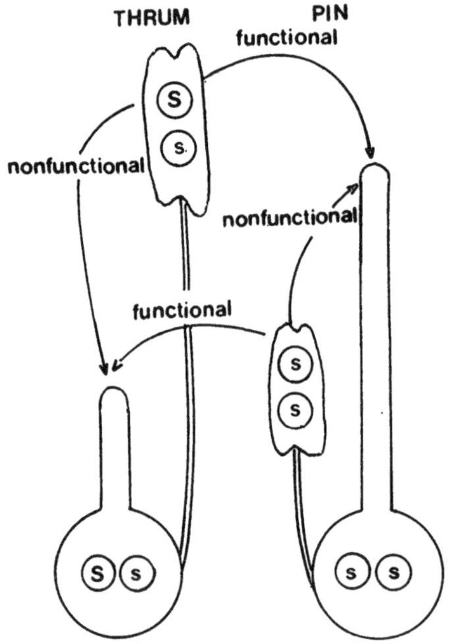

Fig. 5.1 Heteromorphic system of incompatibility. Also sporophytic incompatibility, since S is dominant to s, and both S and s pollen of plants with genotype Ss behave the same. Both flowers are self-sterile, and in crossing produce equal numbers of SS (thrum) and ss (pin) genotypes.
Source: Briggs and Knowles, 1967

Homomorphic incompatibility

Homomorphic incompatibility includes gametophytic and sporophytic incompatibility.

Gametophytic incompatibility: This type of incompatibility was first found in *Nicotiana sanderae*, by East and Mangelsdorf (1925). Later on it was found in red clover, white clover, and several other forage legumes. The ability of pollen to fertilize depends on a series of alleles in a single locus, designated S_1, S_2, S_3, ... S_n. Diploid stylar tissue will have any two alleles of the series but never the same two. A single pollen grain will have only one allele because it is haploid. If the stylar tissue contains an identical allele as the pollen grain then the pollen tube cannot penetrate into the stigma and pollination cannot take place because pollen or tube growth is inhibited. Conversely, pollination may take place if the allele in the pollen grain differs from those of the stigma and the ovary. The fertilization capacity of pollen depends on its own genotype. Therefore, the crossings of plants with different genetic constitutions may be: (1) completely incompatible, (2) one half of the pollen compatible, and (3) fully compatible (Table 5.1).

Sporophytic incompatibility: This type of incompatibility is also conditioned by a series of alleles in a single locus and, unlike in the gametophytic type, the functionality of pollen is determined by the genetic constitution of the plant producing it (Table 5.1).

Although incompatibility creates difficulties because it precludes the possibility of developing inbred lines to exploit the advantages of heterosis, attempts can be made to make use of the incompatibility phenomenon, especially in breeding vegetables and flowers. More will be said about it in the chapter on breeding methods.

Particular attention should be paid to incompatibility in fruit growing where poor pollination may cause yield reduction. It is counteracted by developing varieties which are good pollinators. Since pollinating insects may fail to be active in bad weather conditions, attempts have been made to induce autofertility alleles (Lewis, 1954).

Frequency of genes and genotypes

Populations of open pollinated plants, due to the system of mating, are governed by regularities different from those for self-pollinated plants. Namely, the system of open pollination is based on random mating which offers constant possibilities

Table 5.1 Three levels of cross-compatibility under the gametophytic and sporophytic systems of self-incompatibility

System	Cross Female Male	Male Gametes Functional	Male Gametes Non-functional	Genotypes of Progeny
Gametophytic				
Both plants with the same genotype: no pollen functional	$S_1S_2 \times S_1S_2$	None	All	None
Plants differing in one allele: one-half of pollen functional	$S_1S_2 \times S_1S_3$	S_3	S_1	S_1S_3, S_2S_3
	$S_1S_3 \times S_1S_2$	S_2	S_1	S_1S_2, S_2S_3
Plants differing in both alleles: all pollen functional	$S_1S_2 \times S_3S_4$	S_3, S_4	None	S_1S_3, S_2S_3, S_1S_4, S_2S_4
	$S_3S_4 \times S_1S_2$	S_1, S_2	None	S_1S_3, S_2S_3, S_1S_4, S_2S_4
Sporophytic[a]				
Both plants with the same genotype: no pollen functional	$S_1S_2 \times S_1S_2$	None	All	None
Plants differing in one allele: all or none of pollen functional	$S_1S_2 \times S_2S_3$	None	S_2, S_3	None
	$S_2S_3 \times S_1S_2$	S_1, S_2	None	S_1S_2, S_1S_3, S_2S_2, S_2S_3
Plants differing in both alleles: all pollen functional	$S_1S_2 \times S_3S_4$	S_3, S_4	None	S_1S_3, S_2S_3, S_1S_4, S_2S_4
	$S_3S_4 \times S_1S_2$	S_1, S_2	None	S_1S_3, S_2S_3, S_1S_4, S_2S_4

[a] Assuming the order of decreasing dominance in the pollen to be S_1, S_2, S_3, and S_4, and no dominance in the style.
Source: Briggs and Knowles, 1967

of genetically different gametes being fertilized. As will be explained next, the system of open pollination, in addition to producing homozygosity, constantly maintains the state of heterozygosity while the system of self pollination, as the number of generations increases, leads to almost complete homozygosity. As was explained previously in Chapter 4, this is why the effect of selection on open pollinated populations is different from the effect it has on self pollinated populations, and why the breeding methods for the former must inevitably be different.

A population of an open pollinated plant species comprises homozygous genotypes (*AA* and *aa*) as well as heterozygous genotypes (*Aa*). The proportion between the genotypes may change, depending on the frequency of *A* and *a* genes. Therefore, the populations of open pollinated organisms consist of certain frequencies of genotypes and frequencies of genes.

Hardy (1908) in England and Weinberg (1908) in Germany discovered that an equilibrium between genes and genotypes is achieved in large populations. The equilibrium may be maintained from generation to generation under conditions of random mating and equal fertility of all genotypes present in the population. The frequency of genotypes in a population depends on the frequency of genes in the preceeding generation, not on the frequency of genotypes. After the first generation of open pollination, the frequency of genotypes stabilizes, i.e., an equilibrium is reached for any initial frequency of genes in the case of a single locus. The equilibrium is maintained until the frequency of genes in the population is changed.

If, in a population, gene *A* has the frequency p and gene *a* the frequency q, under the conditions of random mating, i.e., that any individual has equal chance of mating with any other individual in the population, the genotype frequencies in the progeny will be:

♀		♂ p *A*	q *a*
p	*A*	p^2AA	$pqAa$
q	*a*	$pqAa$	q^2aa

The population is in equilibrium at the following frequencies of genes and genotypes:

$$p^2AA + 2pqAa + q^2aa = 1 \text{ or } 100\%.$$

This is the general formula of Hardy-Weinberg's law of equilibrium.

If the frequency of genotypes differs from that of the initial population, it is evident that the latter population was not in equilibrium. In the system of random mating, if the frequency of genes is not changed, an equilibrium in the frequency of genotypes is achieved in the course of one generation regardless of the frequencies of genotypes of the parents.

Changes in frequency of genes in a population

Although a large random mating population is relatively stable, the state of equilibrium is more of an exception than the rule because factors of disequilibrium are constantly at work. The most important factors which change the frequencies of genes and genotypes are migration, mutation and selection. Their action may be individual although it is usually combined.

Migration

Migration is the intrusion of new individuals into an existing population. The phenomenon is less pronounced with plants, because of their stationary character, than with animals and especially insects, which are mobile. But even with plants, wind and animals carry seeds to considerable distances, bringing immigrants to populations settled in an area. Pollen transfer, by wind or by insects, is also a form of migration from one population to another – it is actually a migration of gametes, i.e., genes, which can cause the development of new genotypes in the visited population.

Changes in the frequency of genes (Δq) in the new mixed population depend on the number of immigrants (m) and the difference between the frequencies of the genes of the immigrants (q_m) and the genes of the hosts (q_0), and, according to Falconer (1981) may be calculated as follows:

$$\Delta q = m(q_m - q_0).$$

In practical breeding of open pollinated plants it is necessary to keep in mind any eventual arrival of immigrants to the experimental population. Their frequency may be low but they may also cause considerable changes in the frequencies of genes and genotypes in the experimental population.

Mutation

Spontaneous mutations occur in natural and artificial populations at the frequency of about 10^{-6} per gene per generation, as explained in Chapter 3 when discussing the sources of genetic variability.

Considering the low frequency of mutations, they are a factor which may be disregarded in breeding programmes which last a relatively short period of time. If, however, a lethal or semilethal mutation is encountered (e.g., chlorophyll defects, etc.), the harmful gene should be eliminated by selection. Conversely, if a positive mutation is detected (e.g., disease resistance, etc.), it should be preserved for further work.

Selection

A population comprises different genotypes; some of them survive better and produce more progenies than others. Changes in the frequency of genes which take place in a population as a result of differences in the viability and fertility among its genotypes may in fact be defined as natural selection. More fertile genotypes produce more progenies, thus increasing their gene frequencies and causing the population to be in a disequilibrium. The contribution of genotypes to the next generation is called the *fitness* of and individual, *adaptive value*, or *selective value* (Falconer, 1981).

Selection may occur at the level of phenotype, genotype, zygote and gamete. Thus, diploid and haploid selections are distinguished. Regardless of the level of occurrence, selection unanimously brings about changes in the frequency of genes. The effect of selection depends on the frequency of desirable genes, the effect of genes and the intensity of selection.

Selection for homozygotes: If selection is done for gene A, which is fully dominant, genotypes AA and Aa must be dealt with because they are phenotypically undistinguishable. In the case of open pollinated plants and random mating, the following genotypes will mate – $AA \times AA$, $AA \times Aa$, and $Aa \times Aa$, and the following progenies will be produced – AA, AA, Aa, AA, Aa, Aa, and aa. It is evident that the effect of selection is not as rapid as it is the case with self-pollinated plants. In addition to genotypes AA, there occur heterozygous genotypes Aa and recessive genotypes aa which are not selected for but which keep developing from the crossing of genotypes $Aa \times Aa$. If the initial frequency of gene a is high and the intensity of selection low, the gene will persist in the population for a long time although it will ultimately be eliminated. If its initial frequency is low and the intensity of selection (s) high, the gene will be eliminated fairly quickly.

If a gene is completely eliminated from a population by selection, then $s = 1$ and $q = 0$. In the case of complete selection against recessive genes, the change in the frequency of genes is:

$$\Delta q = -q^2/1 + q.$$

If the initial frequency q is high, the frequency of genes drops down abruptly. If it is low, the nominator $1 + q$ is almost 1 and the changes in each generation are almost equal to $q2$. With low frequencies ($q = 0.01$), therefore, the changes of q per generation are 0.0001, i.e., only 1 out of 10,000 genes are lost through selection. Thus, the selection against recessive genes is a very slow process if their frequency in the population is low.

When breeding open pollinated plants, the question is often asked how many generations are needed to eliminate a lethal recessive gene (e.g., albinism) from a population. If, at the selection intensity of $s = 1$ and hypothesis that there is no mutation, q_0, q_1, q_2 and q_n are put for the frequencies of genes after the 0, 1st, 2nd, and n-th generation, the calculation is:

$$q_1 = q_0/1 + q_0,$$

$$q_2 = q_1/1 + q_1 = q_0/(1 + q_0)/1 + (q_0/1 + q_0) = q_0/1 + 2q_0,$$

$$q_n = q_0/1 + nq_0.$$

The last formula is used for the calculation of the number of generations of selection necessary to achieve a certain change in gene frequency. To change the gene frequency from 0.10 to 0.01, 90 generations are needed at $s = 1$ and 185 generations at $s = 0.50$ (Strickberger, 1968; Falconer 1981).

Selection for favourable heterozygotes: Heterozygotes *Aa* are superior in fitness to homozygotes (*AA* and *aa*) in a large number of natural and artificial populations. If the fitness of genotypes *Aa* is 1, the fitness of homozygotes *AA* is lower by the selection coefficient designated s, and the fitness of homozygotes *aa* is lower by coefficient t. Changes in the frequency of genes occuring in one generation are calculated in the following manner:

	AA	*Aa*	*aa*	Total	Frequency *a*
Initial frequency	p^2	$2pq$	q^2	1	q
Adaptive value	$1 - s$	1	$1 - t$		
Frequency after selection	$p^2(1 - s)$	$2pq$	$q^2(1 - t)$	$1 - p^2s - q^2t$	$\dfrac{q(1 - qt)}{1 - p^2s - q^2t}$

The change in the frequency of genes is then calculated:

$$\Delta q = pq(ps - qt)/1 - p^2s - q^2t.$$

If $\Delta q = 0$, an equilibrium is reached and the frequency of genes does not change. This happens when $ps = qt$. When it does happen, the equilibrium values may be expressed as follows:

$$ps = qt,$$

$$\bar{q} = s/s + t, \quad \bar{p} = t/s + t.$$

These equilibrium values are independent of the initial frequency of genes in a population and are defined only by the selection coefficient against the homozygotes. It means that if s and t are larger than 0 and their values are constant, p and q will attain a stable equilibrium. In that case, both alleles, *A* and *a*, will stay in a population, at the frequencies q and p, as long as the selection favours heterozygotes over homozygotes.

This effect of selection maintains genetic variability in a population which in turn brings about genic polymorphism, a phenomenon of the maintenance of two or more alleles in a single locus in a population (at frequencies of about 1%).

Effect of selection of quantitative traits

The main effect of selection is a change in the frequency of genes in a population; in natural selection, the change is always related with fitness.

Since quantitative traits are controlled by a large number of genes and since they are affected by environmental factors, the effect of selection on changes in gene frequency is expressed in changes of phenotypical values as mean values of the trait in question, its variance and covariances.

There are three basic modes of action of natural selection affecting quantitative traits of a population: stabilizing, directed and disruptive selection (see Falconer, 1981).

Stabilizing selection: Stabilizing selection (also called optimum, balancing, or selection for mean value) is manifested in individuals possessing mean values for quantitative traits but highest fitness. The progenies of these individuals retain the mean values but the variance is reduced because of the loss of a portion of both of the extreme variants (Fig. 5.2).

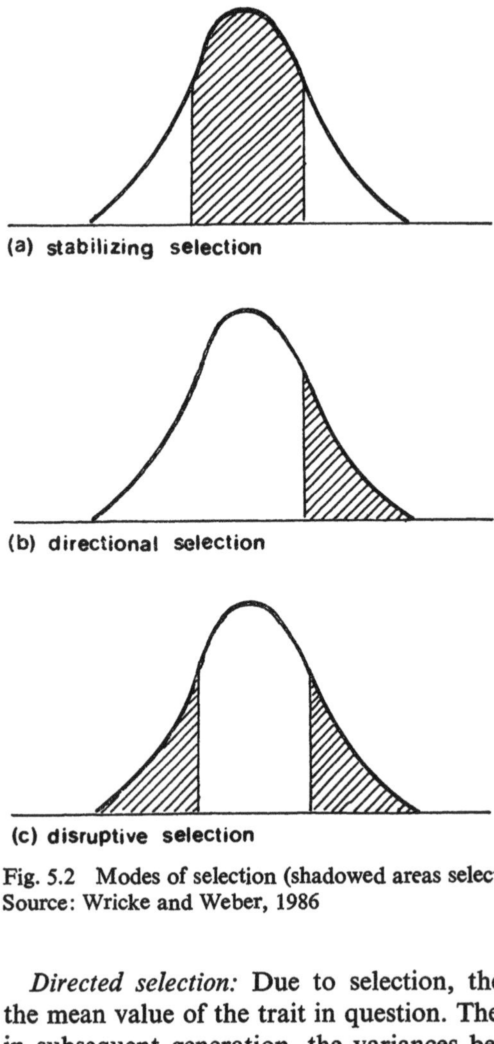

(a) stabilizing selection

(b) directional selection

(c) disruptive selection

Fig. 5.2 Modes of selection (shadowed areas selected).
Source: Wricke and Weber, 1986

Directed selection: Due to selection, the fitness may increase or decrease with the mean value of the trait in question. The directed selection increases mean values in subsequent generation, the variances being changed or remaining unchanged.

Disruptive selection: If extreme variants in a population have a higher adaptive value than the variants around the average mean values, there occurs disruptive selection for plus and minus variants. The mean value may remain unchanged in subsequent generations but the variance increases.

In all these cases the success of selection depends on the heritability of the quantitative trait in question, i.e., to which degree the phenotype reflects the genotype.

In plant breeding, directed selection is most frequently used, in which individuals with the best mean values are selected as parents for subsequent generations. It happens frequently that the selection for an economically important trait is limited by the poor vitality of extremely selected phenotypes. This is the consequence of the reaction of the stabilizing natural selection which counteracts the effects of the artificial selection. The disruptive selection is not frequently used for artificial populations; in natural populations it contributes to genetic polymorphism.

Small population

The phenomena and principles discussed above are valid for a large population (containing several thousand individuals) in which the frequencies of genes and genotypes remain unchanged from one generation to another provided that they are not affected by selection, mutation, and migration. These forces are systematic and constant, acting either in the direction of gene fixation or equilibrium. This is why it is possible to forecast the direction of their action in natural populations.

In breeding work, however, one usually deals with a small number of individuals, i.e., with a small population which contains only a portion of the genetic variability of the source population. The small population is just a sample of the large population and it exhibits many irregularities and specific characteristics in relation to the regularities which govern the large population.

Because of the selection of representative genotypes, i.e., gametes, from the large population, the small population undergoes changes in gene frequency from one generation to another. In other words, the small population is not stable because it is the medium for dispersive processes which lead to:

(1) the division of the population into sub-populations, groups, or lines;
(2) reduction in genetic variability and increase in homozygosity.

Since in sub-populations mating between similar individuals takes place, (i.e., inbreeding), genetic variability decreases within the groups, some genes are fixed and some eliminated, and homozygosity increases. Wright (1951) named this process of dispersal of groups or lines *random drift*.

The development of different local populations of open pollinated plants is an example of the dispersive processes at work in natural populations. Brieger (1950) found a number of local populations of maize in South America. These populations were quite distinguishable among themselves but each population was highly uniform. The same regularity is seen in Yugoslavian local populations of rye, especially those coming from mountain regions. Because of spatial isolation and growing them in karstfields, harvesting by hand, etc., these populations had split into a number of smaller populations in which the inbreeding was intensified resulting in increased homozygosity and uniformity. So, the rye population from the region of Lika is uniform but considerably different from the population from the region of Glamoć and even more so from the population from the region of Nevesinje although all three regions are located within a perimeter of 200 kilometers (Borojević, 1957).

The same or similar dispersive processes take place in other open pollinated species of grasses, legumes and other plants. This is what makes local populations interesting for breeding.

Gene fixation

Dispersive processes in a small population are effective only up to a certain limit since the frequency of genes cannot go beyond 0 and 1. There is no return from the limit values, which, once reached by a line, cannot be changed unless by mutation. When an allele attains the frequency of 1, it is fixed in that population; when it attains the frequency of 0, it is irretrievably lost. When a gene is fixed, all individuals in a line have the same genotype regarding that gene locus. When all loci in a line are fixed, the line becomes genetically uniform, i.e., a new inbred line is made.

Here is an example of gene fixation. If each one of one hundred populations starts with two heterozygotes $Aa \times Aa$, the progenies in the first generation will consist of three genotypes: *AA*, *Aa*, and *aa*, in the ratio 1 : 2 : 1. If only two parents are selected to make the next generation, any combination among the three genotypes may occur. However, the probability of occurrence of a certain combination depends on the frequency of the genotypes, as shown in Table 5.2.

The probability that two parental genotypes *AA* will be selected is 1/16. The same stands for the probability of selecting two parental genotypes *aa*. It means that genes *AA* will be fixed in 6 out of the 100 populations and genes *aa* will be fixed in another set of six populations. The remaining 88 populations will each be made up from combinations derived from the following crosses: the following combinations: $AA \times Aa$, $AA \times aa$, $Aa \times Aa$, and $Aa \times aa$. In the next generation (derived from these 88 populations) combinations made by crossing the three genotypes *Aa*, *Aa*, *aa* may again produce further populations homozygous for both alleles so that there will be 15 of them population or $15/100 = 15\%$. The process will continue in subsequent generations until the number of combinations achieving fixation is 1/2 N (N being the number of parents per generation if the process was started with two parents and 100 populations). It means that all one hundred initial populations, after a large number of generations, will consist of groups which achieved fixation, their frequencies of genes ranging between the groups *AA* and *aa* in proportion to the values of p and q in the initial population.

Thus the rate of gene fixation depends not only on the frequency of genes but also on the number of parents in a population. If N equals 100,000 and the initial frequency of genes is 0.5, only a slight random drift will occur after a long period of time. At the same initial frequency and a small size of population, $N = 1{,}000$, alleles are fixed or eliminated in a short time (Wright, 1951).

Table 5.2 Possible combinations of genotypes in a population where three genotypes *AA*, *Aa* and *aa* are in 1:2:1 ratio

Mating of genotypes	Probability	Frequency of genes *A*	*a*
AA×*AA*	1/4×1/4 = 1/16	1.00	0
AA×*Aa* *Aa*×*AA*	2×1/4×1/2 = 4/16	0.75	0.25
AA×*aa* *aa* ×*AA*	2×1/4×1/4 = 2/16	0.50	0.50
Aa *Aa*	1/2×1/2 = 4/16	0.50	0.50
Aa ×*aa* *aa* ×*Aa*	2×1/2×1/4 = 4/16	0.25	0.75
aa ×*aa*	1/4×1/4 = 1/16	0	1.00

Thus the number of parents is an important parameter for the calculation of random drift and it is expressed as *effective population size*, N_e (Falconer, 1981). The effective population size differs from the actual size because not all members of the populations are parents which leave behind their progenies. In the case when out of 1,000 individuals only 400 pairs mate, the effective population size is 800. If the ratio between sexes is not 1 : 1, the effective population size is again diminished. When there are 10 males to 200 females, the effective population size is larger than 10 but much smaller than 200, presumably because mating is not quite random.

Such phenomena do take place in natural conditions and they should be also taken into account when dealing with laboratory populations and small populations of plants.

The dispersion of a basic population into a large number of small groups (lines) intensifies the process of inbreeding which results in an increase of homozygosity and a decrease of heterozygosity. An increase of homozygosity in a small population may be expressed through variances and inbreeding coefficient (see Wahlund's law, Strichberger, 1968).

In all examples given in previous paragraphs we described the frequency of genotypes for a single gene locus in several lines. These regularities pertain also to a large number of gene loci in a single line. Thus an increase in the number of homozygous gene loci and a proportional decrease in the number of heterozygous loci occur in each line of a small population.

References

Borojević, S., 1957. (Biological and agronomical properties of Yugoslav local varieties of rye (*Secale cereale L.*). Polj. znanstvena smotra, (Zagreb) 16: 5–28.

Brieger, F. J., 1950. The genetic basis of heterosis in maize. Genetics 35 (4).

Briggs, F. N., and Knowles, P. F., 1967. Introduction to Plant Breeding. Reinhold, New York 426 pp.

East, E. M., and Mangelsdorf, P., 1925. A new interpretation of the hereditary behaviour of self-sterile plants. Proc. Nat. Acad. Sci. (USA). 11: 166–171.

Falconer, D. S., 1981. Introduction to Quantitative Genetics. 2nd edition. Longman, London, 340 pp.
Hardy, G. H., 1908. Mendelian proportions in a mixed population. Science, 28: 49–50.
Lewis, D., 1954. Comparative incompatibility in angiosperms and fungi. Adv. Genetics, 6: 235–285.
Strickberger, M. W., 1968. Genetics. Macmillan, New York. 865 pp.
Weinberg, W., 1908. Über den Nachweis der Vererbung beim Menschen. Jahreshefte Verein f. Vaterl. Naturkunde in Württemberg, 64: 369–382. (Translated by C. Stern, 1943. The Hardy-Weinberg Law. Science, 97: 137–138).
Wricke, G., and Weber, W. E., 1986. Quantitative Genetics and Selection in Plant Breeding. Walter de Gruyter, Berlin, 406 pp.
Wright, S., 1951. The genetic structure of populations. Ann. Eugenetics 15: 323–354.

6. TRAIT, GENOTYPE, AND PHENOTYPE

The previous chapters briefly reviewed the fundamentals of plant breeding from the genetic aspect and reviewed the differences between self-pollinated and open pollinated plants. Naturally, the gene is the basic unit of inheritance which determines the direction of a certain process and, ultimately, a certain trait. However, breeding is not aimed at a gene or genes but a specific trait – a phenotype. Since inheritance is most important for breeding, it is necessary to establish connections between a gene and a trait, a genotype and a phenotype, and a genotype and environmental factors.

Trait

In genetics and even more so in the breeding of organisms, the expression "trait" serves to make a distinction between individuals or between cultivars. Some of the traits are flower colour (red or white), plant height (tall or short), disease resistance (resistant or susceptible), yielding capacity (high-yielding or low-yielding), etc.

Thus, a trait is a certain characteristic which defines a phenotype. Each genotype possesses a large number of traits. However, it is not always simple to agree on the definition of a trait because a geneticist considers it from a different angle than does a breeder or a biochemist.

In the background of each trait there is a gene or genes which controls the development of the trait. This is the genetic side of a trait which determines a genotype. However, the formation of a trait is also affected by encironmental factors which introduce variation and modify the trait. As has already been emphasized by Johannsen (1909) a phenotype is the product of the combined action of a genotype and the environment, and the expression of each trait is conditioned by genetic factors and environmental ones. Actual contributions of the two sets of factors to the makeup of a trait differ from one trait to another and they are not always easily distinguished. For breeders, the genetic component is of primary importance. They are interested in what is inherited and what is transmitted to progenies, and this is, especially in the case of the quantitative traits, more or less obscured by uninheritable variability resulting from the action of the environment.

For example, a man is always impressed by the size of spikes, ears, or fruits on the plants in border rows and he cannot resist the temptation of selecting them. Next year, the progenies of these "superior" plants to not surpass the progenies of plants selected from the middle of the plot. It means that a modification had been selected which occurred in consequence of a larger growing area, more solar radiation, etc., i.e., as a consequence of the uninheritable variability which is not transmitted to progenies.

The extent of the use of the genetic component of variability in the selection for a trait depends on the effect and number of genes controlling that trait as well as on the effect of environmental factors.

A trait controlled by major genes, such as flower or fruit colour, form of flower, leaf, fruit, branch, etc., may be visually distinguished and it usually is repeated in progenies. However, if a major gene has a dominant effect, similar phenotypes may not have the same genotype. For example, we select two plants with red flowers, i.e., with the same phenotype. All progenies of one of them may have red flowers and the progenies of another one may have red and white flowers. It means that the first plant was homozygous (*CC*) and the second one was heterozygous (*Cc*) for the red color of the flower.

In the case of a large number of genes controlling a trait, apparently similar phenotypes may contain different gene loci. For example, the round form of pumpkin is controlled by genes *AAbb* in some cultivars and by genes *aaBB* in others. There are many other types of gene interaction for various traits which have been mentioned in Chapter 3 when cources of genetic variability were discussed.

Several gene loci may control the development of a single trait but also a single gene locus may affect several traits. Such gene loci are said to exhibit a pleiotrophic effect. For example, there exist a gene locus which controls the violet colour of the stem and sepals in *Primula sinensis* and this is also the case in some other plants.

The shorter the connection between the primary action of a gene and its expression in a phenotype, the easier it is to perceive the gene – trait relationship. The sicle cell anaemia in humans results from a replacement of a base in the codon (*GAA* by *GUA*), i.e., valine replaces glutamic acid in position 6 of the beta chain which brings about a change in haemoglobin. In that case there exists a direct connection between the gene and the trait. When breeding plants for a large number of traits, however, there is a long way between the primary gene action and the actual expression of the traits in the phenotype. There also occur interactions with other genes among which some affect the traits at one stage of development and other traits at a later stage. When we add here environmental effects which modify gene action, it becomes obvious that it is difficult to perceive a direct connection between genes and traits, especially quantitative traits such as protein content, fruit size, grain yield, etc.

When such quantitative traits are concerned, the trait becomes more of an agronomic or breeding category than of a genetic category. The more important a trait is economically, the more complex is the approach required. The yield in fact should not be considered a single trait but rather a group of traits. Grain yield of small grains, for example, is composed of the following components: number of plants (spikes)/m^2, number of grains per spike, and absolute grain mass. Each component may be considered separately, but when put together, their final product is grain yield. Mather and Jinks (1971) refer to these yield components as sub-traits and to the yield as a super-trait or super-character.

Gene effects and environmental factors bring about a continuous variation of quantitative traits. To distinguish between genetic and non-genetic variability, it is absolutely necessary to test the progenies. Even that may not be enough and it may be that a series of experiments is required in order to recognize genetic interactions and the interactions with the environment within the total phenotype variability.

Phenotype and components of phenotypic variability

Since a quantitative trait of a phenotype is determined by measurement, the obtained value represents the phenotypic value of the analysed individual. This is, in fact, a compound value consisting of the genotypic value of the individual and any deviation due to the action of environmental factors. According to Mather (1949) it may be expressed in the following manner:

P (phenotype) = G (genotype) + E (environmental effect).

Individuals differ in phenotypic values. These differences are conditioned by genetic differences among the individuals, environmental factors, and interactions between the genotype and the environment. Therefore, the phenotypic value is variable and it includes components which may be determined by the analysis of variance. The phenotypic variance consists of the genotypic variance, environmental variance and their interactions:

$$V_P = V_G + V_E + V_{GE},$$

where V_P is phenotypic variance, V_G is genotypic variance, and V_E is environmental variance.

The source of genotypic variance (V_G) is the genetic constitution of the quantitative trait itself. If the genes express an additive effect, the genotypic value of the trait may be increased or decreased by replacing one of them. For example, if A_1A_1 has the value of 6 cm, and A_1A_2 = 7 cm, it means that the presence of the gene A_2 causes a change of 1 cm. On the other hand, some genes express the dominant effect so that the presence of an allele may increase the genotypic value. In that case genotype A_1A_2 will have the value of 8 cm instead of 7 cm. Also, there may occur an interaction among different gene loci which is known as the phenomenon of epistasis. For example, *Aa* expresses the additive effect when it is together with *BB*, but the dominant effect when with *bb*, as shown by genotypic variance which comprises the following components:

$$V_G = V_A + V_D + V_I.$$

It means that the genotypic variance (V_G) consists of the variance resulting from the additive gene effect (V_A), the variance resulting from the dominant gene action (V_D), and the variance resulting from the interaction between the genes (V_I).

The phenotypic variance (V_P) consists of:

$$V_P = V_A + V_D + V_I + V_E + V_{GE}.$$

The values of individual components may be estimated experimentally. If all individuals belong to the same genotype, the variances observed in the experiment may be attributed to enviromental factors. An inbred line, which is practically homozygous, has the same genotype. If the F_1 generation is made by crossing two inbred lines, it will be genetically uniform although heterozygous. Thus the variances of the parents and the F_1 may be taken as a measure of environmental variance (V_E).

The components of genotypic variance may be analysed via the variances in the F_2 and later generations of back crossing, as elaborated by Mather (1949). Since the F_2 is a segregating generation, it contains the variance of each genotype as well as the variance resulting from the action of environmental factors. In the case of a single pair of allelic genes (A_1 and A_2), three genotypes will develop in the F_2 and their ratio will be:

$$1/4\, A_1A_1 : 1/2\, A_1A_2 : 1/4\, A_2A_2.$$

Each genotype has its own genotypic value which represents a deviation from the parental mean,

$$A_1A_1 = -a, \quad A_1A_2 = d, \quad \text{and} \quad A_2A_2 = a \text{ (Fig. 6.1).}$$

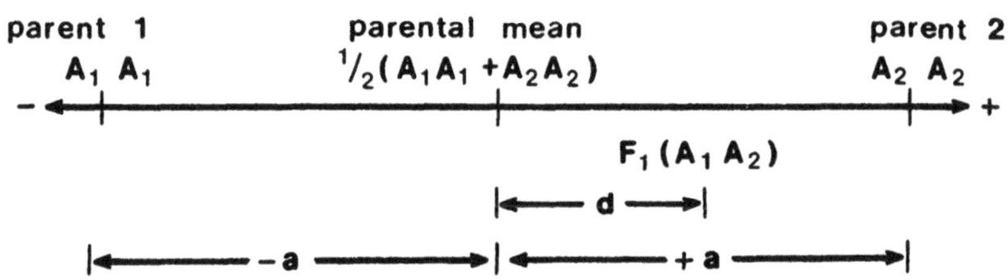

Fig. 6.1 Quantitative values of the parents and deviations from these values brought by additiveness (a) and dominance (d). When dominance is absent, the value of F_1 generation equals the parental mean. When the value of dominance is between 0 and a, the value of F_1 is between the parental mean and the value of one of the parents. When $d = a$, dominance is complete and the value of F_1 equals the phenotypic value of one of the parents.

If these values are put in the given relationship, the mean value of F_2 will be:

$$1/4(-a) + 1/2(d) + 1/4(a) = 1/2d.$$

Since the variance of each genotype is equal to the square of the deviation from the mean value multiplied with its frequency $f(x - x)^2$, the total variance of F_2 is:

$$1/4(-a - 1/2d)^2 + 1/2(d - 1/2d)^2 + 1/4(a - 1/2d)^2 =$$
$$= 1/4(a^2 + ad + 1/4d^2) + 1/2(1/4d^2) + 1/4(a^2 - ad + 1/4d^2) =$$
$$= 1/2a^2 + 1/4d^2.$$

If a^2 is replaced by A and d^2 by D and the components of environmental variance (E) are added, it comes out that the total variance of F_2 is:

$$V_{F2} = 1/2A + 1/4D + E.$$

These components represent in fact the variances resulting from additive (V_A), dominant (V_D), and environmental effect (V_E).

Therefore, the components of variance for the parents (P_1 and P_2), the generations which are usually assessed (F_1 and F_2), back crosses with one parent (B_1) and another parent (B_2) may be expressed as follows:

$V_{P_1} = E$

$V_{P_2} = E$

$V_{F_1} = E$

$V_{F_2} = 1/2A + 1/4D + E$

$V_{B_1} = 1/4A + 1/4D + E$

$V_{B_2} = 1/4A + 1/4D + E$

$V_{B_1} + V_{B_2} = 1/2A + 1/2D + 2E.$

The method of calculation of components of variances will be illustrated by the example of the inheritance of the number of spikelets per spike in crosses between the Mexican wheat variety Siete Cerros and the Soviet variety Bezostaia-1 (Table 6.1).

Table 6.1 Mean values ($\bar{x}$) and variances (σ^2) for the number of spikelets per spike in the crosses of two wheat cultivars

Parents and hybrids	No. of plants analysed	No. of spikelets per spike $\bar{x}$	σ^2
P_1 (Siete Cerros-*SC*)	40	19.4	0.70
P_2 (Bezostaja-1 — *B*)	40	20.8	0.53
F_1 ($SC \times B$)	40	19.9	0.58
B_1 ($F_1 \times SC$)	40	19.3	1.30
			0.98
B_2 ($F_1 \times B$)	40	19.7	0.72
F_2 ($F_1 \times F_1$)	500	20.2	1.34

Source: Malek, 1976

First it is necessary to calculate the environmental variance which is contained in the variances of the parents and the F_1 generation:

$E = (V_{P_1} + V_{P_2} + V_{F_1})/3 =$

$= (0.70 + 0.53 + 0.58)/3 = 0.60.$

If the value of $E = 0.60$ is subtracted from the total variance of $F_2 = 1.34$ and $2E$ (2×0.60) from the average variance of the back crosses (2×0.98), there remain only the variances resulting from additive and dominant effects:

$V_{F_2} - E = 1/2A + 1/4D + E - E = 1/2A + 1/4D = 1.34 - 0.60 = 0.74,$

$V_{B_1} + V_{B_2} - 2E = 1/2A + 1/2D + 2E - 2E = 1/2A + 1/2D =$

$= 2 \times 0.98 - 2 \times 0.60 = 0.76.$

These two equations may be solved by multiplying the first one with 2 and subtracting the second one from it:

$$1/2A + 1/4D = 0.74/\times 2$$
$$A + 1/2D = 1.48$$
$$-1/2A + 1/2D = 076$$

$$1/2A \quad = 0.72$$
$$A \quad = 1.44$$

When the value of A is included in any equation, the value of D is calculated as follows:

$$A + 1/2D = 1.48$$
$$1.44 + 1/2D = 1.48/\times 2$$
$$2.88 + D \quad = 2.96$$
$$D \quad = 2.96 - 2.88 = 0.08.$$

Thus, the total variability of F_2 consists of the following components:

$V_A = 1/2A$	$= 0.72$	or	53.7%
$V_D = 1/4D$	$= 0.02$	or	1.5%
$V_E = E$	$= 0.60$	or	44.8%
V_{F_2}	$= 1.34$	or	100.0%

This analysis shows that in the F_2 generation there existed a higher genetic variability for the number of spikelets per spike (53.7 + 1.5%) than environmental variability (44.8%), which is due to the genetic differences between the varieties Siete Cerros and Bezostaia-1 (Table 6.1). Furthermore, the major part of the genetic variability was due to the additive gene action (53.7%) and only a small portion was due to the dominant gene action (1.5%).

The above is an example of the simplest calculation of components of genetic variance, omitting the variances due to interallelic interactions (epistasis) which are frequent with quantitative traits. Such calculations call for more complex formulae of biometrical genetics, after the models of Mather and Jinks (1971), Falconer (1981), or other authors to which the reader may refer.

The components of genetic variability may also be calculated from diallel crosses by the methods of Jinks (1954), Haymann (1954), and Mather and Jinks (1971), which allow us to perceive the interactions between the additive and the dominant gene effects. However, the values obtained pertain to all participating cross combinations in the diallel and these are, as such, fairly useless because what is important for a breeder is individual genetic variance of each cross combination.

Heritability

Since selection is in fact based on phenotypic values, it is important to know what is the probability of the selected phenotype giving rise to the same progeny. If the genetic variance for a certain trait is high and the environmental variance is low, it is to be expected that the progeny will be very much alike the selected phenotype. And vice versa, if the genetic variance is low and the environmental variance is high, the progeny may be very different from the selected phenotype.

A similarity between a parent and its progenies depends largely on the components of genetic variance ($V_A + V_D$). In the case of the additive variance (V_A), the parental phenotype is a reliable indicator of its genotype, i.e., it will produce a progeny similar to itself; in the case of the dominant variance (V_D), the progeny will differ from the parental phenotype according to the nature of the intra-allelic interaction.

The relationship between genotypic variance and total phenotypic variance is called the heritability (H or h^2) of a trait of a certain population and it is designated as:

$H = V_G/V_P$, or

$h^2 = \sigma^2{}_G/\sigma^2{}_P$.

This is the heritability in the broad sense or the degree of genetic determination (Falconer, 1981). Heritability in the narrow sense is a relationship between the additive variance and total phenotypic variance:

$H = V_A/V_P$, or

$h^2 = \sigma^2{}_A/\sigma^2{}_P$.

For the previous example, the heritability for the number of spikelets per spike is:

$H = 0.74/1.34 = 0.552 = 55.2\%$ (in the broad sense)

$H = 0.72/1.34 = 0.537 = 53.7\%$ (in the narrow sense).

The heritability values are relatively high, indicating that there existed large genetic differences between the parents and that in the F_2 and later generations it may be possible to select genotypes with a large number of spikelets (derived from Bezostaia-1) and combine them with a large number of grains (derived from Siete Cerros). Since the environmental variance is fairly high too, it may obscure the actual value of the genotype and mislead the breeder into selecting modifications which will not produce a superior progeny in subsequent generations.

It has already been mentioned that the additive variance determines the success of selection. Consequently, it is also called the breeding value. Falconer (1981) states that heritability is a convenient expression of the phenotypic value which serves as a guide to the breeding value or as a degree of agreement between the phenotypic and the breeding values.

Other methods for determining heritability

Heritability after hybridization is calculated most completely according to the formulae of Mather (1949).

Heritability in the broad sense may also be calculated from the F_2 generation providing that the environment had the same effect on the parents and the F_2 generation. The difference between the average value of the parents' variance and the F_2s' variance is the genotypic variance. Then heritability is calculated according to Mahmud and Kramer (1951) by the formula:

$$h^2 = (\sigma^2 F_2 - (\sqrt{\sigma^2 P_1 \times \sigma^2 P_2})/\sigma^2 F_2) \times 100.$$

This is the calculation of heritability in the broad sense; when translated to the number of spikelets per spike, it is:

$$h^2 = (1.34 - (\sqrt{0.70 \times 0.53})/1.34) \times 100 = 54.5\%,$$

and for plant height given in Table 4.6 it is:

$$h^2 = (10.45^2 - (\sqrt{4.83^2 \times 3.27^2})/10.45^2) \times 100 = 85.5\%$$

If the F_1 generation is grown in parallel with the F_2 generation and the parental material, in the same year, the variances of the F_1 and the parents are taken as the environmental variance which is then subtracted from the variance of the F_2 (Allard, 1960; Beil and Atkins, 1965). The use of the variance of the F_1 is avoided in a way because of the effect of superdominance and its interaction with the environment, which is not always found in the F_2 generation.

Heritability may also be calculated as a regression of the breeding value over the phenotypic value:

$$h^2 = b_{AP}$$

where A is the additive variance or the breeding value and P is the phenotypic value. The formula of the last parameter is:

$$P = A + R$$

where R is the remainder which includes deviation due to dominance, interactions and environmental effects.

The sum of products $AP = >A(A + R) = >A^2 + AR$.
Since A and R are not in correlation, $AR = 0$, and the covariance is:

$$cov_{P/P} = \Sigma A^2,$$

$$cov_{A/P} = V_A.$$

Since the regression coefficient is:

$$b = W_{XY}\sigma^2_X,$$

the regression coefficient of the breeding value for the phenotypic value is:

$$b_{A/P} = cov._{A/P}/o^2_P = V_A/V_P.$$

Heritability may also be calculated as the coefficient of correlation between the breeding value and the phenotypic value:

$$r = \sqrt{\frac{s[y(x-\bar{x})]}{(x-\bar{x}^2)} \times \frac{[x(y-\bar{y})]}{(y-\bar{y}^2)}}$$

and, further,

$$r_{AP} = b_{AP} \times (\sigma_P/\sigma_A),$$

$$r_{AP} = b_{AP} \times (1/h).$$

Since $b_{AP} = h^2$, it means that

$$r_{AP} = h^2 \times (1/h)$$

$$r_{AP} = h$$

i.e., the coefficient of correlation between the breeding value (A) and the phenotypic value (P) is the value of heritability. Therefore:

$$h^2 = V_A/V_P, \quad h^2 = b_{AP}, \quad h = r_{AP}.$$

(For the derivation of the above formulae, see the textbooks of Falconer (1960, 1981), Mather and Jinks (1971), etc.).

Since the heritability values of individual traits are very important for the genetic gain from selection, it will be discussed in more detail in the chapter on breeding methods.

Interaction genotype-environment in the process of breeding

So far we have considered components of genetic variance and the relationship between genetic and environmental variance in the expression of a trait. However, as certain traits may interact with environmental factors (V_{GE}), so may a genotype as a whole (especially in relation to the yield) interact with environmental factors; the latter interaction must be taken into account in the course of a breeding process.

Generally the process of developing new cultivars takes a long time. Breeding material is thus exposed to the effect of environmental factors over a period of several years. In the case of annual plants, it takes about 10 years on average to develop a new variety and get it into commercial production; in the case of many perennial plants, the process is much longer because they may take much longer to mature before their agronomic value can be assessed, or which may take several years before they will produce progeny.

Those phenotypes which are expected to possess gene recombinations and to have desired agronomical traits start to be selected in the F_2 generation. Because of large environmental variations from one year to another, it happens that one year is suitable for testing drought resistance, the next for testing winter hardiness, the third for testing disease resistance, etc. After five to six years, it may be expected that a breeding material which has passed all these tests will exhibit a wide adaptability which may protect it from a pronounced environmental stress. This material, however, can hardly be expected to use a large number of favorable environmental factors advantageously and its yield may not be very high. On the other hand, it may happen that the preliminary tests coincide with generations of segregation, when the larger portion of the material is still heterozygous. And later on when lines are being selected, low temperatures, drought or disease attacks simply do not take place. In that case, the selected lines will exhibit their shortcomings in commercial production.

Therefore, in order to avoid being dependent on the limiting factors of the natural environment, artificial environments are created in greenhouses, climatic chambers and laboratories. In these, the materials in the segregating generations, (at first selected plants and later on, lines) are tested for winter hardiness, drought resistance, disease resistance, etc. Another method in which the effects of climate and pathogens are fully examined is that some breeding centres grow the materials in segregating generations and select desired plants in at least two locations and this is a good substitute for extra experimental years. All these tests lower the risks of any unfavorable interaction between genotype and environment.

The interaction between genotype and environment will be discussed in more detail in the chapter on varietal adaptability and stability.

References

Allard, R. W., 1960. Principles of Plant Breeding. John Wiley & New York, 485 pp.

Beil, G. M., and Atkins, R. E., 1965. Inheritance of quantitative characters in grain sorghum. Iowa State Sci. 39: 345–358.

Falconer, D. WS., 1960, 1981. Introduction to Quantitative Genetics. Longman, London, 340 pp.

Hayman, B. I., 1954. The theory and analysis of diallel crosses. Genetics 39: 787–809.

Johannsen, W. L., 1909. Elemente der exacten Erblichkeitslehre. Gustav Fischer, Jena.

Jinks, J. L., 1954. The analysis of continuous variation in a diallel cross of *Nicotiana rustica* varieties. Genetics, 39: 767–789.

Mahmud, I., and Kramer, H. H., 1951. Segregation for yield, height and maturity following a soybean cross. Agron. J. 43: 605–609.

Malek, M. A., 1976. Inheritance of final yield components in diallel crosses of vulgare wheat. Doctorate thesis, University Novi Sad (Yugoslavia).

Mather, K., 1949. Biometrical Genetics. Methuen, London, 162 pp.

Mather, K., and Jinks, J. L., 1971. Biometrical Genetics. Second Ed. Chapman and Hall, London.

7. METHODOLOGIES OF PLANT BREEDING

Having discussed the principles of genetics and selection which are the foundation of plant breeding, we shall go into breeding procedures and the methods of developing new varieties.

In accordance with the overall development of science in this century, a number of new methods supplemented the classical breeding procedures such as the selection from natural populations and intraspecific hybridization. These new methods, used exclusively or in combination with hybridization, facilitate the development of various new varieites.

The main avenues of plant breeding are:

(1) selection from natural and local populations,
(2) intraspecific hybridization,
(3) use of heterosis,
(4) chromosome engineering,
(5) induced mutations,
(6) grafting,
(7) tissue and cell culture
(8) protoplast fusion (parasexual interspecific hybridization)
(9) genetic engineering.

In this chapter only the first two methods will be discussed. Natural populations are used as a material for the development of new varieties for those species which have not been subjected to much plant breeding and also wherever natural populations may serve as a source of desirable genes, i.e., traits. Local or domesticated populations are natural populations which had been selected once or several times and then were grown as local varieties.

Intraspecific hybridization is the main avenue to produce new varieties. Since this method is discussed practically in all chapters, its importance will only be emphasized here. The other methods of plant breeding, with respect to their particular characteristics, will be dealt with in separate chapters or, if they are not widely used they will be mentioned in the other chapters.

Selection from natural and local populations

Natural populations were the first plant materials used for selection. The first written records about plant breeding date back to the 18th century. Van Mons in Belgium, Knight in England, and Cooper in the United States of America (according to Tavćar, 1959) were the first to describe mass selection methods in various agricultural crops. At the beginning of the 19th century, John de Couteur, an American wheat grower, observed a variation in a wheat field and decide to select individual spikes. He planted them in separate rows and found a large uniformity in some rows as well as large differences between rows. He multiplied some of the lines and thus produced the first selected varieties which subsequently became quite popular. At the same time, P. Shireeff in Scotland applied the method of individual selection to wheat and oats and produced several varieties which spread into commercial production.

The work of T. Hallett in England around the middle of the 19th century was particularly important for the further development of plant breeding. He studied the mechanisms of the tranfer of traits from selected plants to their progeny. He started from the hypothesis that the best selected plants will give the best progeny. Of course, this was not always the case since he also selected non-heritable variations. Still, Hallett came out with a considerable number of very good varieties. Particularly successful were his internationally recognized barley varieties named Chevalier. At the same time in France, important advances in plant breeding were made by Louis de Vilmorin, the founder of a famous seed company which still exists and operates. Using the method of individual selection, he not only made new wheat varieties from local populations but also followed the changes resulting from continuous selection over a long period of time. De Vilmorin made notable success in sugar beet breeding increasing the content of sugar from 7 to 15%.

International recognition for the breeding of agricultural crops has also been earned by the Swedish Seed Association established in 1886 in Svaloff. Hjalmar Nilsson, its first director, introduced the method of individual selection from local populations by which improved varieties of agricultural crops have been developed.

Selection from natural and local populations has continued in the Twentieth century. For example, some natural wheat populations from Kenya have been used as a source of genotypes resistant to some physiological races of stem rust (*Puccinia graminis* f.sp. *tritici*). These genotypes were subsequently crossed with susceptible commercial wheat varieties in the United States of America, Canada and other countries.

In developed countries, natural or local populations of the main self-pollinated plant species, such as wheat, barley, oats, soyabean, etc., have been used up and presently they do not serve as source materials for the selection of new varieties. Exceptionally, if they possess some valuable traits (disease resistance, high content of proteins or lysine, etc.), they are included in hybridization programmes as gene donors for these traits.

However, a multitude of natural and local populations of grasses, grain and forage legumes, medicinal herbs, and other plant species exist in both developed and developing countries. These populations are used and will continue to be used as a source of materials for selection, especially of the open pollinated species whose populations possess high genetic variability which, in the process of selection, keeps renewing itself. It is mentioned in Chapter 2 that a possibility and necessity of domesticating new plant species still exists. Similarly it is feasible to breed new crops with valuable traits from natural wild populations.

Intraspecific hybridization

The discovery of the principles of the mechanisms which control inheritance of traits by G. Mendel in 1865 and confirmed and further developed by numerous geneticists and breeders since then, has allowed the breeding of organisms to grow to be an established science. The knowledge of the possibility of crossing parents with different traits in order to obtain progenies possessing a recombination of parental traits led to planned hybridization, the forecast of gene recombinations and the programmed development of varieties possessing certain traits.

High genetic variability is present in natural populations of open pollinated organisms because open pollination allows continuous random mating of various genotypes. By man-directed hybridization, especially of open pollinated organisms, it is possible to create a higher genetic variability than encountered in natural populations. Numerous crossings of varieties and species have already been made which, due to incompatibility, mechanical barriers, or geographical isolation, would have never developed in natural conditions.

In Chapter 3 it was explained how genetic recombinations may be effected by hybridizing genetically different parents. Such hybridization produces a pool of genetic variability which cannot be achieved by any other method. Furthermore, it is a part of almost all methods of plant breeding. This is why hybridization or recombination breeding has been and will continue to be the main tool for the development of new plant varieties and animal races.

Table 7.1 Pedigree of wheat cultivars San Pastore and Libellula

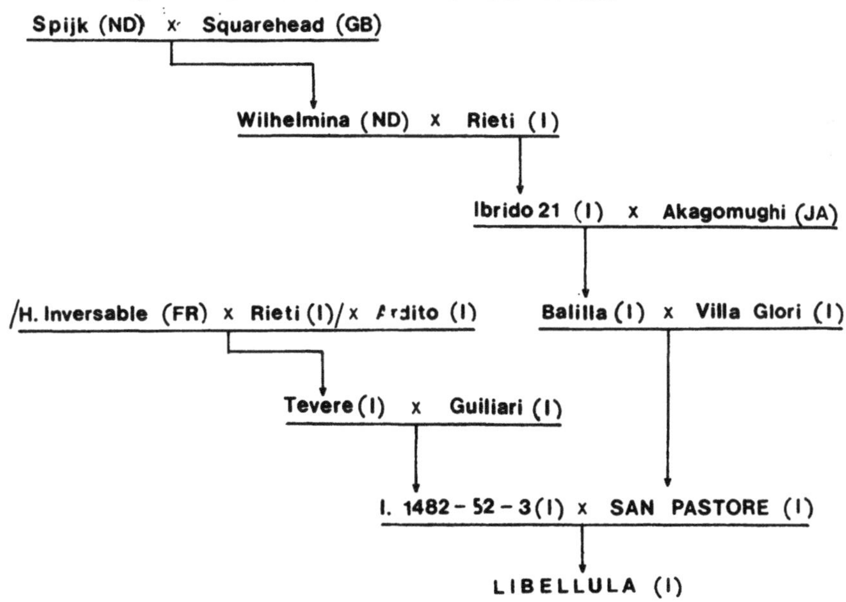

Source: Rabinovič, 1972

Since all chapters of this book deal with various aspects of hybridization, its characteristics and importance for plant breeding will not be explained in much detail here. Instead of a conclusion, here are examples of the pedigrees of the Italian wheat variety San Pastore (Table 7.1), the Soviet variety Bezostaia-1 (Table 7.2) and the Yugoslav variety Sava (Table 7.3). A pedigree of a variety includes not only the direct parents but all predecessors which had participated in the crossing programme. San Pastore contains genes which come not only from its direct parents, Balilla and Villa Glori, but also some genes of the varieties Spijk from the Netherlands and Squarehead from England which had been selected far back at around the turn of this century (Table 7.1).

When descending generations are followed, it is interesting to see that quite different varieties, developed in different countries and in different programmes,

Table 7.2 Pedigree of wheat cultivars Bezostaia-1

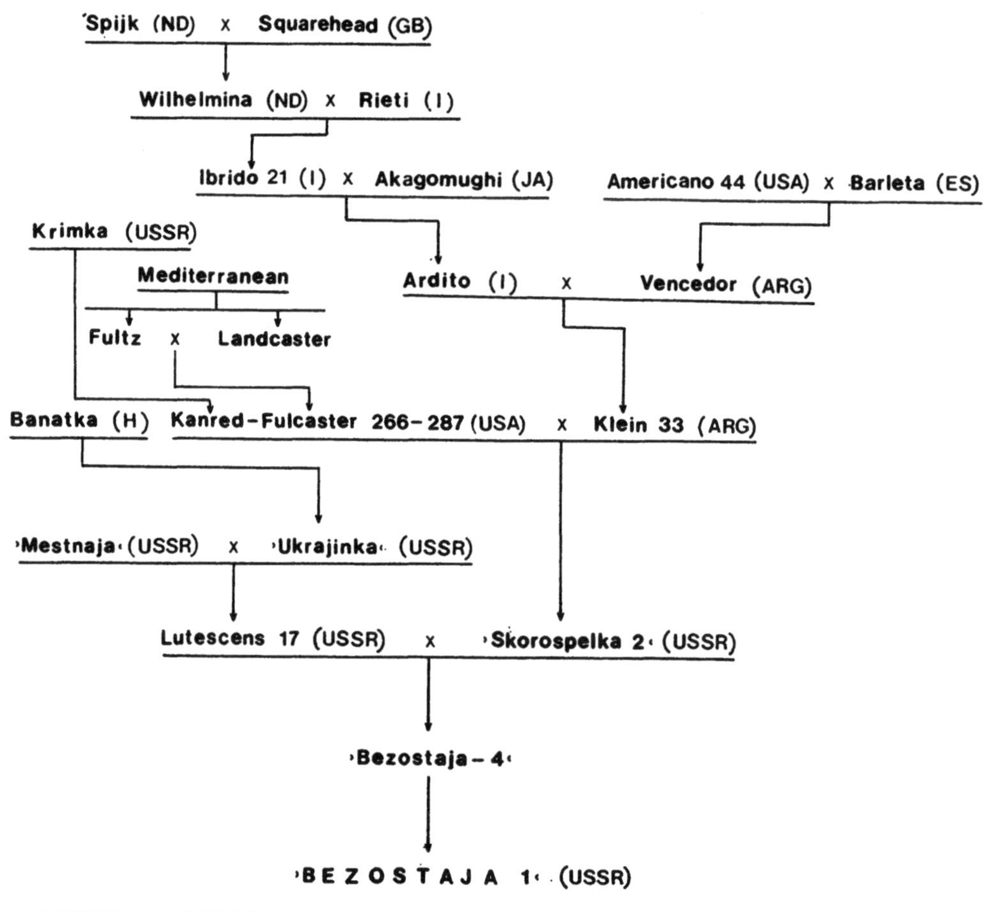

Source: Rabinovič, 1972.

may have a line of common ancestors. For example, the varieties San Pastore, Bezostaia-1, and Sava have in their pedigrees as many as six common ancestors – Spijk, Squarehead, Wilhelmina, Rieti, Akagomughi, and Fultz (Table 7.1, 7.2, and 7.3). Nevertheless, the three varieties differ in a large number of traits resulting from gene recombinations taking place in their common ancestors and the ancestors which they had not shared. It may be stated that the three varieties possess considerably more versatile gene recombinations than their ancestral local varieties whose pedigrees do not include as many parents.

The hybridization or recombination breeding method has been far more successful and productive than the selection from natural or domestic populations, as convincingly illustrated in Figure 7.1 for wheat breeding in Sweden (MacKey, 1980).

Table 7.3 Pedigree of wheat cultivars Sava

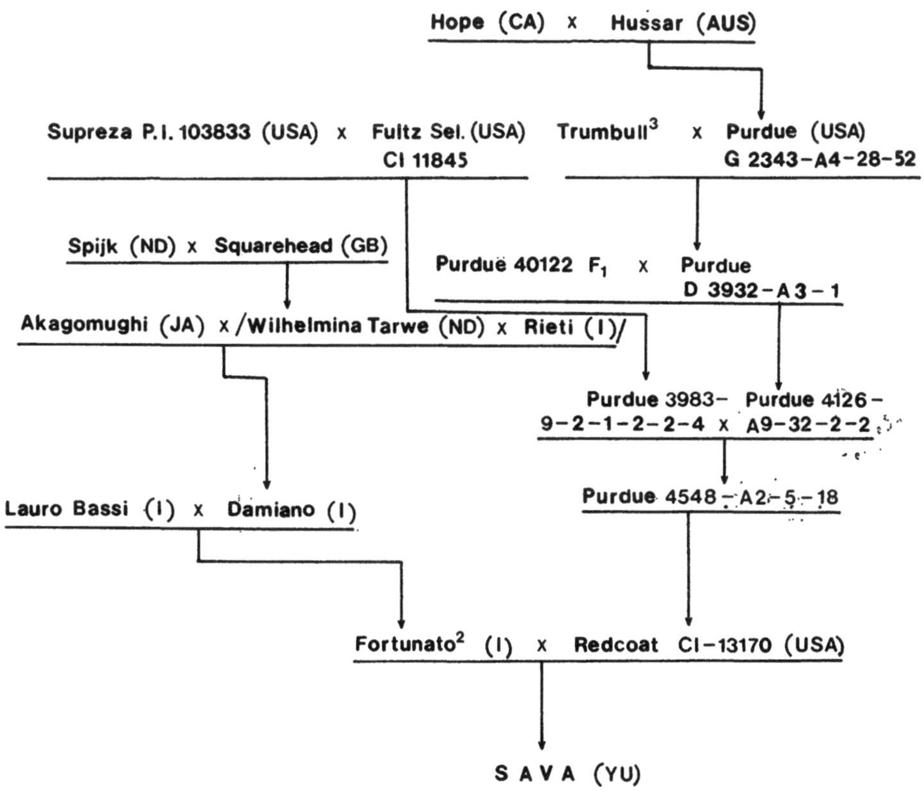

Source: Borojević, 1981

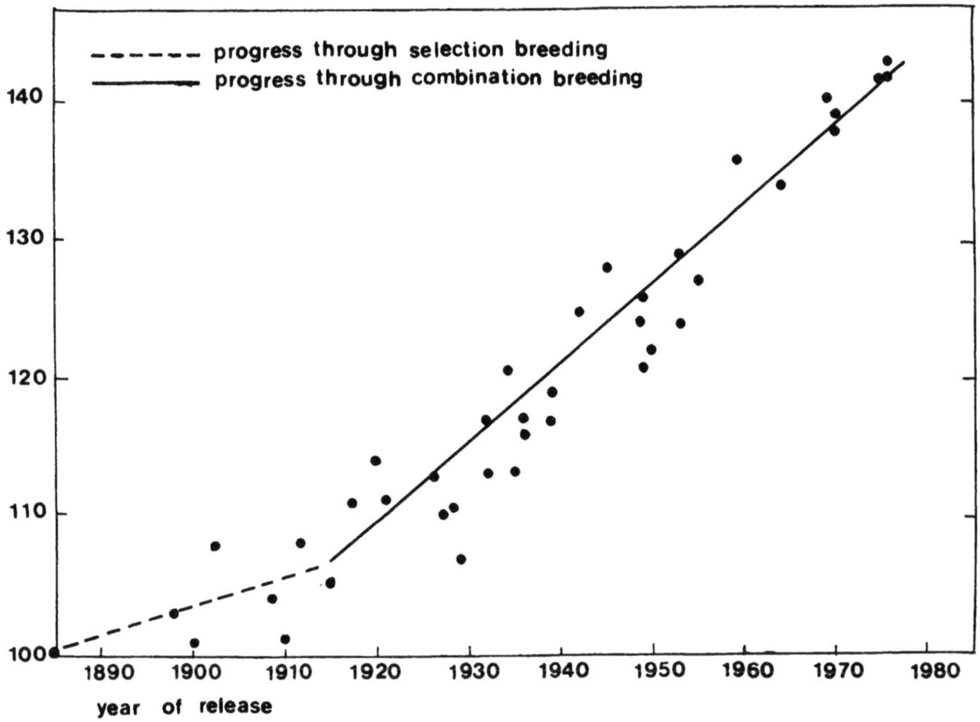

Fig. 7.1 Progress in Swedish breeding of winter wheat. Relative grain yield for all released varieties with inbred cultivar Sammet = 100.
Source: Mac Key, 1980

References

Borojević, S., 1981. (Principles and Methods of Plant Breeding). (In Serbocroatian). Cirpanov, Novi Sad, 386 pp.

Mac Key, J., 1980. Genetic potentials for improved yield. Proc. Workshop Agric. Potentiallity Directed by Nutritional Needs (Ed. S. Rajki). Akademiai Kiado, Budapest, pp. 121–143.

Rabinovič, S. V., 1972. Sovremenie sorta pšenici i ih rodoslovie. Izd. Urožai, Kiev, 325 pp.

Tavčar, A., 1959. Oplemenjivanje bilja. Polj. nakladni zavod, Zagreb, 275 pp.

8. A CONCEPT OF THE DEVELOPMENT OF NEW CULTIVARS

Since plant breeding is a lengthy process which encounters many uncertainties and which depends on a number of biotic, abiotic, and economic factors, improvements in the efficiency of plant breeding may be expected if a breeding concept is established and a programme of realization of the concept is designed.

Breeding objectives

First of all it is necessary to define breeding objectives, i.e., which problem is tried to be solved by developing new cultivars of a certain crop.

This decision depends on the land area available and other conditions. If it is feasible to increase the acreage of the crop in question, by expanding it to different agroecological regions, it will be necessary to develop different types of cultivars adapted to different cultural practices and climates. If the acreages for a crop cannot be increased and the demand for food cannot be met, it is necessary to develop cultivars possessing high yield potentials capable of returning the investments in research and of increasing the average yield per unit area.

Breeding objectives depend also on the intensity of production and the level of cultural practices that may be provided. An intensive production, which implies a quality soil tillage, high mineral fertilization, full protection against pests and weeds, mechanized harvesting, etc., calls for a different set of cultivars than those suitable for less intensive production.

In view of the human population explosion, a need arises to cultivate various abnormal and defective soils, such as acid, saline, swamped, and pronouncedly sandy soils, etc. In addition to using agricultural practices aimed at the normalization of such soils, there should be concurrent work on the development of cultivars of various species capable of tolerating such adverse soil conditions. For example, barley varieties have already been developed which bring satisfactory yields when irrigated with sea water. Also, there is a new *Triticale* synthetic species which may successfully be grown on highly alkaline soils and soils with high concentrations of aluminium.

It is of particular importance that cultivars are produced which tolerate various climatic stresses, e.g., low temperatures without a snow cover, late frosts, dry soil, drought at the stage of seed filling, etc. Such varieties could expand the areas under crop towards the poles as well as to higher altitudes.

Another aspect that has to be taken into account when developing new cultivars are the requirements of the market. Of course, new cultivars must have a high nutritional value but the traits such as flavour, fragrance or transportability are of no lesser importance. It should not be overlooked either, when defining a breeding concept, that the requirements of domestic markets frequently differ from those of foreign markets. For example, Mexican wheat varieties succeeded in entering the markets of India and Pakistan only after their original red grain colour was converted to white, because the local people customarily make flat bread (chapaties) exclusively from white wheat flour. The demands of market are still more pronounced in vegetable and fruit production which brings about a large diversification in many breeding programmes.

In any case, it is necessary to establish an interaction between the market and breeding programmes. Breeding should be guided by market demands but, on the other hand, it should insist on developing new and different cultivars which could change the tastes and desires of the customers and thus create new demands on the market.

Finally, the economy of production cannot be disregarded when defining a breeding concept. Farmers may hardly be expected to accept new cultivars if these call for increased production costs due to special methods of harvest, transportation, or protection which are not offset by increased profits.

Designing a model of a cultivar

In developed countries, plant breeding has reached a point at which the chances of a major breakthrough are more and more difficult. Thus in order to develop cultivars superior to the ones used at present, (which is possible because the genetic limit has not been reached in any crop), efforts should be made to design a detailed breeding programme. This would identify what should be done at certain phases and how results relate to the ultimate ideotype which is intended to be developed. The drawing given in Figure 8.1 is very instructive in that respect. It appears as if somebody has made a drawing of a wild pig, saw its shortcomings, made a drawing of the type of pig that he wished to have, and then made it. The differences between a wild pig and the Swedish landrace pig are such that neither a large population nor a wide reaction norm of wild pig cultivated in favorable conditions would have helped in producing the landrace. It took a series of transitional genotypes, hybridization and the selection of new gene recombinations in domesticated environmental conditions, and a long line of generations to achieve the final goal.

The theoretical basis for the development of a model cereal cultivar relating to specified environments has been worked out by the Swedish scientist MacKey (1966). Searching for efficient criteria for breeding for yield, MacKey started with photosynthesis as the first factor which determines the yield of grain. Since currently used methods do not allow for the practical application of physiological parameters (such as net assimilation rate, assimilate translocation in grain, etc.) in breeding, MacKey pointed out that morphologically distinctive traits which may be relevant for yield realization in different environments should be taken into consideration when designing a model of cultivar. The cereals possess several yield components which develop successively in the course of ontogenesis (such as tillering, number of spikes,

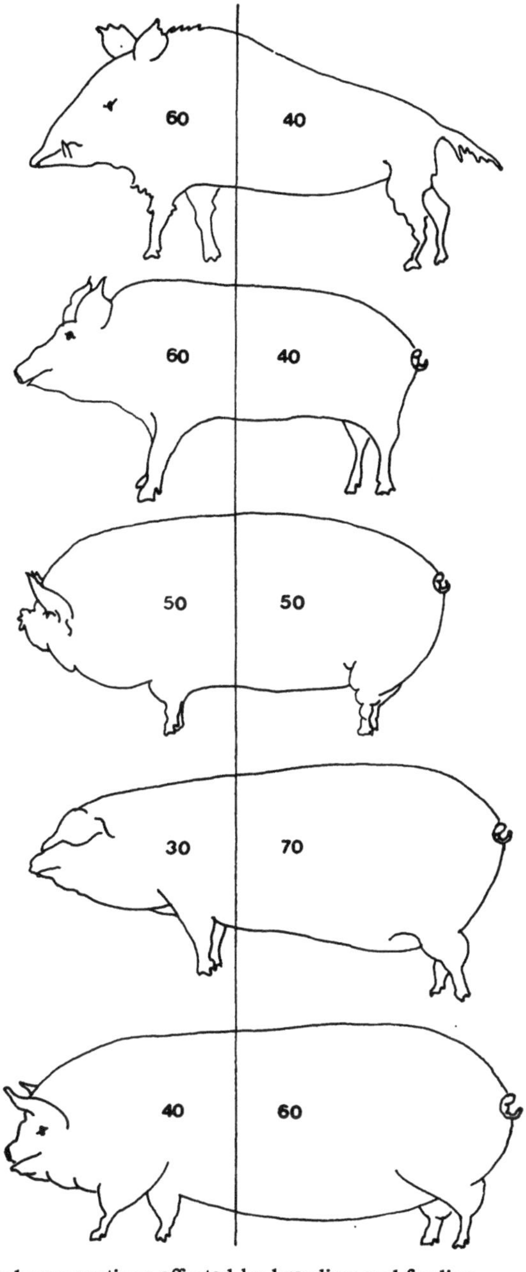

Fig. 8.1 Changes in body proportions effected by breeding and feeding.
Source: Our drawing initiated by the idea of an Argentine drawing published in "Pig Production" by D. Ilančić, Agr. Press, Zagreb, 1958

number and mass of grains per spike) and it is valuable for a breeder to estimate the effect of environmental factors on the development of these traits.

At the 3rd International Wheat Genetics Symposium held in Australia in 1968, Dr. J. Donald presented a model of a dwarf wheat type with erect leaves, long awned spike, and monoculm stem, i.e., a wheat plant which does not produce tillers (Fig. 8.2).

In contrast to ideotype modeling, Finlay (1968) voiced the opinion that it is not

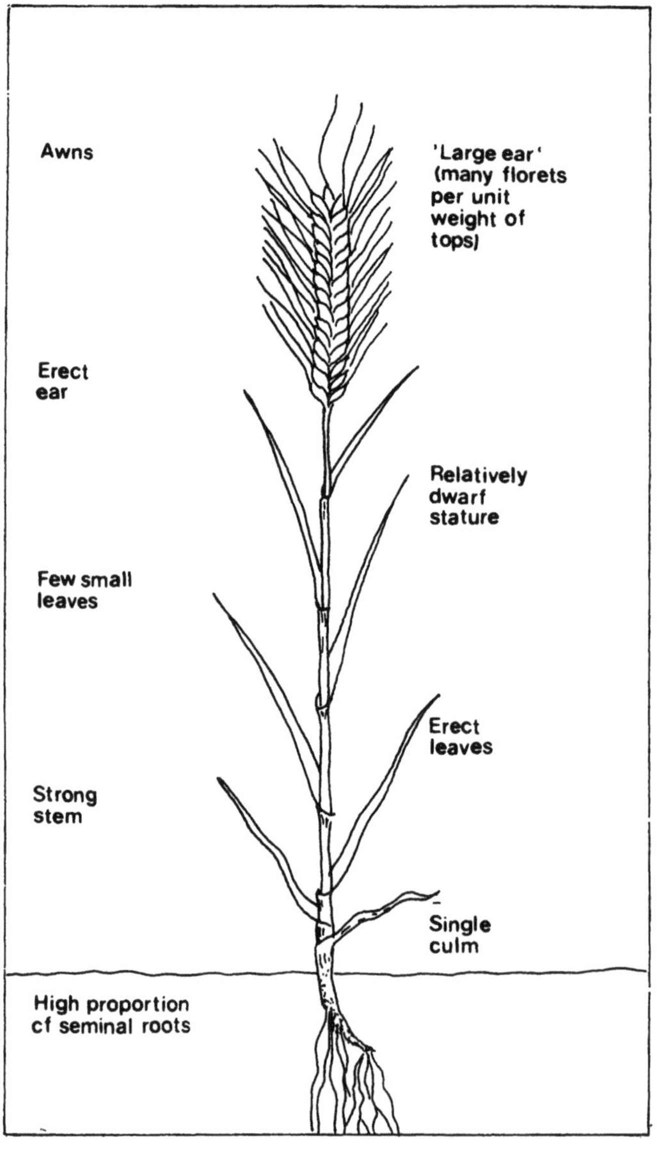

Fig. 8.2 A design for a wheat ideotype.
Source: Donald, 1968

desirable to define a model of cultivar because such a procedure may rapidly reduce genetic variability. He maintained that a high yield and high adaptability may be achieved in different ways. Seconding the latter opinion, Martinić (1973) emphasized a need for the development of cultivars possessing a wide genetic adaptability. Evans (1973) and Blixt and Vose (1984) suggest that breeders should give preference to the concept of adaptability rather than to the concept of ideotype.

Their attitude accrues from the standpoint that new cultivars should perform satisfactority in a wide range of environments. Of course, such cultivars are needed too, especially for the large territories of Australia, the Soviet Union, the United States, Argentina, etc. However, it is difficult to accept that a cultivar with a high genetic yield potential will also be highly adaptable. A grower, especially a small one, is interested in a cultivar which will yield better than any other on his land, in his region, regardless of the performance of the cultivar in another environment. If a high average yield is set as an objective, adaptability to a wide range of environments may best be covered by growing several cultivars possessing different traits and not by a single variety.

Genetic variability of high-yielding cultivars should not be estimated only on the basis of their phenotype. Cultivars identical in plant height, growing season, or winter hardiness may differ in a number of other parameters, e.g., physiological traits, disease resistance, quality characteristics, etc. Simmonds (1981) states: "Two cultivars of similar yield may get there by different routes" (p. 46). However, it does not mean that breeding should be left to chance or that it is not necessary to plan how to breed new cultivars. There is not a single breeder who believes that it is possible to make the ideal cultivar, but there are many breeders who design models of new cultivars. It is not a matter of fashion but a necessity to work out a breeding programme in detail and take into account all relevant factors. Model designing is detrimental ony if it is taken too literaly, if only one type of cultivar is designed, if only some traits are bred for, without taking care of environmental conditions and market requirements.

Understandably, it is not enough to outline in the model only the outward appearance of a phenotype; its genetic structure should be specified and the environment for which it is made should be defined. When designing an ideotype it should be started from the well-known fact that a phenotype is the realization of a genotype in certain environmental conditions. In doing this, it is useful to adhere to the following principles:

– *Determination of traits which result from the relationship of a cultivar and its environment.* This approach was initiated by Donald (1968) when he designed his wheat ideotype. First of all, limiting factors should be assessed – low or high temperatures, drought, pests, etc. – in order to see what traits a new cultivar should possess to be able to counteract these limiting factors. Favourable environmental factors, those which already exist and those that may be improved – soil quality, distribution and amount of rainfall, irrigation, mineral fertilization, sowing date, etc. – should also be taken into account in order to allow for the full expression of genes for certain traits and to use the maximum of the genetic potential for yield of the newly developed cultivar.

– *Determination of traits which result from the relationship of plants within a cultivar.* This principle has also been noted by Donald (1968) when he explained the advantages of erect leaves. This is the question of stand density which should be determined in dependence of the architecture of leaves or branches, plant height, the level of cultural practices, the risk of pathogens, etc. Since an intensive production calls for monoculture, it is desirable to reduce the competion among plants in a plot to a minimum.

In the case of companion cropping, such as maize – soyabean, rye – vetch, etc. it is necessary to take into account the height of each crop, the position of leaves of the taller crop, and the arrangement of plants in the plot in order to avoid shading which reduces the photosynthetic activity of plants. In the case of mixed cropping or intercropping, attempts should be made to make the competition between the species compensating (cooperating) rather than detrimental.

– *Observance of market demands.* Depending on the crop in question, it is necessary to itemize how a new cultivar should stand with respect to quality, protein or vitamin content, its behaviour in storage or during transportation, etc.

After considering the above principles, bearing in mind that it is difficult to breed for a large number of traits, it may be helpful to define the main parameters of a breeding project:

(1) Breeding for high harvest index or total biomass

The increases in the harvest index of a number of grain and fruit crops, for example, rice, wheat, maize, sunflower, pea, etc., may be consider the major breakthrough in plant breeding achieved in recent decades. With the old wheat cultivars, 20–30% of the total biomass was the grain yield; the new short cultivars (semi-dwarf and dwarf) have a harvest index over 40%. Since the optimum partitioning has not been attained yet with a large number of cultivars and even crops, such as for example the soyabean, breeding for a harvest index of 50% or more should be set as a target.

As regards species grown for their vegetative mass, e.g., forage crops, some vegetables, etc., breeding efforts should be directed towards increases in total biomass, placing emphasis on certain plant parts according to the crop in question.

(2) Breeding for traits which secure optimum canopy

As the number of plants per unit area is the primary yield component, it is essential to breed for those traits which will ensure it, e.g., resistance to low temperatures, lodging resistance, etc.

(3) Breeding for high photosynthetic activity

So far, the approach to this parameter has been to breed for resistance to pathogens which attack the photosynthetic apparatus, or breeding for a suitable vegetative period, etc. In order to increase the efficiency of photosynthesis, which is a key to further increases of genetic potential for yield, it is necessary to include in breeding programmes those parameters which may be selected phenotypically: harvest index, erect leaves, effective leaf area duration, etc.

(4) Breeding for rational use of nutrients

The breeding for resistance to lodging, to ensure higher canopy, has been a roundabout way aiming at the economy of nutrient use. It is time to try a new approach, wherever it is applicable, which is to breed for a more efficient root system.

(5) Breeding for higher genetic potential for yield via final yield components

Eventual improvements in the parameters mentioned above do not automatically ensure the attainment of high yields. It is thus necessary to breed for final yield components such as the number and mass of grains, fruits, or other end products (Grafius, 1964). It is important to keep it in mind because in a large number of crops whose reproductive organs are in fact the economic product the sink capacity (a spike, an ear) limits the yield more than the source capacity (photosynthetic apparatus) (Borojević et al., 1980).

Since breeding is based on phenotype, it may be helpful to define a desired phenotype by the main traits and yield components which may be followed successively in the course of ontogenetic development (MacKey, 1966; Borojević, 1966, 1971). Of course, the traits and the components should be maximized and the final expression should be the desired genetic potential for yield. Such a model of high-yielding wheat cultivars adapted to the environmental conditions of the continental part of Yugoslavia, i.e., central Europe, is shown in Figure 8.3.

Understandably, a desired genotype should be composed by incorporating into it genes for those traits which are intended to be improved.

Believing that designing a model of cultivars is a useful guideline in breeding, Swaminathan (1978) and Sinha and Swaminathan (1984) point out that, considering the new technologies and new demands, it is necessary to introduce new parameters in plant breeding, e.g., efficiency of food production per unit of cultural and solar energy invested, photo- and thermo-insensitive cultivars to suit different weather conditions, plant types for different cropping systems, etc.

At the end of designing a concept of breeding new cultivars, the following points should be considered:

(1) Will the available materials allow the development of genetic variability necessary for the realization of the model? The answer should indicate the direction of breeding effort (hybridization, production of heterosis, induced mutation, gene transfer, etc.).

(2) Will the genetic population intended to be made contain sufficient genetic variability to ensure successful recombining of the desired traits? The choice of parents and methods of crossing should be carefully made.

(3) Which methods of selection from the genetic population will ensure the most efficient extraction of superior genotypes and subsequent development of new cultivars? It is necessary to decide which methods of selection to use (pedigree, recurrent, etc.).

(4) What should be the genetic composition of new cultivars to suit certain environmental conditions and market requirements? Define the genetic composition of the new cultivars to be made (pure line, synthetic, population, etc.).

(5) How to test the selected lines, hybrids and populations in order to reliably select superior genotypes? Define experiment design, level of cultural practices, locations, duration of experiment, etc.

These and related problems will be discussed in subsequent chapters.

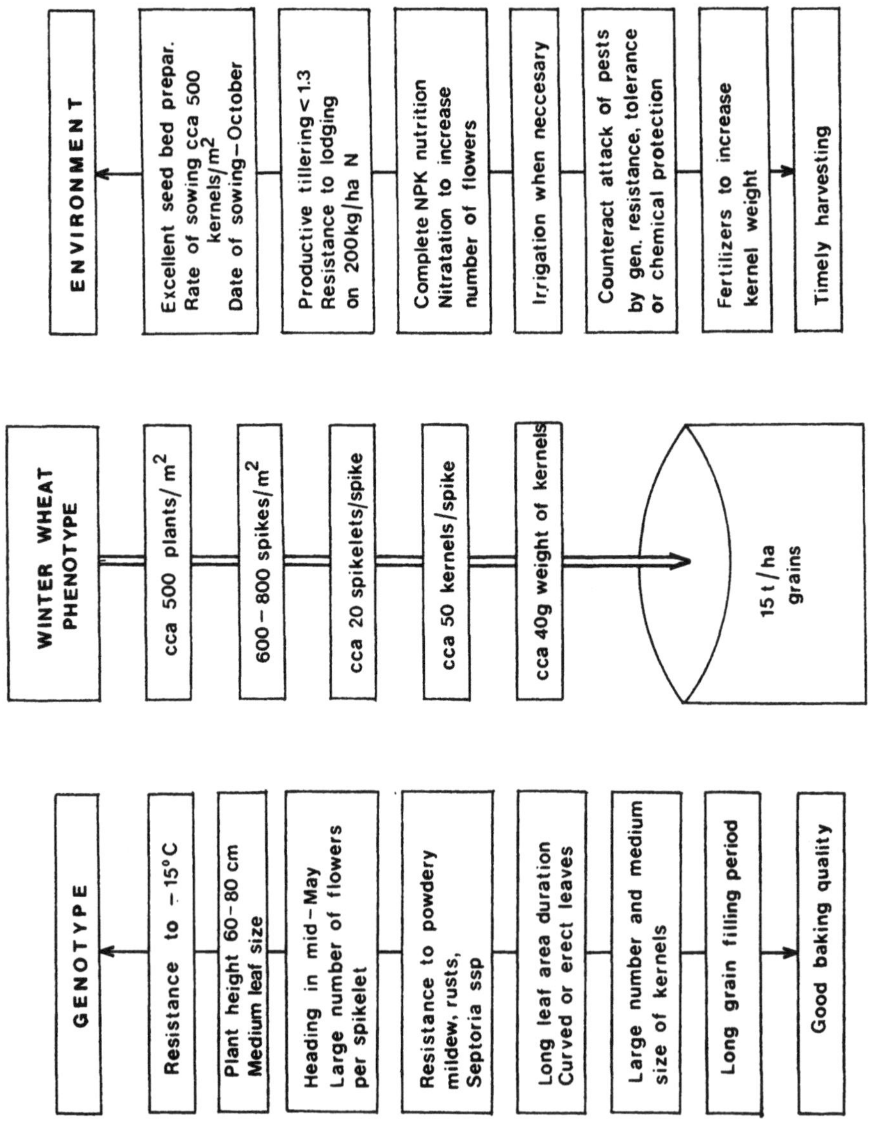

Fig. 8.3 A model for the development of high-yielding wheat varieties for the conditions of Yugoslavia.
Source: Borojević, 1971, 1978, 1985

References

Blixt, S. G., and Vose, P. B., 1984. Breeding towards an ideotype – aiming at a moving target? In: P. B. Vose and S. G. Blixt (Editors), Crop Breeding – contemporary basis. Pergamon Press, Oxford, pp. 414–426.

Borojević, S., and Potočanac, J., 1966. Development of the Yugoslav Programme for Creating High-yielding Wheat Varieties. Savremena poljoprivreda (Novi Sad), XIV: 7–35.

Borojević, S., 1971. Elaborating a model of high-yielding wheat varieties. Savremena poljoprivreda, (Novi Sad), XIX: 33–47.

Borojević, S., 1978. Genetics, environment and modeling in breeding of organisms. Savremena poljoprivreda, (Novi Sad), XXVI: 5–27.

Borojević, S., Čupina, T., and Krsmanović, M., 1980. Green area parameters in relation to grain yield of different wheat genotypes. Zeit f. Pflanzenzüchtung, 84: 265–283.

Davies, D. R., 1977. Creation of new models for crop plants and their use in plant breeding. Applied Biology, 11: 87–127.

Donald, C. M., 1968. The design of a wheat ideotype. Proc. 3rd. Int. Wheat Genet. Symp., Canberra, p. 377–387.

Evans, A. M., 1973. Commentary on plant architecture and physiological efficiency in the field bean. In: M. W. Adams (Editor), Potentials of field beans and other food legumes in Latin America. Seminar series No. 2E, CIAT, Columbia, pp. 279–286.

Finlay, K. W., 1968. The significance of adaptation in wheat breeding. Proc. 3rd. Inter. Wheat Genet. Symp., Canberra, p. 403–409.

Grafius, J. E., 1964. A geometry for plant breeding. Crop Sci. 4(3): 241–246.

Mac Key, J., 1966. The wheat plant as a model in adaptation to high productivity in different environments. Savremena poljoprivreda (Novi Sad), XIV: 29–39.

Martinić, Z., 1973. Wide-general vs. narrow-specific adaptation of common wheat varieties. Proc. 4th Inter. Wheat Genet. Symp., Missouri Agr. Exp. Stat., Columbia, p. 501–508.

Simmonds, N. W., 1981. Principles of Crop Improvement. Longman, London, 408 pp.

Sinha, S. K., and Swaminathan, M. S., 1984. New parameters and selection criteria in plant breeding. In: P. B. Vose and C. C. Blixt (Editors), Crop Breeding – a contemporary basis. Pergamon Press, Oxford, p. 1–32.

Swaminathan, M. S., 1978. Recent trends in crop improvement in India. Proc. 5th Int. Wheat Genetics Symp. New Delhi, p. 1267–1298.

9. SELECTION OF MATERIAL FOR BREEDING

The selection of material to work on is decisive for the success of any plant breeding programme. For hybridization, which is the most frequently used method of obtaining new cultivars, the choice of parental pairs is of the utmost importance. If the selected parents do not possess those genes which are intended, by the breeder, to be recombined in new cultivars, then there is no chance of success regardless of the model designed and the type of cultivar planned to be developed.

Choice of parents

There are in fact three concepts in the choice of parents for crossing: the concept of variety, the concept of character, and the concept of the gene (Borojević and Potočanac, 1966a).

The concept of variety. It covers the selection of a large number of varieties for crossing, under the assumption that desired combinations of positive traits will occur in the new cultivars. This concept was characteric for the early breeding programmes by hybridization, when knowledge of the genetic basis of the traits intended to be recombined from different parents was rather limited. Nowadays it is seldom used, generally only when starting new breeding programmes and when there is no time to study in detail the traits of the available varieties. The success of this concept depends on a large number of cross combinations among which some may be promising for the development of prospective lines and varieties. However, a breeding programme based on this concept is burdened with risks and uncertainties and it may happen that whole series of combinations must be discarded although considerable efforts and time had been invested in them.

The concept of character. This concept implies the choice of parents for crossing on the basis of the knowledge of their traits which are intended to be combined in new cultivars.

Phenotypically different varieties may possess the same or similar winter-hardiness, earliness, plant height, grain mass, etc. The choice of parents on the basis of their traits means that the breeder must look for differences in the traits to be combined. Thus the crossing programme starts with a larger genetic diversity among the parents, and with a clearer picture of prospective combinations than it is the case with programmes based on the concept of variety. A smaller number of cross combinations is required while the chance of obtaining prospective lines and new cultivars is higher.

In the case of equal genetic contributions from both parents, it is important to select parents with maximum number of positive traits and minimum number of negative traits. If the method of complete back crossing is used, it is only necessary for the recipient or the recurrent parent to have many positive traits. Regarding the choice of parents, the concept of character is presently the most frequent in breeding programmes.

The concept of gene. This concept implies the choice of parents for crossing on the basis of the knowledge of the genetic constitution of desired traits.

Genetic studies conducted so far offer information about the genetic basis of a large number of traits of many varieties and plant species. For example, it has been found that wheat resistance to stem rust (*Puccinia graminis* f.sp. *tritici*) is controlled by more than 20 major genes (Sr1, Sr2, etc.), that the resistance to leaf rust (*P. recondita*) is controlled by more than 35 major genes (Lr1, Lr2, etc.). Some varieties possess a single gene for resistance, some a combination of several genes, etc.

Similar data have been gathered for the resistance to leaf stripe (*Helminthosporium* species) and similar pathogens of barley, maize, sorghum and other crops. The genetic basis of resistance of the sugar beet, tomato, pea and other crops has also been intensively studied.

Winter-hardiness, for example, is a complex character which involves the resistance to low temperature, the duration of the stage of vernalization, etc. Winter hardiness has been found to be controlled by major and minor genes which may have dominant or additive effect. Wheat chromosomes that carry the genes for winter hardiness have been successfully determined by the method of monosomic series. It has been found that some winter wheat cultivars possess the genes for spring habit and that some spring wheat cultivars possess the genes for winter hardiness (Maystrenko, 1978).

When the polyploid nature of wheat, oats, lucerne and other crops is considered, it should not be thought strange when spring type varieties are obtained by crossing winter type varieties and vice versa.

In the following example, if winter hardiness is controlled by a single major gene on one chromosome of each wheat genome (*Triticum aestivum* subsp. *vulgare*), then, when, two varieties are crossed we have the situation:

	Genome	*A*	*B*	*D*
Variety 1	Gene	W_1W_1	W_2W_2	w_3w_3
Variety 2	Gene	w_1w_1	w_2w_2	W_3W_3

In the case of full dominance of the gene for winter hardiness (WW), both varieties will be equally winter hardy. The F_1 plants will be winter hardy and the F_2 generation will segregate a certain percentage of spring type plants (w_1w_1 w_2w_2 w_3w_3). In the case of the additive gene effect, variety 1 will be more winter hardy than variety 2, and transgressive segregation in both directions may occur in the F_2 generation producing genotypes more resistant than either parent (W_1W_1 W_2W_2 W_3W_3) as well as spring genotypes (w_1w_1 w_2w_2 w_3w_3).

Under the assumptions that a trait is controlled by a number of minor genes (which is regularly the case with the quantitative traits) that the effect of gene W in genome A

is different from its effects in genomes *B* or *D* and that one gene may modify or supress another gene, it is evident that there exist ample chances for differences even within just one trait in varieties offering large opportunities of gene recombinations in their progenies.

Recent genetic studies have shown that a number of quantitative traits are controlled not only by minor genes but also by major genes. As already mentioned in Chapter 4, major genes for reduction of plant height in wheat (Rht genes), for vernalization (Vrn), photoperiod (Ppd), etc. have already been located on certain chromosomes. Major genes are preferred in plant breeding because they are more reliably manipulated and gene recombinations may be forecasted with more certainty than in the case with the minor genes.

The more that is known about the genetic constitution of a trait, then the easier it is to select varieties – carriers of controlling genes, there will be more planing in the selection of parents for crossing, and the chances of success will be higher. Frankel (1947), a well-known Australian geneticist and breeder, pointed out already in the late forties that the breeding work should not rely any more on observations but that it was necessary to manipulate genes for quantitative traits.

Therefore, wherever it is possible, parental pairs for crossing should be selected on the basis of the concept of the gene. Selection is not difficult regarding the traits controlled by major genes. Regarding the quantitative traits, such as the number of grains, fruit size, protein content, etc., they are controlled by sets of minor genes whose genetic constitutions are not sufficiently known, especially not in many cultivars and species. In that case, the concept of trait is a safer approach to the selection of parental pairs than is the concept of the gene.

Genetic diversity versus geographical distance

When discussing the choice of parents for crossing, the concept of geographical distance of parental pairs should be mentioned. The concept was introduced and used by a Soviet fruit breeder Michurin (1939) who claimed that superior cultivars are produced by crossing or grafting geographically distant cultivars or even species. Later on, the concept was adopted and advocated by field crop breeders (Cicin, 1963; Lukyanenko, 1970, etc.).

In the first four decades of this century, in fact, the concept of geographical distance stood for the concept of genetic diversity. Various cultivars developed in breeding centres in different countries were genetically divergent from cultivars developed in breeding centres in other countries, as confirmed when these materials were exchanged and compared.

After World War II, when due to the expansion of trade and traffic, open borders, and so on, there occurred an exchange of information and materials on a scale that had not been seen at any time before, the geographical distance of cultivars and other plant materials lost its significance.

The wheat cultivar Sava does not possess its traits because of the geographical distance between the cultivars Fortunato and Redcoat (See Table 7.3) but because of genetic differences between the two cultivars. Fortunato does not possess "geographical" characteristics peculiar to Italian cultivars since it is a cross between a Japanese cultivar Akagomughi and a Dutch cultivar Wilhelmina Tarve. The pedigree of Redcoat is similar (Table 7.3). A certain geographical specificity is imprinted by the site of selection, but the site cannot introduce genes of resistance to low temperatures if they are not present in one of the parents used for crossing. Furthermore, different genotypes may be developed in the same place, their traits depending on the genetic divergence of their parents and the direction of selection chosen by a breeder.

Thus, it appears that the concept of genetic diversity is more useful for the selection of parental pairs than is the concept of geographical distance. The latter concept is more meaningful where effect of natural selection is concerned.

It should be pointed out that the selection of materials is important not only for hybridization programmes but also for any action intended to make a certain change. The programmes of induced mutations by irradiation, chemicals, or other agents imply too the selection of materials suitable for treatment, i.e., such materials which, having undergone mutations, will possess traits of interest for genetic studies or plant breeding. Similarly, if tissue culture is used in a breeding programme, it is important to select a truly superior genotype and produce the largest possible number of its progenies. To obtain haploids or doubled haploids from anther culture, the best lines or hybrid combinations must be selected in order to obtain materials suitable for breeding. In the case of the recombinant DNA technique, breeders must be certain that the gene which is transferred is really the one they wanted to use (resistance to a pathogen, herbicide, etc.).

Combining ability

The notion of good combining ability implies the capacity of a parent to produce superior progenies when combined with another parent.

It has been observed that two good cultivars do not necessarily produce progenies superior to them, either in individual traits or on the whole. Also, not all crosses between inbred lines show hybrid vigour. Thus one talks of good and poor combining ability exhibited between certain parents. It is not possible to foresee the combining ability of two parents merely on the basis of the knowledge of their traits. Only the recombination of parental traits may be expected with a degree of certainty. For example, if one parent is early and susceptible to diseases and the other is late and resistant, it may be expected that they will produce a hybrid or a cultivar which will be early and resistant. However, it is impossible to foresee whether these new lines will in addition to these positive recombinations, possess a higher yield potential than the better parent, or higher protein content, etc.

Combining ability should thus be examined always when the objective is the development of superior progenies, i.e., when heterosis is practically exploited or, to a degree, when hybridization is used to develop new cultivars. There are general and "specific" combining abilities.

General combining ability (*GCA*) is the average value of a parent (line) estimated on the basis of its performance when crossed with other parents (lines).

Specific combining ability (*SCA*) is the behaviour of a parent (X) when crossed with another parent (Y). The average value of that specific combination may diverge from the average general combining ability of these two parents.

The mean value of lines X and Y is thus calculated:

$$M_{XY} = GCA_X + GCA_Y + SCA_{XY}.$$

The question what makes a line a good or a poor combiner is answered by the method of diallel crosses, i.e., by crossing each line of a given set with all other lines of the set (Table 9.1). The number of crosses, taking into calculation the reciprocal ones, is $n(n-1)$. If a set consists of ten lines, it is necessary to make $10(10-1)=90$ cross combinations; if a set consists of 150 lines, the number of diallel crosses is 22,350. As it is practically impossible to handle such a large number of crosses, general combining ability is evaluated by testing all lines against a common parent, i.e., a tester. In the case of maize, a best hybrid or a standard variety are used as testers. Only those lines which performed well with the tester are selected for diallel crossing for specific combining ability. Lines which exhibit heterosis in particular combinations are then used as components for the development of corn hybrids.

With plants such as lucerne, rye, etc., in which it is difficult to control pollination, combining ability is evaluated by the polycross method. All lines are first sown in a random arrangement where it is probable that each line will be pollinated by all other lines. As the male parent remains unknown, differences between the lines, in yield and other traits, are attributed to the females and such differences are used for the assessment of their general and specific combining abilities.

When two or more parents are hybridized not merely for the use in the F_1 generation but to develop a new line or a cultivar, it is important to know the combining ability of the parent in order to make use of the phenomenon of epistasis.

When organisms which multiply by seed are crossed, uniform lines are obtained only in late generations, starting from the F_5, because of segregation of traits in early generations. It therefore takes a long time to wait and see whether a cross combination will or will not produce superior lines. Breeders thus look for reliable indicators of good combining ability which would be visible already in the F_1 or F_2 generation. Weiss et al. (1947), Fowler et al. (1955), Lupton (1961), Borojević (1966) arrived at the conclusion that:

a) parents with good general combining ability will probably render good lines in later generations;

b) good combining ability is expressed as heterosis in a large number of cases;

c) heterosis for earliness is a fairly reliable indicator of good combining ability;

d) there exists a positive correlation between earliness, plant height, and some traits of the ear in the F_2 generation on the one side and performances of lines in later generations on the other.

These conclusions on combining ability are certainly valuable but there are many exceptions which question the reliability of these conclusions.

The genetic basis of combining ability has not been fully elucidated as yet. In view of the theory of diallel crosses, Jinks (1955) and Hayman (1958) considered general combining ability to be a result of additive gene action, and specific combining ability a result of epistasis, i.e., interallelic interaction. Falconer (1967) associated differences in general combining ability to the additive component of genetic variance and to interaction of alleles and differences in specific combining ability to the non-additive variance, i.e., dominance and epistasis, and their interactions.

Statistical analysis of combining ability

A mathematical model for the analysis of combining ability elaborated by Griffing (1956) is based on the analysis of variance of a diallel series of crossings. Depending on the available data, one of the four experimental methods may be used:

Method 1: involves parents, F_1 or F_2, and reciprocal crosses [all p^2 combinations].

Method 2: involves parents and F_1 or F_2 generation (ignoring reciprocals) [$1/2p$ $(p+1)$ combinations].

Method 3: involves F_1 or F_2 and reciprocals (without parents) [$p(p-1)$ combinations].

Method 4: involves only mean values of F_1 generations, without parents and reciprocals [$1/2p(p-1)$ combinations].

Each method requires a different analysis. Here is the mathematical formula for the combining ability analysis according to model I, method 2, which includes the following calculations:

$$X_{ij} = \mu + g_i + g_j + s_{ij} + 1/bc\, \Sigma_k \Sigma_l\, e_{ijkl}$$

where

X_{ij} = the mean value between crosses of i and j parents,

μ = population mean effect

g_i = the g.c.a. effect for the ith parent

g_j = the g.c.a. effect for the jth parent

s_{ij} = the s.c.a. effect for the cross between ith and jth parents

e_{ijkl} = the environmental effect associated with the $ijkl$th individual observations.

The analysis of variance for general (GCA) and specific (SCA) combining ability is given in Table 9.1.

The analysis of combining ability for the number of grains per spike will be illustrated by an example of diallel crosses of six wheat varieties conducted in Novi Sad.

The basic sample for the analysis comprised 40 plants for each parent, 15 F_1 combinations and 30 back crosses, and 500 plants for each of 15 F_2 combinations.

Although the experiment was conducted in a randomized block design in four

Table 9.1 Analysis of variance for method 2 giving expectations of mean squares for the assumptions of model I

Source	Degrees of freedom	Sum of squares*	Mean squares	Expectation of mean squares Model I
General combining ability GCA	$p-1$	Sg	Mg	$\sigma^2 + (p+2)\frac{1}{p-1}\Sigma gi^2$
Specific combining ability – SCA	$p(p-1)/2$	Ss	Ms	$\sigma^2 + \frac{2}{p(p-1)}\sum_{i\leq j}\sum sij$
Error	m	Se	$M\acute{e}$	σ^2

* Where

$$Sg = \frac{1}{p+2}\left[\sum_i (Xi. + xii)^2 - \frac{4}{p}X..^2\right],$$

$$S_s : \sum_{i\leq j}\sum x_{ij}2 - \frac{1}{p+2i}\Sigma(X_i. + x_{ii})^2 + \frac{2}{(p+1)\cdot(p+2)}X..^2$$

The following F ratios may be used to test g.c.a. and s.c.a.

effects:

for differences among g.c.a. effects

$F(p-1, m) = Mg/M\acute{e}$

for differences among s.c.a. effects:

$F[p(p-1)/2, m] = Ms/M\acute{e}$

The effects may be estimated as follows:

$$\hat{g}_i = \frac{1}{p+2}\left[X_i. + x_{ii} - \frac{2}{p}X..\right],$$

$$s_y = x_{ij} - \frac{1}{(p+2)}X_{i.} + x_i + X_{j.} + x_{jj} + \frac{2}{(p+1)\cdot(p+2)}X..$$

where

p = number of parents

m = degrees of freedom for error of the analysis of variance in a randomized block design

$M\acute{e}$ = the mean square of error divided by the number of replications

$(X_{i.} + x_{ii})$ = the total value for rows + mean value of parents

$X..$ = the sum of individual value of the parents and the crosses

Source: Griffing, 1956

Table 9.2 Results of diallel crosses in the F_1 generation for the number of grains per spike of six wheat cultivars

Cultivar	1. UPI-301	2. Caj.-71	3. S. Cerros	4. Libellula	5. Sava	6. BZ-1	Total	$X_i + \bar{x}_{ii}$
1. UPI-301	*54.1*	53.4	63.2	56.6	57.3	56.2	340.8	394.9
2. Cajeme-71	53.4	*47.4*	62.2	58.1	61.7	53.0	335.8	383.2
3. Siete Cerros	63.2	62.2	*65.2*	62.4	71.0	60.4	384.4	449.6
4. Libellula	56.6	58.1	62.4	*50.6*	58.0	47.7	333.4	384.0
5. Sava	57.3	61.7	71.0	58.0	*61.5*	54.5	364.0	425.5
6. Bezostaja-1	56.2	53.0	60.4	47.7	54.5	*43.9*	315.7	359.6
	340.8	335.8	384.4	333.4	364.0	315.7	2074.1	
$\bar{x}$	56.8	56.0	64.1	55.6	60.7	52.6		

LSD za 5% = 2.00
1% = 2.67

Source: Malek, 1976

replications and in the same year, the analysis of variance was made for each generation separately because of differences in the number of plants per generation. The F_1 generation will be taken for illustration of the procedure of the analysis of variance for combining ability (Table 9.2).

The calculation on the basis of the data presented in Table 9.2:

$X\ldots$ (table's total) $= 54.1 + (54.3 + 63.2 + \ldots 54.5 + 43.9) = 1\,198.4;$

$(X\ldots)^2 = 1\,198.4^2 = 1{,}436{,}162.6;$

$\Sigma X_{ij}^2 = (53.4^2 + 63.2^2 + \ldots 54.5^2) + (54.1^2 + 47.4^2 + \ldots 43.9^2) = 69{,}238.3.$

The sum of squares for general combining ability (S_g or S_{GCA}) is calculated as follows:

$S_{GCA} = 1/6 + 2(394.9^2 + \ldots 359.6^2) - 4/6 \times 1{,}436{,}162.6 = 663.14.$

The sum of squares for specific combining ability (S_s or S_{SCA}) is calculated as follows:

$S_{SCA} = 69{,}238.3 - (1/6 + 2) \times (394.9^2 + \ldots 359.6^2) +$
$+ 2/(6 + 1)(6 + 2) \times 1{,}436{,}162.6 = 186.48.$

Using the same procedure and the data for the number of grains in the F_2 generation, the analysis of variance for general and specific combining ability is made for these generations.

Now a table of the analysis of variance for combining ability of all tested generations can be put together (Table 9.4).

The analysis shows that the cultivars included in diallel crosses had highly significant differences in GCA and SCA. As the mean of squares for GCA is 10–28 times larger than the mean of squares for SCA, it may be stated that the additive gene action played a more important role in the inheritance of the number of grains per spike than the non-additive gene action (dominance and epistasis).

Table 9.3 Results of diallel crosses for the number of grains per spike in wheat

	F_2 generation					
Cultivars	1	2	3	4	5	6
1. UPI-301	*54.1*	49.9	58.8	46.4	53.0	46.7
2. Cajeme-71		*47.4*	57.5	50.6	52.8	49.4
3. Siete Cerros			*65.2*	53.2	61.5	54.0
4. Libellula				*50.6*	51.7	44.9
5. Sava					*61.5*	51.3
6. Bezostaja-1						*43.9*

LSD za 5% = 2.02, za 1% = 2.68

	Back crosses					
1. UPI-301	*54.1*	49.2	57.9	50.5	54.9	48.6
2. Cajeme-71		*47.4*	55.6	50.3	52.0	48.6
3. Siete Cerros			*65.2*	51.4	61.2	54.4
4. Libellula				*50.6*	52.3	43.9
5. Sava					*61.5*	50.9
6. Beuostaia-1						*43.9*

LSD za 5% = 1.83, za 1% = 2.44

Source: Malek, 1976

Table 9.4 Analysis of variance for combining ability for the number of grains per spike

Source of variation	Degrees of freedom	Mean squares[1]		
		F_1	F_2	Back crosses
General combining ability – GCA	5	132.63**	115.70**	108.43**
Specific combining ability – SCA	15	12.44**	4.60**	3.85**
Error	60[2]	0.50	0.51	0.42

** F test for 1% significance.

[1] Mean square values obtained by dividing the value of sum squares by the degrees of freedom.

[2] Degrees of freedom from the table of analysis of variance after the randomized block design.

Since general and specific combining ability of individual cultivars or lines are of interest in breeding, these parameters may be calculated from diallel crosses by the following formulae:

GCA effect for each cultivar in the F_1 generation:

$$g_1 = \frac{1}{6+2}(394.9 - \frac{2}{6}1{,}198.4) = \frac{1}{8}(394.9 - 399.4) = -0.56$$

$$g_2 = \frac{1}{8}(383.2 - 399.4) = -2.05$$

$$\vdots$$

$$g_6 = \frac{1}{8}(359.6 - 399.4) = -4.97$$

Standard error (SE) of the difference between any two GCA values is calculated by the formula:

$$SE = \sqrt{\frac{2}{P+2} \times M\acute{e}} = \sqrt{\frac{2}{6+2} \times 0.50} = 0.35$$

LSD for 5% = 0.35×2.00 = 0.70

for 1% = 0.35×2.66 = 0.93

GCA effect for the F_2 generation and back crosses is calculated according to the same procedure. The data obtained are then included in a table for GCA (Table 9.5).

Table 9.5 Estimate of general combining ability (GCA) for the number of grains per spike in the diallel crosses of six wheat cultivars

Parents	F_1	rank	F_2	rank	Back crosses	rank
1. UPI-301	−0.56	3	−0.65	3	+0.11	3
2. Cajeme-71	−2.05	5	−1.65	4	−2.13	4
3. Siete Cerros	+6.27	1	+5.90	1	+5.31	1
4. Libellula	−1.92	4	−2.52	5	−2.36	5
5. Sava	+3.26	2	+3.14	2	+3.35	2
6. Bezostaia-1	−4.97	6	−4.26	6	−4.28	6
SE(gi)	0.35		0.36		0.33	
LSD za 5%	0.70		0.72		0.66	
1%	0.93		0.96		0.88	

The cultivars Siete Cerros and Sava exhibited highly significant GCA effects in all three generations which allows for the conclusion that they are good general combiners for the number of grains per spike. The other cultivars, especially Bezostaia-1 and Libellula, exhibited negative GCA effects and may be considered rather poor general combiners for the examined trait. It should be mentioned that there existed a good agreement among the GCA values in all three generations which made the conclusions more reliable.

To have the whole picture of combining ability, it is necessary to calculate SCA values. SCA effect of each cross combination in the F_1 generation is calculated as follows:

$$S_{12} = 53.4 - \frac{1}{6+2}(394.9 + 383.2) + \frac{2}{(6+1)(6+2)} \times 1{,}198.4$$

$$= 53.4 - 97.3 + 42.8 = -1.10$$

$$S_{13} = 63.2 - \frac{1}{}(394.0 + 449.6) + \frac{2}{56} \times 1{,}198.4$$

$$= 63.2 - 105.6 + 42.8 = 0.40$$

$$S_{56} = 54.5 - \frac{1}{8}(425.5 + 359.6) + \frac{2}{56} \times 1{,}198.4$$

$$= 54.5 - 98.1 + 42.8 = -0.80$$

Standard error (SE) of the difference between any two SCA values is calculated by the following formula:

$$SE = \sqrt{\frac{2p}{p+2} \times M\acute{e}} = \sqrt{\frac{2 \times 6}{6 \times 2} \times 0.50} = 0.85$$

LSD for 5% $= 0.85 \times 2.00 = 1.70$
for 1% $= 0.85 \times 2.66 = 2.26$

SCA effect for the F_2 generation and back crosses is calculated according to the same procedure.

All data are then gathered in a table of SCA effects (Table 9.6).

Of the 15 cross combinations in the F_1 generation, ten combinations exhibited positive SCA effects for the number of grains per spike. The best general combiners, Siete Cerros and Sava, produced highly significant SCA values (S. Cerros × Sava) (Table 9.6). Positive SCA effects were obtained also in the crosses between good and poor combiners, e.g., Cajeme-71 × Sava, S. Cerros × Libellula, S. Cerros × Bezostaia-1. Some combinations of poor parents produced nevertheless positive SCA values – UPI-301 × Libellula, UPI-301 × Bezostaia-1, Cajeme-71 × Libellula, Cajeme-71 × Bezostaia-1 (Table 9.6).

The obtained SCA values were not as consistent among the three generations as the GCA values. For example, only four F_2 cross combinations and five back crosses displayed positive SCA values (Table 9.6).

Considering their performance in all three generations, the following cross combinations were found to be the best specific combiners for the number of grains per spike – Cajeme-71 × Libellula (both with poor GCA), Cajeme-71 × Bezostaia-1 (both with poor GCA), S. Cerros × Sava (both with good GCA), S. Cerros × Bezostaia-1 (one with good GCA, another with poor GCA), Cajeme-71 × S. Cerros (one with poor GCA, another with good GCA), UPI-301 × Bezostaia-1 (both with poor GCA) (Table 9.5).

Table 9.6 Estimate of specific combining ability (SCA) for the number of grains per spike in diallel crosses of six wheat cultivars

Cross combination	Specific combining ability (SCA) F_1	F_2	Back crosses
UPI-301 × Cajeme-71	−1.06	−0.45	−1.41
UPI-301 × Siete Cerros	+0.44	−0.90	−0.15
UPI-301 × Libellula	+2.04	−3.08	+0.12
UPI-301 × Sava	−2.45	−2.14	−1.18
UPI-301 × Bezostaia-1	+4.69	−1.04	+0.14
Cajeme-71 × Siete Cerros	+0.90	+0.60	−0.21
Cajeme-71 × Libellula	+5.00	+2.12	+2.15
Cajeme-71 × Sava	+3.41	−1.34	−0.95
Cajeme-71 × Bezostaia-1	+2.95	+2.66	+2.38
Siete Cerros × Libellula	+1.00	−2.83	−4.18
Siete Cerros × Sava	+4.41	−0.19	−0.08
Siete Cerros × Bezostaia-1	+2.05	−0.29	+0.74
Libellula × Sava	−0.39	−1.56	−1.32
Libellula × Bezostaia-1	−2.45	−0.96	−2.10
Sava × Bezostaia-1	−0.84	−0.23	−0.80
SE(*sij*)	0.85	0.88	0.79
LSD za 5%	1.70	1.76	1.59
1%	2.26	2.34	2.11

On the basis of the data obtained, it does not seem possible to draw general conclusions on the combining ability of individual varieties. First of all it is important to know if the GCA data for a certain variety allows for the estimation of SCA of that variety. The best general combiner, the cultivar S. Cerros, had four positive SCA value in five cross combinations; the second best general combiner, the cultivar Sava, had only two positive SCA values (Table 9.5). The poorest general combiner for the number of grains per spike, the cultivar Bezostaia-1, produced three combinations with positive SCA values and two with negative SCA values. Another poor general combiner, the cultivar Cajeme-71, produced four combinations with positive and two with highly positive SCA values (Table 9.5).

This is an indication that general combining ability is not a reliable basis for the estimation of specific combining ability of a variety. The reliability may increase in a diallel which includes a much larger number of parents, so that each parent is checked in a large number of combinations. Nevertheless, it is probable for parents with good general combining ability to have good specific combining ability more frequently than parents with poor general combining ability.

Numerous researchers (Singh and Gupta, 1969; Bhatt, 1971; Gill et al., 1972; Kraljević-Balalić, 1974; Halloran, 1975; etc.) found that highly positive SCA values are frequently obtained when a parent with a high GCA and a parent with a poor GCA are crossed. The same phenomenon was observed in the presented example (Malek, 1976). However, it is important information that a parent with a poor GCA may render very positive SCA values.

All studies of combining ability in diallel crosses conducted so far have indicated that parents with the highest mean values for the tested trait are the best general combiners and parents with lowest mean values for the trait are the poorest general combiners. This is a further confirmation of an old empirical observation that parents with many positive and few negative traits should be selected for crossing. The fact that the highest SCA values were exhibited by the crosses between a good and a poor general combiner support another old experience which says that at least one parent must have a high mean value for the trait which is tried to be improved. Since the improvement in a single trait is seldom set as a breeding objective, but at least two or more traits, it is desirable for the other parent to possess a high mean value for another trait. Only then a breeder can expect recombinations of two or more positive traits in a new cultivar.

These are the reasons which preclude the application of the method of diallel crosses for the testing of combining ability in self-pollinated plant species. Instead, parents possessing different mean values of desired traits are crossed directly.

If lines are not used directly as commercial cultivars but for the production of hybrids, as is the case with maize, sunflower, etc., their combining ability should be tested in order to estimate which lines may produce good heterotic hybrids (see Chapter 16).

References

Bhatt, G. M., 1971. Heterotic performance and combining ability in a diallel cross among spring wheats. Australian J. Agric. Res. 22: 359–368.

Borojević, S., 1966. Combining ability in wheat crosses. Proc. 2nd. Int. Wheat Genetics Symp., Lund 1963, Hereditas, suppl. vol. 2, 102–118.

Borojević, S., and Potočanac, J., 1966. The development of the Yugoslav programme for creating high-yielding wheat varieties. Savremena poljoprivreda (Novi Sad), XIV, 11–12: 7–36.

Cicin, H. B. (Ed.), 1963. Gibridi otdalenih skreščivanij i poliploidi. Izdat. A.N. SSSR, Moskva, 204 pp.

Falconer, D. S., 1964, 1967. Introduction to Quantitative Genetics. R. MacLehose and Co. Ltd., Glazgow, Ronald Press, New York, 3rd edition, 365 pp.

Fowler, W. S., and Heyne, F. G., 1955. Evaluation of bulk hybrid tests for predicting performance of pure line selections in hard red winter wheat. Agron. J. 47: 430–434.

Frankel, O. H., 1947. The theory of plant breeding for yield. Heredity, 1: 109–120.

Gill, K. S., Dhillon, S. S., and Bains, K. S., 1972. Combining ability and inheritance of yield components in crosses involving Indian and exotic wheat germplasm. Indian J. Genet. & Pl. Breeding. 32: 421–430.

Griffing, B., 1956. A generalised treatment of the use of diallel crosses in quantitative inheritance. Heredity 10: 31–50.

Halloran, G. M., 1975. Genetic analysis of yield in wheat. Z. Pflanzenzüchtung, 74: 298–321.

Hayman, B. I., 1958. The theory and analysis of diallel crosses. II. Genetics, 43: 64–85.

Jinks, J. L., 1955. A survey of the genetical basis of heterosis in a variety of diallel crosses. Heredity 9: 223–238.

Kraljević-Balalić, Marija, 1974. Nasledjivanje veličine lisne površine i sadržaja hlorofila kod vulgare pšenice. Doktorska disertacija, Novi Sad.

Lupton, F. G. H., 1961. Studies in the breeding of self-pollinating cereals. 3. Further studies in cross prediction. Euphytica, 10: 209–224.

Lukyanenko, P. P., 1970. Hybridization of distant ecogeographical forms of winter wheat. Proc.

Soviet-Swedish Symp. on Selection of Self-Pollinated Crops. Krasnodar 1967. Izd. VASHNIL, Moskva, p. 10–23.
Maystrenko, O. I., 1978. Cytogenetic study of the growth habit and spike emergence time in wheat *Triticum aestivum L.* XIV. Inter. Congress Genetics, Moskva.
Malek, M. A., 1976. Inheritance of final yield components in diallel crosses of vulgare wheat. Phd. thesis, University of Novi Sad, (Yugoslavia).
Michurin, I. V., 1939. Sočinenija. Tom I. Ogiz. Seljhozgiz, Moskva, 655 pp.
Singh, K. B., and Gupta, V. P., 1969. Combining ability in wheat. Indian J. Genet. and Pl. Breed., 29: 227–232.
Weiss, M. G., Weber, C. R., and Kalton, R. R., 1947. Early generations testing in soyabeans. J. American Soc. Agron. 39: 791–811.

10. IMPORTANCE OF CROSSING METHODS

Crossing methods affect the success of breeding programmes to the extent in which they contribute to the enlargement of genetic variability. The actual choice of a hybridization method depends on the knowledge of whether the available parents are capable of ensuring the recombination of positive traits in new cultivars.

With regard to the different methods of producing the genetic population from which genotypes are selected for the development of new cultivars, there exist three ways of plant breeding: 1) selection breeding, 2) combination or divergent breeding, and 3) convergent breeding.

Selection breeding

This approach is based on the selection of certain phenotypes from a population formed through open pollination or random mating, without interference in the genetic structure of the population by the breeder. It includes the selection from natural populations of those cross-pollinated species which are difficult to hybridize artificially, whose sources of cytoplasmic male sterility have not been identified, and whose inbred lines have not been made yet, e.g., some forage legumes and grasses. Phenotypes are thus selected from open pollinated populations, intercrossed by the polycross method, and further selected to develop commercial cultivars.

Divergent breeding

Various procedures exist which make use of genetically diverse parents in order to recombine their desirable traits. These methods also rely on gene interactions to obtain cultivars with desirable traits. Such methods are known as recombination or divergent breeding.

The single cross method – $A \times B$

This is the method of crossing of two parents, for example:
wheat cultivars: Bezostaia-1 × Siete Cerros,
maize lines: B73 × C103,
tomato cultivars: Zaria × Cometa,
apple cultivars: Jonathan × Delicious.

If parents are carefully selected to possess the maximum of positive traits and the minimum of negative traits, progenies possessing positive traits of both parents will be fairly frequent in the F_2 generation. Therefore, this method is relatively quick and reliable. It has produced innumerable cultivars of wheat (Tables 7.1–7.3), barley, rice and many other crops.

The single cross method, $A \times B$, is most successful in cases when F_1 hybrids are commercially used. The F_1 generation contains the largest number of desirable genes of both parents and this frequently results in the expression of hybrid vigour. Single cross hybrids are used in the production of maize, sorghum, cucumber, tomato, onion and other crops.

If one parent is an advanced cultivar and the other a primitive cultivar, it is not probable that lines superior to the better parent will results from any cross-breeding, even though the primitive cultivar possesses a trait which would further improve the advanced cultivar. For example, the wheat cultivar Bankuty1205 may be used instead of Bezostaia-1 as a donor of high quality grain. However, Bankuty-1205 also has a tall stem, it is susceptible to lodging, less resistant to rusts, and its production per spike is lower than that of Bezostaia-1. Therefore, in the F_2 and subsequent generations, most plants of a single cross including Bankuty-1205 will be intermediate in all traits in respect to the parents and there will be a low frequency of plants possessing the desired genes of both parents.

If a sufficient number of cultivars with desirable traits is not available, the breeder must include primitive cultivars with the desired traits in his breeding programme. Then it is advisable to make one back cross with the advanced cultivar to introduce 75% of its germplasm into the single cross and so increase the frequency of the desired genes and promote the success of selection:

(1) A (advanced cultivar) $\times$ B (primitive cultivar)
germplasm 50% 50%
intermediacy

(2) $F_1(A \times B) \times A$
germplasm 25% 25% 50%

BC_1F_1 75% germ plasm from A
higher frequency of desired genes

An example of the use of one back cross is the pedigree of the cultivar Sava which was made by crossing an advanced Italian cultivar, Fortunato, with less advanced American cultivar, Redcoat, i.e., (Fortunato $\times$ Redcoat) $\times$ Fortunato (see Table 7.3).

The method of the three-way cross – $(A \times B) \times C$

When it is not possible to find two parents that possess all desired traits, it is necessary to select a third parent and cross it with the F_1 generation:

$$
\begin{array}{l} A \times B \\ \downarrow \\ F_1 \times C. \end{array}
$$

If one of the parents must be less advanced cultivar, i.e., a cultivar with several negative traits or undesired genes, the other two cultivars must be advanced, possessing a large number of positive traits. It makes quite a difference which parents will be used in the first crossing and which in the second:

(1) A (advanced) $\times$ B (advanced) $\times$ C (less advanced)
germplasm 25% 25% 50%

(2) (A $\times$ C) $\times$ B
germplasma 25% 25% 50%

In the first example, the advanced cultivars (A and B) incorporate 50% of the germplasm in the three-way cross and the less advanced cultivar (C) incorporates 50% of the germ plasm. It reduces the probability of finding desired genes together in certain progenies which lowers the chance of success. In the second example, the advanced cultivars incorporate 75% of their germ plasm and the less advanced cultivar only 25% which greatly increases the frequency of desired genes.

The crucial point for three-way crosses, as pointed out MacKey (1962), is the selection of the third parent which is used at the last step of crossing. If it is a genotype well-adapted to the agroecological conditions of the region for which the new cultivar is made, the success will be higher.

With some crops, a hybrid of three-way crosses may be used directly for production. With maize, the F_1 generation of two inbred lines is sometimes crossed with a third inbred line or a cultivar (top cross); the resulting hybrid, if it is heterotic, is then grown commercially. The expression of heterosis in that case is determihed by the choice of the third line or cultivar used in the second step of crossing. Three-way crosses are not genetically homogeneous, as are F_1 of single -cross hybrids, and they segregate in traits. There are not many successful three-way combinations and their use, in comparison with single- and double-cross hybrids, is limited.

The method of successive crosses – [(A $\times$ B) $\times$ C] $\times$ D

If all the traits intended to be incorporated in a new cultivar cannot be found in three parents, it becomes necessary to introduce a fourth and sometimes even a fifth parent.

(1) $A \times B$
↓
(2) $F_1 \times C$
↓
(3) $F_1 \times D$
↓
selection

Their portions of the germ plasm are:

[(A $\times$ B) $\times$ C] $\times$ D
12.5% 12.5% 25% 50%

It is essential for the method of successive crosses to use in the last step of crossing a cultivar with the maximum number of positive traits because its germplasm makes up 50% and the germplasm of all other parents makes up the other 50 percent.

This method may be fairly successful in combining genes from many parents. It has been used in Yugoslav wheat breeding programmes when switching from the hybridization of less advanced cultivars to the hybridization of advanced cultivars.

However, the method of successive crossing takes a long time to produce a desired genetic population; after that, an additional few years are necessary to select suitable homozygous lines from the segregating generations. The method is thus used to improve existing cross combinations, when parent components are found to be incapable of ensuring the desired recombination of traits. The time to produce an F_1 may be shortened by growing two generations of the crossing programme per year in a greenhouse.

The method of double crosses – $(A \times B) \times (C \times D)$

This method is counted among complex crossing methods which are resorted to when two parents are not sufficient to provide all the desired traits. When the method of double crosses is used, each of the four parents equally contribute their germplasm to the new hybrid:

	$(A \times$	$B) \times$	$(C \times$	$D)$
germplasm	25%	25%	25%	25%

The frequency of desirable genes in an F_2 generation of an $F_1 \times F_1$ cross is considerably lower than that in an F_2 generation of a single cross, especially if one or two parents of a double cross possess a large number of negative traits. Therefore, as stated by Sneep (1977) and Sneep et al. (1979), the complex methods of crossing cannot be used for the combining of more than ten desirable alleles without a risk of losing a large portion of them.

Otherwise, the method of double crossing is quite efficient for the production of $F_1 \times F_1$ hybrids, as is the case with the production of double-cross maize hybrids aimed at the exploitation of heterosis.

The method of diallel crosses

Plant breeding institutes keep collections of varieties originating from various countries. As it is impossible to know beforehand the combining ability of a certain variety, the method of diallel crosses is sometimes used to pinpoint parents capable of rendering a large number of superior lines or those capable of producing heterotic F_1 hybrids.

It is not possible to estimate in advance whether a cross $A \times B$ will be superior to a cross $A \times C$ or $D \times E$. The following crossing sheme must therefore be established:

1. $A \times B$				
2. $A \times C$	6. $B \times C$			
3. $A \times D$	7. $B \times D$	10. $C \times D$		
4. $A \times E$	8. $B \times E$	11. $C \times E$	13. $D \times E$	
5. $A \times F$	9. $B \times F$	12. $C \times F$	14. $D \times F$	15. $E \times F$

The number of combinations may be calculated by the formula:

$n - (n - 1)/2,$

where n is the number of varieties in the crossing programme.

Fifteen cross combinations can be made from a set of six varieties, and 190 combinations from a set of 20 cultivars in a diallel cross. Essentially, a large number of cross combinations must be performed with a limited number of varieties, knowing beforehand that a portion of the combinations will be unsatisfactory. This is why the method of diallel crosses is not used in breeding programmes of self-pollinated plants but only for the testing of general and specific combining ability of inbred lines to find heterotic combinations (maize, etc.). Otherwise, the diallel cross method is used exclusively for genetic studies since it offers information on the effect of a number of genes, combining, ability, and gene interactions.

Convergent breeding

If emphasis is placed in a breeding programme on the improvement of a single trait or a couple of traits in an otherwise good cultivar, taking care to maintain the other positive traits of that cultivar, methods of convergent breeding are applied. This can be done by the backcrossmethod and various versions of convergent crossing.

The backcross method – $(A \times B) \times A)$ 2—6

In the backcross method, an F_1 generation is crossed several times with one of the parents.

The initial crossing	A	$\times$	B
	recurrent parent		donor

1st backcross (50% genes coming from A)	$F_1 \times A$
2nd backcross (75% A)	$BC_1F_1 \times A$
3rd backcross (87.5% A)	$BC_2F_1 \times A$
4th backcross (93.7% A)	$BC_3F_1 \times A$
5th backcross (96.9% A)	$BC_4F_1 \times A$
6th backcross (98.4% A)	$BC_5F_1 \times A$

The method was first used and recommended by Harlan and Pope (1922) and Briggs (1922) in the United States of America. Then it was used widely in the course of the 'twenties and the 'thirties for the incorporation of disease resistance in standard cultivars (Briggs and Allard, 1953). The backcross method is very efficient for the introduction of one or two positive traits into a good cultivar, maintaining at the same time its other positive traits. The cultivar which brings this new trait is called a *donor* and the cultivar which receives it a recipient or a *recurrent* parent.

The intention of the backcross method is to incorporate a trait, most frequently the resistance to a pathogen, into a recurrent parent without changing its other traits. It is achieved through a series of back crosses of progenies which exhibit the donor's desired trait with the recurrent parent. After six series of back crossing, the progenies will posses 98.4% of the germplasm of the recurrent parent as well as the desired trait from the donor. If six or more back crosses are performed, the procedure is called the method of *complete backcrossing* because the recurrent parent has been completely restituted, differing theoretically from the original recurrent parent in a single trait. The method of complete back crossing is resorted to when a donor is a cultivar possessing a large number of negative traits but one highly valued positive trait.

As the method of complete backcrossing returns the recurrent cultivar to its original type, a small number of new genes being introduced into it after a long process of hybridization, the method is considered to be conservative and its use is recommended only when it is estimated that the original cultivar may hardly be excelled, and that it should be kept as it is but improved in only one trait.

If, on the other hand, a donor of a certain trait possesses a number of other positive traits, it is not necessary to completely restitute the recurrent parent. In that case, the method of incomplete backcrossings is recommended, i.e., only one or two backcrosses with the recurrent parent need be made (Borojević, 1965). It allows gene recombination for other positive traits too, resulting in a number of genetically diverse lines.

The backcross method may be considered a method of selection because individuals possessing a trait obtained from the donor are being selected throughout the entire process of hybridization. More about it will be said when reviewing methods of selection.

Other methods of convergent crossing

When breeding reaches a high level, as is the case in some developed countries, and when it becomes difficult to develop better cultivars than those already grown, it may be advisable to start using different versions of convergent and divergent crossing. Mac Key (1962) recommended several versions, such as: (1) convergent crossing according to the principle of maximum recombination, (2) convergent crossing according to the principle of transgressive recombination, (3) convergent crossing according to the combined principle of transgressive recombination and incomplete back crossing, and (4) convergent back crossing (Table 10.1). A difference among

Table 10.1 Different versions of the convergent cross

1. Convergent cross according to the principle of maximum recombination:

1st year	$A\times B$ $\quad C\times D$ $\quad E\times F$ $\quad G\times H$	50% A-type
2nd year	$\times \qquad\qquad \times$	25% A-type
3rd year	$\times$ ↓	13% A-type

2. Convergent cross according to the principle of transgressive recombination:

1st year	$A\times B$ $\quad A\times C$ $\quad A\times D$ $\quad A\times E$	50% A-type
2nd year	$\times \qquad\qquad \times$	50% A-type
3rd year	$\times$ ↓	50% A-type

3. Convergent cross according to the combined principle of transgressive recombination and incomplete backcross:

1st year	$A\times B$ $\quad A\times C$ $\quad A\times D$ $\quad A\times E$	50% A-type
2nd year	$\times A$ $\quad \times A$ $\quad \times A$ $\quad \times A$	75% A-type
3rd year	$\times \qquad\qquad \times$	75% A-type
4th year	$\times$ ↓	75% A-type

4. Convergent cross according to the combined principle of transgressive accumulation and backcross (convergent backcross):

1st year	$A\times B$ $\quad A\times C$ $\quad A\times D$ $\quad A\times E$	50% A-type
2nd year	$A^2\times B$ $\quad A^2\times C$ $\quad A^2\times D$ $\quad A^2\times E$	75% A-type
3rd year	$A^3\times B$ $\quad A^3\times C$ $\quad A^3\times D$ $\quad A^3\times E$	88% A-type
4th year	$A^4\times B$ $\quad A^4\times C$ $\quad A^4\times D$ $\quad A^4\times E$	94% A-type
5th year	$\times \qquad\qquad \times$	94% A-type
6th year	$\times$ ↓	

Source: Mac Key, 1962.

these methods is in the percentage of germplasm of the recurrent parent (A-type) planned to be retained in new lines. The percentage ranges from 12.5 in complex methods of divergent crossing to 93.7 in nearly complete backcrossing.

There also exist other methods of divergent and convergent crossing which are used in plant breeding. For example, the composite or multiple cross (Harlan et al., 1940) or the synthetic cross (Sprague, 1946) use 8, 16 or even more parents to produce in pairs their F_1 generations in several cycles and ending up with the final F_1 where genes of all parents will be included. The idea was to maximize multiple-parent recombination of genes, but the question is how much gene recombination of so many parents will take place in a final population (Frey, 1983). As analysed by Ackerman and Mac Key (1948) it would require an enormous number of crosses and sizes of population if only one desirable gene from each parent should be present in a recombined plant.

It may be concluded that the probability of success in breeding is not increased by the use of complex methods of crossing but by a careful selection of parents which should possess many positive traits and hence many desirable genes.

References

Ackerman, A., and Mac Key, J., 1948. The breeding of self-fertilized plants by crossing. In: A. Ackerman, O. Tedin and K. Froier (Editors), Plant Breeding Svalöf 1896–1946. K. Bloms Boktr., Lund, p. 46–71.

Borojević, S., 1965. (On the efficiency of the incomplete backcross method in wheat breeding. Savremena poljoprivreda, (Novi Sad), 5: 403–415.

Briggs, F. N., and Allard, R. W., 1953. The current status of the backcross method of plant breeding. Agron. J. 45: 131–138.

Frey, K. J., 1983. Plant population management and breeding. In: D. R. Wood (Editor), Crop Breeding. American Society of Agronomy, Madison, USA, p. 55–88.

Harlan, H. V., Martini, M. L., and Harland, S., 1940. A study of methods in barley breeding. United States Department of Agriculture Tech. Bull. 720.

Harlan, H. V., and Pope, M. N., 1922. The use and value of backcross in small grain breeding. J. Hered. 13: 319–322.

Harrington, J. B., 1952. Cereal Breeding Procedures. FAO, Rome, 122 pp.

Mac Key, J., 1962. The 75 years development of Swedish plant breeding. Hodowla Roslin Aklimatyzacja i Nasiennctwo. (Warsaw), Tom 6, Zeszyt 4–5.

Sneep, J., 1977. Selection for yield in early generations of self-fertilizing crops. Euphytica 26: 27–30.

Sneep, J., Murty, B. R., and Utz, H. F., 1979. Current breeding methods. In: J. Sneep and A. J. T. Hendricken (Editors), Plant Breeding Perspectives. Pudoc, Wageningen, p. 104–233.

Sprague, G. F., 1946. Early testing on inbred lines of corn. J. American Soc. Agron. 38: 108–117.

11. NUMBER OF CROSS COMBINATIONS AND SIZES OF F_2 GENERATIONS

It has already been said that success in plant breeding depends on the choice of parental pairs and methods of crossing. In addition to that, the number of cross combinations in the F_1 generation and the number of plants grown in the F_2 generation are important constituents.

Number of cross combinations in the F_1 generation

When selecting materials for crossing, questions arise about which cultivars or lines to include in a crossing programme and how many different cross combinations, $A \times B$, $A \times F$, $C \times D$, $B \times E$, $K \times N$, etc., should be made annually or within the time at disposal to carry out a programme.

The number of cross combinations for the production of F_1 seeds depends on:

(1) the structure of the sexual organs of the plants and the feasibility of carrying out the crossing,

(2) the concept of the choice of parental pairs,

(3) the available manpower and facilites necessary for the work.

In the case of bisexual plants, e.g., cereals, the anthers are fairly large and they are easily removed by forceps while they are still green. Furthermore, these emasculated flowers are pollinated relatively simply by pollen gathered from nonemasculated plants. This is why most breeding stations dealing with bisexual plant species perform each year several hundred cross combinations.

On the other hand, there are plant species whose male flower structures are not easily removed because their anthers are small and are imbedded deep in the flower, e.g., soyabean and some other legumes. With those species, only a small number of cross combinations may be performed each year.

Gametocides have been introduced in recent years for bisexual plants (see Chapter 16). Gametocides destroy anthers and pollen without damaging the stigma. After the treatment, pollen of a variety selected for crossing as a male component is easily applied.

With monoecious and dioecious plants, in which male sexual organs are separated from female sexual organs, emasculation is not necessary. The female organs are simply isolated (for example the ear and silk of maize) from natural open pollination before pollen is deliberately applied from another plant. In crossing programmes involving such plant species, a few hundred cross combinations may be performed anually at one institute (station).

There are large differences in the number of cross combinations performed in different breeding institutes and by individual breeders. Some breeders make a limited number of cross combinations, under one hundred, but they pay special attention to the selection of parental pairs for crossing. They usually select them on the basis of the concept of the gene or the concept of character taking care that both parents possess many positive traits and the smallest possible number of negative traits.

Most stations which breed wheat and other self-pollinated plants have the practice of making one hundred to five hundred cross combinations a year. Parental pairs are selected on the basis of the concept of character but, since the combining abilities of all cultivars intended to be used in crossing programmes are not known, an increased number of cross combinations is necessary.

More than one thousand cross combinations of wheat, rice, barley, maize, etc., are performed yearly at some stations, particularly in international plant breeding centres. Knowing neither the genetic constitution of all available varieties nor their combining abilities, but wishing to save time, they include in crossing programmes all new materials that come to their collections. In stations which pursue this practice, only the best F_1 cross combinations are selected for the F_2 generation.

It is difficult to decide what is better, to make a small number of carefully planned cross combinations or to go for a large number of combinations hoping that some of them will give superior lines after the selection in the F_2 and subsequent generations. Both ways may be more or less successful. Nevertheless, it has been realized that a very small number of cross combinations a year (10–20), no matter how carefully planned, is not a guarantee of success. A small total number of cross combinations will inevitably produce a small number of combinations with good specific combining ability. Conversely, the crossing of more than two thousand cross combinations a year is a work at random and it is certain that many cross combinations will have to be discarded already in the F_1 and F_2 generation because of the lack of desired gene recombinations.

It would be useful if by a statistical analysis one could estimate the optimum number of cross combinations to obtain by high probability a large number of superior lines. Baker (1984) tried to make such a calculation, but since it deals with a single quantitative trait and does not take into account epistasis and linkage (which are always present), it is uncertain to rely on such formulae in a breeding programme. However, Baker concluded that 50 or so crosses may be required for maximum response to selection for a quantitative trait.

If we take it that breeding is usually done for multiple traits instead of a single trait, it may be said that at least a few hundred cross combinations per year per station per crop should be performed. This has been confirmed by empirical experience also.

Yonezawa and Yamagata (1982) tried to settle the dispute by offering a mathematical model which predicts a cross combination in which a larger number of promising genotypes could be expected. In conclusion, these authors emphasize that the number of cross combinations should be increased rather than the number of plants per a cross combination.

It should be pointed out that the number of cross combinations may vary from one year to another. When new programmes are established or when breeding objectives are changed, several hundreds or thousands of cross combinations are usually performed during a few years. On the other hand, when new parental pairs are not available, the number of cross combinations may be considerably reduced.

These considerations of the number of cross combinations in the F_1 generation pertain mostly to self-pollinated species whose F_1 plants cannot be used commercially and must be selected through segregating generation to produce homogeneous lines. With crops whose hybrid seed may be used for commercial production of F_1 plants, as with maize, rye, onion, cucumber, etc., it is definitely recommendable to carry out several hundred cross combinations each year in order to detect heterotic combinations.

Finally, it is frequently asked how many flowers, spikes, or ears to cross per combination. This depends on the number of plants planned to be grown in the F_2 generation. If the breeder wishes to test about 4,000 wheat or barley plants in the F_2 generation, it would be enough to cross only three spikes. Supposing that 20 seeds will develop per spike, there will be sixty plants in the F_1 generation. Each plant may easily render one hundred seeds, thanks to tillering, so there will be more than 6,000 plants in the F_2 generation.

To obtain 1,000 F_2 soyabean plants, which is a lot considering the difficulties that are encountered in crossing, it may be enough to obtain five to ten F_1 seeds per cross combination.

In the case of tomato, only one or two flowers need to be crossed because one to two hundred seeds develop in each fruit.

With maize, for which only the F_1 generation is tested, three to five ears should be pollinated (about one hundred seeds are obtained per ear) to produce a sufficient amount of seeds for the replicated testing of F_1 plants.

The size of the F_2 population

For self-pollinated and open-pollinated plant species whose F_1 plants cannot be used directly for commercial production, the F_2 and further generations must be grown in order to select new genotypes which recombine the parents' genes. The success of selection depends to a large extent on the number of plants in the F_2 generation per cross combination.

When two varieties or pure lines are crossed, the largest number of different gene combinations is achieved in the F_2 generation. In the case of self-pollinated plant species, the frequency of homozygous genotypes increases in the F_3 and subsequents generations and only the F_2 generation contains all new combinations that might be expected. The number of gene recombination differing from their parents depends on:

(1) the number of gene loci in which the parents are different,
(2) the linkage of gene loci,
(3) the number of alleles in each gene locus.

If two parents differ in two pairs of allelic genes, four different kinds of gametes will develop in the F_1 generation. Thus, the F_2 generation should consist of at least 16 plants for all gene combinations to appear ($4\times4 = 16$) (Table 11.1). If two parents differ in five gene pairs, the minimum number of plants in F_2 generation is 1,024 in the case of full dominance.

The purpose of crossing programmes is to produce a large number of gene recombinations from the parents used. If parents differ in ten alleles, the number of different gametes in the F_1 generation will be 1,024 and the number of different genotypes in the F_2 generation will be 59,049. For all those gene recombinations to be expressed in the F_2 generation, it would be necessary to grow not less than 10,500,000 plants. If parents differ in 50 pairs of genes, the entire surface of the Earth would not be sufficient for all gene recombinations to be expressed in the F_2 generation (Table 11.1).

Table 11.1 Variability in F_2 generation between parents differing in n allelic gene pairs

Number of allelic gene pairs	Number of different gametes in F_1	Number of genotypes in F_2	Number of phenotypes in F_2 assuming full dominance	no dominance no epistasis	Smallest needed population in F_2
1	2	3	2	3	4
2	4	9	4	9	16
3	8	27	8	27	64
4	16	81	16	81	256
5	32	243	32	243	1,024
10	1,024	59,049	1,024	59,049	1,084,576
20	1,084,576	$3,484\times10^{9}$	1,084,576	$3,483\times10^{9}$	$1,102\times10^{9}$
50	$1,122\times10^{12}$	$7,161\times10^{20}$	$1,122\times10^{12}$	$7,161\times10^{12}$	64×10^{22}
n	2^n	3^n	2^n	3^n	4^n

If a gene possesses several alleles, as in the case of multiple alleles or polyploids, the number of gene recombinations and new genotypes increase considerably, requiring an increased number of plants in the F_2 generation.

When parents for crossing are selected on the basis of differences in a number of gene loci, it is inevitable that a certain number of them will be linked, i.e., will be located on the same chromosome. The linkage reduces the frequency of gene recombinations which would have been higher if these genes had been independent. The extent of reduction depends on the distance between the linked gene loci. If they are close, the number of crossing-over gametes will be small and the number of parental type gene combinations will be correspondingly larger.

The effect of gene linkage on the number of desired gene recombinations in the F_2 generation varies from the case of *coupling phase* (when both dominant gene loci are present in one parent and both recessive gene loci in another) to the case of *repulsion phase* (one dominant and one recessive gene loci in one parent, the opposite gene loci in another parent). An example for two linked gene loci for resistance to smut in wheat was given in Chapter 4, Table 4.5.

If two varieties are crossed, $RRTT \times rrtt$ (coupling phase), 81.9% of F_2 plants will be resistant to smut because of the presence of one, the other, or both dominant genes of resistance. Of the total number of resistant plants, 83.1% will possess both genes of resistance and 22% of those will be homozygous for both genes. If two other varieties are crossed, $RRtt \times rrTT$ (repulsion phase), about 99.5% of F_2 plants will be resistant, about 50% will possess both genes, but only 0.56% will be homozygous for both genes. If the two genes had been independent, 93.75% of F_2 plants would have been resistant, about 60% would have possessed both resistance genes, and only 6.25% would have been homozygous for both genes (see Table 4.5).

Therefore, it is of the utmost importance for the F_2 generation to be as large as possible. Of course, it is not possible to grow several hundred thousand or a million plants per cross combination although theoretically it would be necessary when parents differ in a large number of gene loci. The ruling principle is to grow as many plants as necessary to ensure a high probability of obtaining those genetic recombinations and traits set as the breeding objective. The size of the F_2 population and consequently of the F_3 progenies is a topic of perpetual discussions among breeders and geneticists.

Among several authors who studied the problem, the contributions of Sneep (1977) and Baker (1984) should be particularly mentioned. Using bread wheat as an example, on the assumption of 21 independent gene pairs, Sneep (1977) calculated that the chances of having one plant with all desirable alleles (either homozygous or heterozygous) are 1 in every 421 plants for the F_2, 1 on every 19,343 plants for the F_3, 1 on every 176,778 plants for the F_4, etc. To be practically certain that such a plant could be found, one should grow at least four times as many plants, i.e., 1684 F_2 plants (421×4).

It is practically impossible to find a plant with all desirable alleles in later generations. Therefore, Sneep suggested that F_3 lines of sufficiently large size should be grown. Assuming that the F_2 selected plant is homozygous for 7 desirable loci and heterozygous for 14 desirable loci, the F_3 derived line should consist of at least 57 plants to warrant the one chance to obtain again a desirable plant. To be safe, at least four times as many plants (57×4) should be grown, i.e., 228 per F_3 line.

Baker (1984) calculated a minimum population size depending on different numbers of genes under the assumption that they are independent and with equal effect (Table 11.2). He used the term *expected genotypic maximum value*, which means the maximum value that will be observed in at least 50% of the samples of the quoted size. As is shown in the table, samples of 100 to 500 plants should be sufficient for traits determined by 5 to 10 genes. In the case of 20 or more genes the expected maximum value is similar, whether the sample size is 500, 1,000, or 5,000. Thus, Baker concluded that population sizes of between 500 and 1,000 should be adequate for short-term response in self-pollinated crops. and somewhat smaller populations in medium to long-term recurrent selection programmes.

Our experience in regard to wheat and other cereals is that the optimum population size of the F_2 generation should range between 2,000 and 5,000 plants when dealing with a few hundred cross combination per year.

If less than one hundred cross combinations are performed annually, it is recom-

Table 11.2 Relative expected maximum genotypic values in various sample sizes

Sample size	Number of genes				
	5	10	20	50	100
20	0.89	0.79	0.71	0.63	0.59
50	0.96	0.84	0.75	0.66	0.61
100	0.99	0.88	0.77	0.68	0.62
500	1.00	0.94	0.83	0.71	0.65
1,000	1.00	0.96	0.85	0.73	0.66
5,000	1.00	1.00	0.89	0.76	0.68
∞	1.00	1.00	1.00	1.00	1.00

Source: Baker, 1984

mended that the F_2 generation should include more than 5,000 plants per cross combination. If the number of cross combinations is over one thousand a year, the number of plants per cross combinations could be reduced in accordance with the abilities to carry out the work properly. Naturally, a differential approach to breeding material may be applied. Those cross combinations whose parents possess a complex of positive traits, whose combining abilities are known, or which performed well in the F_1 generation, should be represented by more plants in the F_2 generation than those cross combinations whose parents are not good performers in the F_1 generation.

In any case, the size of the F_2 generation is decisive for success in breeding self-pollinated plant species because this generation contains the maximum number of gene combinations among which desired recombinations should be searched for, brought to the homozygous state, and new genotypes developed from them.

In the case of open pollinated plant species, where parents are heterozygous, new gene recombinations are found already in the F_1 generation. Some additional gene recombinations and genotypes may occur in the F_2 generation. However, the desired genotypes cannot be fixed, i.e., brought to the homozygous state, because of open pollination. To be able to reproduce the desired genotypes, there remains the need to develop inbred lines wherever possible, and to use F_1 hybrids (maize, sorghum, onion, etc.) or to reproduce valuable genotypes vegetatively (potato, grape-vine, fruit trees, poplar trees, etc.).

References

Baker, R. J., 1984. Quantitative genetic principles in plant breeding. In: J. P. Gustafson (Editor), Gene Manipulation in Plant Improvement. Plenum Press, New York, p. 147–176.

Sneep, J., 1977. Selection for yield in early generations of self-fertilizing crops. Euphytica 26: 27–30.

Yonezawa, K., and Yamagata, H., 1982. Selection strategy in breeding of self-fertilizing crops. IV. The sizes of F_2 and F_3 populations in consideration of selection. Japanese J. Breed. 32 (1): 35–44.

12. METHODS OF SELECTION

It is extremely important in the process of developing new cultivars to know which method of selection should be used to choose plants for further work. Basically, a selection programme may be started when a heterogeneous population is available, i.e., when variability is observed in a population. Heterogeneous populations are either found in nature or they are made by hybridizing genetically different varieties.

In principle, there are two main methods of selection – mass selection and individual selection. All other methods are variations or combinations of these two basic methods.

Mass selection

Mass selection implies the selection of plants from a population on the basis of their phenotype, bulk sowing of their seeds and growing of the next generation in order to obtain new cultivars or to maintain the varietal purity of existing cultivars.

This is undoubtedly the oldest method of plant breeding which started soon after man began to cultivate the land. Improvements have been introduced into the method especially in the direction of what is selected rather than how the selected material is handled in later generations.

Mass selection is applied to self- and cross-pollinated plant species alike but, due to the differences between the two systems of mating, it has different effects.

The effect of mass selection depends on:

(1) effect of a gene for a desirable trait,
(2) heritability of a trait,
(3) interaction genotype/environment,
(4) sample size.

If the effect of genes for desirable traits is additive, the success of selection will be higher than if it is dominant. In the former case, repeated selection the next year will not be needed; all that is required is to continue multiplying the plants. In the latter case it will be necessary to repeat the selection.

As mass selection is based on the choice of phenotype, its success depends to a large extent on the heritability of desirable traits. If the heritability is high, the progenies in subsequent generations will be similar to the selected plants. Conversely, if the heritability is low, which is the case with such traits as fruit size, grain yield, etc., the progenies may diverge from the selected plants. Multiple mass selection is then recommended.

If the interaction genotype/environments is high, which occurs with traits with a low heritability, selection will not be very successful.

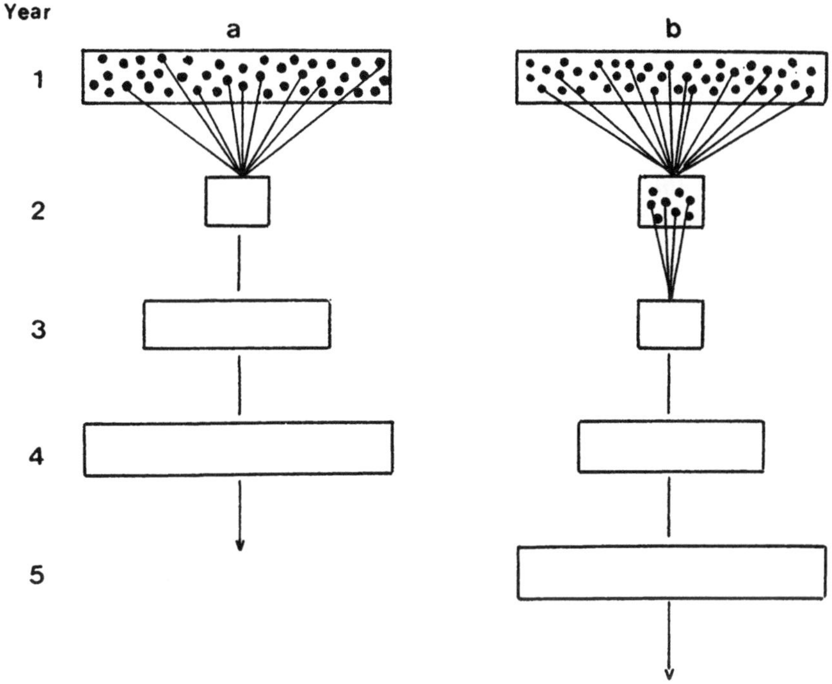

Fig. 12.1 Mass selection from natural and local populations of self-pollinated and vegetatively propagated plants: (a) simple, (b) multiple

Sample size is another important factor in mass selection, especially with open pollinated plants when the sample should be as large as possible to avoid inbreeding depression which frequently causes yield reduction.

In the case of cross-pollinated plant species, new gene recombinations which had not existed in the initial population occur during the course of selection. New genetic variability is thus added by each cycle of selection, making the mass selection method effective.

As there is no crossing between genotypes in populations of self-pollinated plants (except for a low percentage of open pollination), mass selection leads to the separation of homozygous lines, their number depending on the diversity of the source population.

Nowadays, mass selection from natural populations is applied to those self-pollinated species which have not been selected intensively (grasses, forage legumes, etc.) or it is applied when more efficient methods, e.g., individual selection, cannot be afforded.

In some cases, the method of mass selection is successfully used for populations of self-pollinated plants originally developed by hybridization. Romero and Frey (1966) reported a case of successful mass selection for a short stemmed oat derived from a hybrid population. By selecting short plants in the course of five generations, they managed to reduce the stem height by 5 cm in respect to the source population.

Frey (1967) offered another example of successful mass selection for grain mass where the increase was 9% in relation to the source population. The breeding objective was approached indirectly, through mass selection for grain size. Knowing that there was a high positive genetic correlation between the two traits, it was easier to sieve the grains for a certain size rather than to weigh individual grains. This experiment also showed that the correlation between grain mass and grain yield cannot be presented graphically by a straight line but by a curve since it was necessary to grade the grains and only keep the fraction consisting of the medium-sized grains in order to obtain the best genotypes (Frey and Huay, 1969).

One of the oldest examples of the use of mass selection for cross-pollinated plants is the selection for the ear of maize. In regions where maize varieties have not been replaced by hybrids, farmers still select the largest and heaviest ears and use their seeds for sowing. Many maize varieties were developed that way, e.g. Reid Yellow Dent, King Yellow Dent, etc. in the U.S.A. and Vukovar Yellow Dent, Novi Sad Yellow Dent, Zaječar White, etc. in Yugoslavia.

Hallauer and Sears (1969) conducted a detailed study of the effect of mass selection on maize. They could not find any significant yield increase, which confirmed previous empirical observations. The reason being that there is a higher environmental variability for ear size than there is genetic variability and this leads to a low heritability. In addition to that, perpetual selection from a small sample inevitably leads to inbreeding depression. This is why this method of selecting maize and other open pollinated crops gave way in developed agricultural countries to the production of synthetics or hybrids.

Mass selection is a useful tool for the maintenance of purity in commercial cultivars. Several hundred or thousand plants typical for a cultivar are selected from a well-managed commercial plot, harvested, bulked, and sown the next year in a single plot. In the course of the next growing season, the plot is inspected at regular intervals to remove off-types which had passed unobserved in the previous year or which had developed by spontaneous hybridization. The remaining plants are harvested and used for the production of elite seed of the cultivar.

Individual selection

The methods of individual selection which will be discussed next are more efficient than the methods of mass selection for developing new cultivars as well as for the maintenance of commercial cultivars.

a) Individual selection from local populations of self-pollinated plant species.
b) Individual selection from populations of open pollinated plant species.
   i) Individual selection without isolation.
   ii) Individual selection with isolation.
   iii) Individual selection with left over seed.
c) Individual selection of biennial plants.
d) Individual selection of perennial plants.
e) Individual selection of dioecious plants.

a) Individual selection from local populations of self-pollinated plant species

Local populations of self-pollinated species like oats, barley, some forage grasses, vetches, peas and beans may still be found. These populations are more or less heterogeneous on account of the mixing of different genotypes, spontaneous hybridization and natural mutations. However, due to the process of self-pollination, these genotypes are mostly homozygous so that each population contains in fact several homozygous lines.

If plants possessing disease resistance, earliness, resistance to low temperature or drought are observed in such a local population, it is worthwhile to start a cycle of selection.

The individual selection consists of the following steps:

(1) selection of individual plants from a population,
(2) sowing of seed of individual plants in separate rows or small plots,
(3) selection of superior progenies, i.e., lines, on the basis of observations made in the course of the growing season and tests for desirable traits,
(4) testing yield and other traits of the selected lines and seed multiplication.

The procedure is given in Figure 12.2.

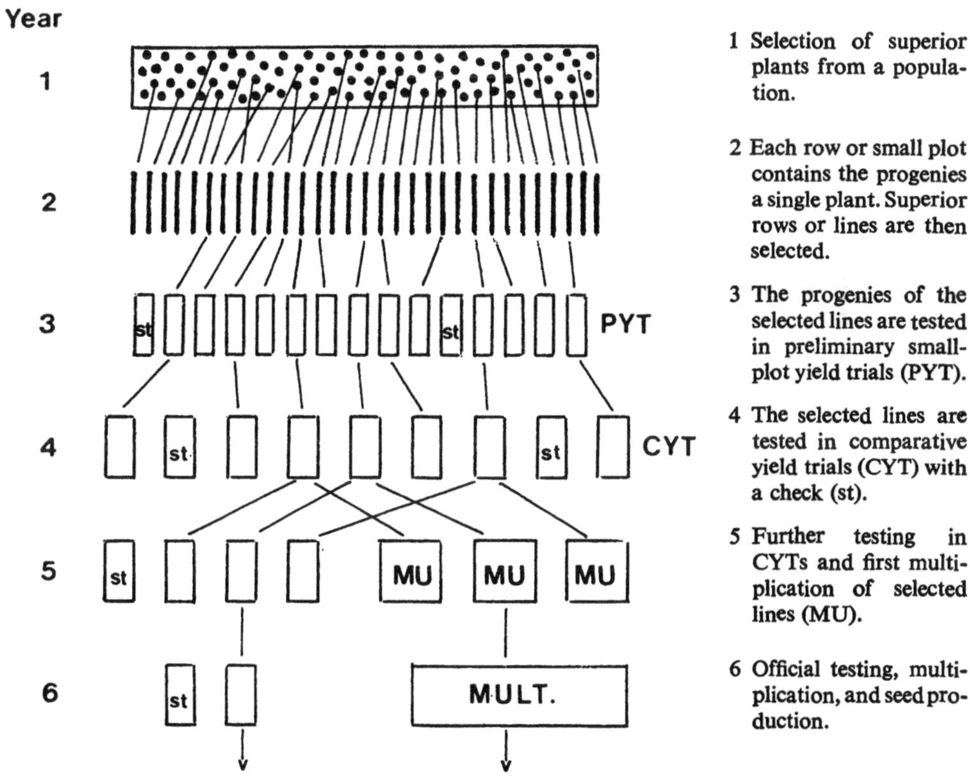

Fig. 12.2 Individual selection from natural or local populations of self-pollinated and vegetatively propagated plant species

The method of individual selection applied to self-pollinated plant species may be highly successful because it is based not only on the selection of phenotype but also of genotype. In the first year, plants are selected on the basis of phenotype. In subsequent years, when progenies of individual plants are grown separately, it is possible to assess the progenies against the original phenotype. If the difference is small, the phenotype is a true reflection of the genotype. If the difference is large, the selection must have been based on environmental variability or non-heritable modifications. Differences may also occur due to random mating in the source populations but this will be indicated by the segregation among the progenies in height, earliness, or other traits.

The method of individual selection applied to a single population of a self-pollinated plant species may produce several lines or cultivars, their number depending on the genetic variability of the source population.

Individual selection is also regularly used for the maintenance of commercial cultivars of self-pollinating crops. Several hundred or thousand plants are selected from a well-managed commercial plot and their seeds are kept separately and visually inspected before sowing in separate rows or small plots. The rows are inspected in the course of the growing season in order to select the best ones. Since the objective is the maintenance of purity and yield potential and not the development of new cultivars, the seeds from the selected rows are bulked and multiplied in subsequent years as elite (breeder seed), original (foundation seed) and further reproductions (commercial seed). The difference between the methods of mass selection and individual selection is that the former method, in the second year, allows only for the removal of off-type plants but not of off-type progenies. The method of individual selection is therefore more efficient for both, the development of new cultivars and the maintenance of commercial cultivars.

b) Individual selection from populations of open pollinated plant species

In the case of open pollinated plant species where it is not possible to make use of heterosis in the F_1 generation and to produce hybrid seed, the method of individual selection is used for the development of new cultivars from either local populations, populations developed by planned hybridization, or population mixtures.

i) Individual selection without isolation. Superior plants are selected from a heterogeneous population of bisexual open pollinated species such as rye, lucerne, etc. The seeds of individual plants are collected and sown in a separate row or small plot. The rows are inspected at regular intervals during the growing season. Undesirable plants should be removed before the stage of flowering to prevent them from pollinating the promising plants. Superior progenies are selected for further work. Since there is no isolation, the selected and the undesirable progenies intermate and the phenotypic value is assessed as the contribution of the female. The contribution of the male remains unknown, i.e., it may be assessed as the average contribution of all progenies. The progenies selected after the second year are bulked and multiplied in the third year as a new cultivar. Such a cultivar contains high variability, its rate depending on the degree of heterozygosity of the originally selected plants and the intensity of selection in later generations. Attention must be paid to work with a sufficiently large population to avoid the risk of inbreeding depression.

ii) Individual selection with isolation. The progenies of selected plants are isolated by canvas, plastic screens, or by planting rows of a barrier species such as hemp etc. If the number of progenies is not too large, a spatial isolation of at least 200 meters is an alternative. If an insect-attracting species is to be selected, individual progenies may be isolated in special cages with a certain number of insects confined in them. The contribution of the male is better assessed with this method but the risk of inbreeding depression is higher and this must be taken into account when deciding on the size of the sample.

iii) Individual selection with left-over seed. The seeds of individually selected plants are divided in two lots, one half is used for sowing and the other half is left as a reserve. Field observations are made during the growing season and the yield is assessed after the harvest. Superior progenies are selected for further work but their seeds are discarded because they have also been pollinated by the inferior progenies. Instead, the reserve seed is used for sowing. The procedure is repeated in the second and the third year. In subsequent years, the best progenies are multiplied and their seeds bulked together as one or as groups, depending on their variability and the effect of inbreeding on the tested species (Fig. 12.3).

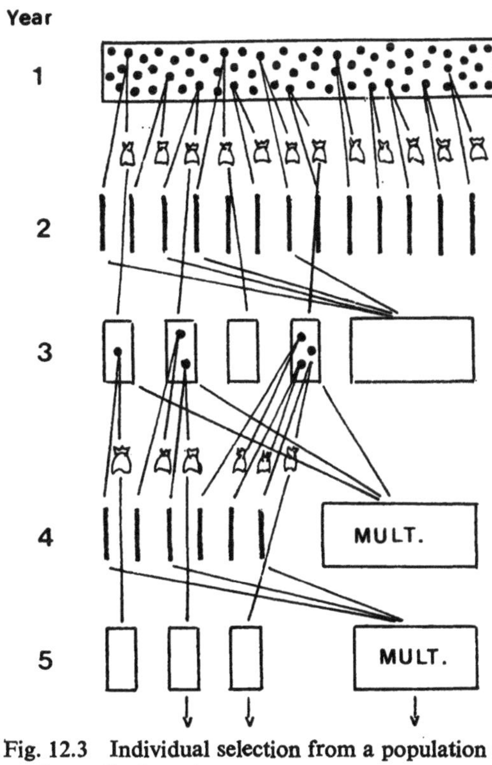

Year

1 Superior plants or ears are selected. A half of seed left for a reserve, the other half sown in individual rows.

2 The best rows are selected, their seed taken for multiplication, and the reserve lot used for sowing the next year.

3 Selection of superior plants may be repeated if necessary.

4 Sowing in rows of the best plants and multiplication of the best progenies.

5 Multiplication of reserve seed of the best rows and further multiplication in large plots.

Fig. 12.3 Individual selection from a population of an open pollinated species with left-over seed, the method ear-to-row

c) Individual selection of biennial plants

The selection of biennial plants, such as sugar beet and other root crops is more complex and it takes longer than the selection of annual plants. Good roots are selected from a variety or a hybrid population in the first year. As each root is the progenitor of an entire family, it is called a "mother plant". The selected roots are then assessed for mass, sugar content, dry matter content, etc. The screening is rigorous and a large number of selected roots are discarded. The remaining ones are cloned, i.e., cut up into three or four pieces which are then planted in separate plots. These plots are usually isolated by hemp rows or plastic foil. Seed plants are obtained from the cloned roots (Fig. 12.4). The reason for cloning is that beets are open pollinated plant species which produce a very small amount of seed when grown in isolation. If, however, a single root is cloned and several plants are obtained from it, then there will be more seeds obtained from such isolation.

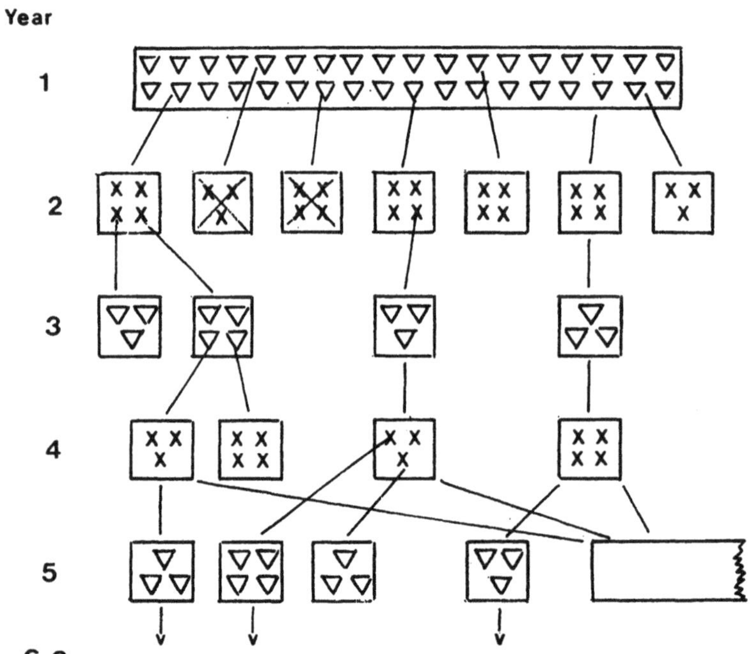

Year

1 Selection of female plants (roots) and cloning.

2 Cultivation of seed plants from clones, in isolation.

3 Cultivation of root plants and selection of superior roots.

4 Cultivation of seed plants in isolation.

5 Continuation of the selection of roots and seed plants on the basis of the assessment of the progenies. Multiplication of elite seed.

6–8 Continue

Fig. 12.4 Individual selection of open pollinated biennial plants (e.g., sugar beet)

Seed plants which are well-developed, healthy, and prolific are taken for further work and the rest is discarded. In the third year, the seeds of the selected seed plants are sown in separate plots to obtain root plants. Roots of the best progenies are selected for the growing of the next generation of seed plants in the fourth year. These plants must be grown in spatial isolation or must be isolated by plastic foil. The best seed plants are again selected for the growing of root plants in the fifth year.

The process continues until satisfactory uniformity is achieved and enough seeds for multiplication are gathered (Fig. 12.4).

In recent years, in sugar beet and in similar crops, tissue culture or other techniques are used for fast clonal propagation of superior genotypes obtained by hybridization.

d) Individual selection of perennial plants

Seed and vegetative multiplication may be combined with some open-pollinated perennial plants, e.g., lucerne, clovers, and some grass species.

In the first year, the best plants are selected from a population. These plants are cloned to obtain a large number of progenies. In the second and third year, the clones are assessed for different traits and yield performance. The seeds of the best clones are then sown in separate plots and the resulting plants are cloned again. In the fourth and fifth year, the clones are assessed for superior agronomic traits and the best clones are multiplied vegetatively and by seed. The materials selected after that are included in comparative yield trials and the seeds of the best progenies are multiplied for commercial production.

If this method of individual selection is applied without isolation, the selected material is fairly heterozygous in spite of several years of selection. It cannot be called a line but a cultivar at best. Homozygosity for quantitative characters and a large number of qualitative characters is difficult to achieve with open pollinated plants such as lucerne. This is why a fair amount of non-uniformity is tolerated in such cultivars and attention is focused to achieve the maximum possible uniformity in the main traits, e.g., vegetative period, the size of stems and leaves, etc.

e) Individual selection of dioecious plants

Special methods are inevitable in the individual selection of dioecious plants such as hemp. The best female plants are selected from a cultivar or a hybrid population and seeds of individual plants are taken for sowing in separate plots in the second year. There will be about 50% of male and about 50% of female progenies in these plots. The plots are inspected in the course of the growing season and inferior male plants are removed before flowering to prevent any undesirable pollination. After maturation, the best female plants are selected again. The procedure is continued in subsequent years until uniform progenies are obtained. Their seeds are then multiplied for commercial production.

The methods of individual selection described in the previous pages are applied to the major open pollinated plant species which differ in their systems of mating and reproduction. Many plant species are specific in their mode of reproduction and this must be taken into account when selecting them from populations.

Methods of selection of self-pollinated plant species from a population made by hybridization

A population made by a planned hybridization of different varieties is the best genetic population from which to select new cultivars. The selection from a hybrid population does not start from homozygous plants present in the population as is the case in natural and local populations of self-pollinated species. Rather, it starts from homozygous and heterozygous plants in the F_2 generation which developed as a result of random mating of gametes of F_1 plants.

Hybrid populations too are subject to mass selection, individual selection, or combinations of the two. However, since each segregating generation is different from the one that preceded it and the one that will succeed it, attention must be paid in each generation to the genetic constitution of individual plants, the number of plants grown and the number of selected genotypes, i.e., to the intensity of selection.

Bulk method

This method, called either the population method or the bulk method, was introduced by Nilsson-Ehle in Sweden at the beginning of this century. The method was adopted and theoretically grounded by Harlan, Martini, and Stevens (1940) in the United States of America. It has been used by many breeders of barley, wheat, oats, soyabean and other self-pollinated plant species.

With self-pollinated plants, gene recombinations in the heterozygous state cannot be used because the process of self-pollination leads to homozygosity which keeps increasing from the F_2 generation to reach 96.9% in the F_6 generation and 99.8% in the F_{10} generation (see Table 4.3). Therefore, the material is grown in bulk to the F_6 generation when the actual selection starts. Practically, it is as follows:

(1) After crossing varieties, $A \times B$, $C \times D$, etc., F_1 seeds of each cross combination are grown in bulk to produce sufficient amounts of seed for the F_2 generation (Fig. 12.5).

(2) Several hundred or thousand F_2 plants of each cross combination are grown together, as mentioned in Chapter 11, and harvested in bulk. It is not possible to use the entire F_2 seed lot for the sowing of the F_3 generation because the area required would be too large. Therefore, only a sample is used, taking care that several hundred or thousand plants per cross combination will be grown in the next generation. The procedure is repeated in the F_4 and the F_5 generation.

Harlan et al. (1940) performed yield testing of the F_2 bulk progenies of barley crosses and left for further selection only those crosses which gave the highest yields hoping that from them superior lines can be obtained. Atkins and Murphy (1949) made the same experiment using oats and found little relationship between F_2 bulk yield and derived lines from later generations.

(3) In the F_6, when 96.9% of homozygosity for all traits has been achieved, individual selection of prospective phenotypes possessing desirable recombinations of traits from their parents can be started. The best lines (rows) are selected for preliminary yield trials (PYT) next year.

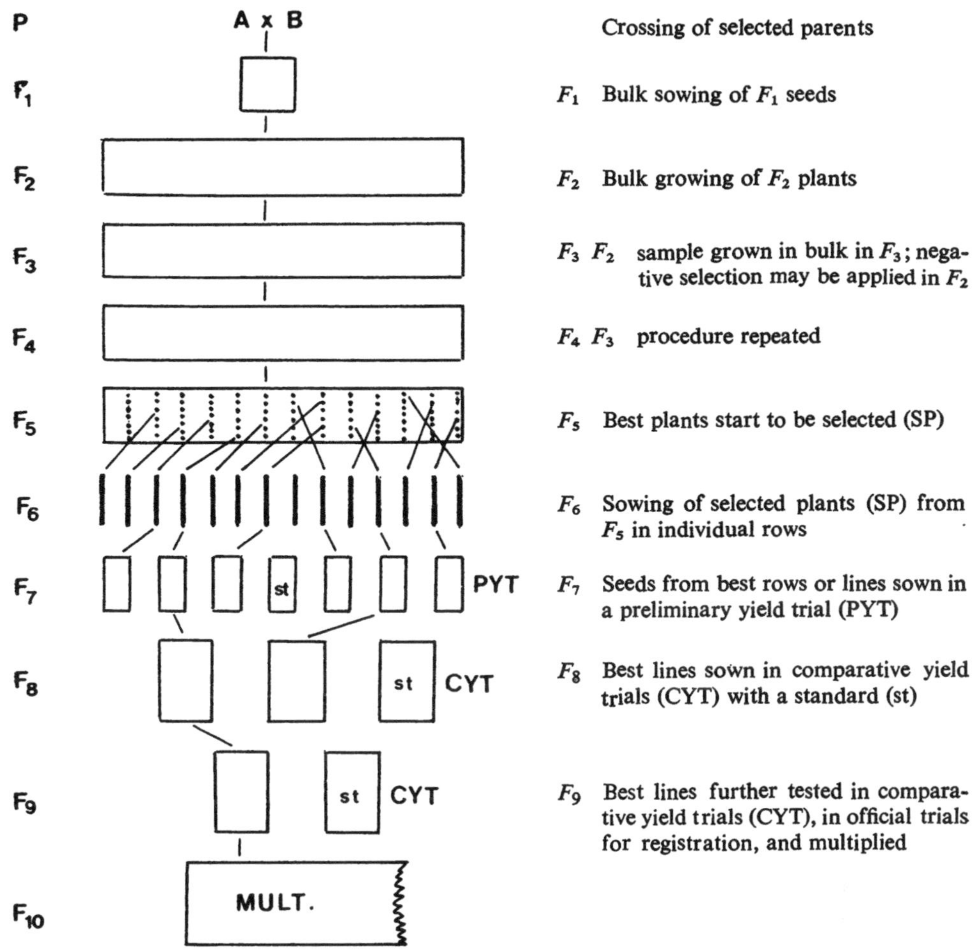

Fig. 12.5 Bulk method of selection after hybridization of self-pollinated plant species

(4) In the F_7 and further generations, the selection is based on progeny performances. The best lines are tested in comparative yield trials (CYT) and finally, the best few are selected for registration as new cultivars (Fig. 12.5).

An advantage of bulk selection is that a large number of cross combinations is grown at a low cost, without large expenditures for manpower and without much observing and selecting. One shortcoming of the method is that valuable genotypes may be lost by being unobserved during the experiments. Another shortcoming is a passive attitude of the breeder towards the breeding material. The latter is particularly unfavourable in junior breeders who should become closely acquainted with the breeding materials they use, the processes that take place within a population and the effect of selection in individual generations.

However, the value of the bulk method is most frequently discussed from the point of the effect of natural selection from F_1 to F_6 occurring due to competition between different genotypes in a population.

Natural selection takes place in a hybrid population grown in bulk because the population consists of genotypes which differ in productivity and in their reaction to the environment. Some genotypes are favoured above the others whose frequency decreases or which may gradually be completely eliminated. If two genotypes are in competition, their survival depends on:

(1) the number of seeds produced by each genotype, and
(2) the number of seeds produced by their progenies.

According to Allard (1960), the theoretical curve of survival of an inferior genotype may be calculated by the following formula:

$$A_n = a \cdot s^{n-1},$$

where A_n is the proportion of the inferior genotype in the n generation, a is the initial proportion of the inferior genotype, and s is the selection index which takes into account the number of produced seeds and the number of progenies from these seeds.

If, for example, two genotypes are mixed in the proportion 50% vs. 50% ($A_1 = 0.5$), and the values of survival capacity of the superior and the inferior genotypes are 1 and 0.9, respectively, the selection index is 0.9. The expected proportion of the inferior genotype, in the F_5 generation for example, will be:

$$A_5 = (0.5) \cdot (0.9)^4 = 0.3645 = 36.45\%$$

It means that the portion of the inferior genotype dropped in five generation from 50% to 36.45% while the superior genotype which produced more seeds and more progenies increased to 63.55%.

Very instructive results concerning competition were reported by Suneson (1949) who conducted an experiment on four barley cultivars. From the initial frequencies of 25% of each cultivar, after fifteen years, the cultivar Atlas took 88%, Club Mariout 11%, Hero 1%, while Vaughn was completely eliminated. However, when these cultivars were grown in pure stands, the cultivar Vaughn brought higher seed yields than Atlas.

If genotypes whose frequency increases over generations are also superior in grain yield then natural selection helps the breeder. Suneson (1956) reported of the effect of natural selection on 379 cross combinations of barley grown in bulk over 28 generations. Only after 20 generations did the population become superior in yield in relation to the standard cultivar Atlas. Calling this the "*evolution method of plant breeding*", Suneson interpreted the yield increase in the population as the result of natural selection which favoured high-yielding genotypes. Such genotypes may be selected as pure lines and multiplied for the establishment of new cultivars.

Gonzales (1976) studied the effect of natural selection on a bulk population of F_2 seeds of 250 cross combinations of oat cultivars, bulk samples of which were grown to F_{12} in three locations in Iowa (USA). Thirty-two lines from each of the

F_3, F_5, F_7, and F_{11} generations were randomly selected in the three locations. The lines were then tested in 15 trials differing in soil fertility and doses of mineral fertilizers. Yield increases from the F_3 to the F_{11} generations were registered in all three locations which may be interpreted as a higher survival rate of the superior genotypes.

A so-called frequency dependent selection should be mentioned at this point. Sometimes it happens that a certain rare genotype has a higher adaptive value than the other genotypes in the population and so it keeps increasing its frequency. Harding and Allard (1969) reported that in *Phaseolus lunatus*, after a hybridization of $SS \times ss$ genotypes, the average adaptive values from the 6th to the 12th generation were 1.00 for the *SS*, 1.32 for the *Ss*, and 0.86 for the *ss* genotypes. This means that the selection favoured the heterozygous genotypes on account of their increased adaptive value. This also explains the higher frequencies of heterozygous genotypes found in later generations of some inbred lines or pure line varieties than should be expected theoretically.

It can also be noted that in insects, as well as in humans, there are cases of "advantage of rare phenotype" leading to genetic polymorphism (Wallace, 1981).

However, the effect of natural selection in a bulk population may also be negative from the standpoint of breeding.

Table 12.1 Expected and realized percentage of short plants in segregating generations of rice

Generation	Expected	Realized %	
	%	No nitrogen	High doses of nitrogen
F_2	25	25	25
F_3	38	31	22
F_4	44	34	19
F_5	47	28	14
F_6	49	24	9

Source: Jennings and Herrera, 1968

Jennings and Herrera (1968) calculated the effect of natural selection in the F_2 to the F_6 generations of a bulk population of crosses between a tall rice cultivar (Peta) and a short cultivar (TN1). Plants were grown in the sowing arrangement 30×30 cm, one part of the population received no nitrogen and the other one was fertilized with a high dose of nitrogen. The result of the competition between tall and short genotypes in the population was that the portion of short plants kept decreasing from one generation to another although theoretically it should have remained at the level of 49% in the F_6 generation (Table 12.1). The tall plants, in the arrangement used, tillered more intensively than the short plants thereby gaining advantage in the competition for light. On the other hand, when 36 tall and 36 short plants were randomly selected from the F_6 generation and grown as a pure crop in a trial with 80 kg/ha of nitrogen, the short plants brought a considerably higher yield than the tall ones (Table 12.2).

Therefore, the effect of natural selection in a bulk population varies according to the prevailing conditions and to the traits which are to be improved. If a large population is established and the plant-to-plant distance is large so as to abate competition between the plants, then natural selection will not cause large changes in the proportion of genotypes. If a high effect of natural selection is desirable, conditions must be created for its full expression.

Table 12.2 Productivity of plants randomly selected from F_6 tall and short lines of rice

Variety or line	Height of plant in cm	Yield of grains in t/ha	
		Range	Mean
Peta	174	1.7–2.7	2.3
36 tall lines	164	1.4–4.1	2.7
5 best tall lines	170	3.1–4.1	3.7
TN_1	105	4.2–5.1	4.7
36 short lines	102	2.8–5.9	4.7
5 best short lines	103	5.5–5.9	5.7

Source: Jennings and Herrera, 1968

The improved bulk method. To increase the efficiency of the bulk method, improvements are introduced in the time and mode of intervention in the population.

To accelerate the selection of new lines, hybrid generations are bulk grown only to the F_4 instead of the F_6 generation. Individual selection of desirable plants starts in the F_4 generation and the pedigree method is used in subsequent generation to select new lines. In this way the time necessary to produce new lines is shortened by one or two generations. In situations when new cultivars are urgently needed, this time saving may be highly advantageous.

The bulk method with positive or negative selection. In a bulk population which is left without any intervention, an average sample of each generation contains a number of genotypes that will never be selected in the F_5 or F_6 generations (Fig. 12.5), e.g., tall and late genotypes in a programme aimed at short and early genotypes.

To increase the number of plants with desirable combinations of traits in the sample, *positive selection* may be applied starting from the F_3 generation. This practice usually involves traits which may be selected visually. For example, all short and medium-short plants are selected and bulk grown in the next generation. The majority of tall plants will thus be eliminated from further work.

In the case of selection for the length of vegetative period, *negative selection* is readily applicable. It is difficult to distinguish early from late plants at the time of harvest, i.e., as the time when this difference really matters, but it is a simple procedure to remove all plants that flower late leaving the other plants to be harvested together and bulk grown in the next generation.

Nowadays, different versions of the bulk method are applied to hybrid generations to breed new cultivars of wheat, barley, oats, soyabean, vetches, peas and other self-pollinated plant species.

Pedigree method

This is a method of individual selection of plants in segregating generations and observing the pedigree of the selected plants until homozygous lines are obtained. It deserves to be described in detail because it is used more frequently in the selection of self-pollinated plant species than is the bulk method (Borojević and Potočanac, 1966a).

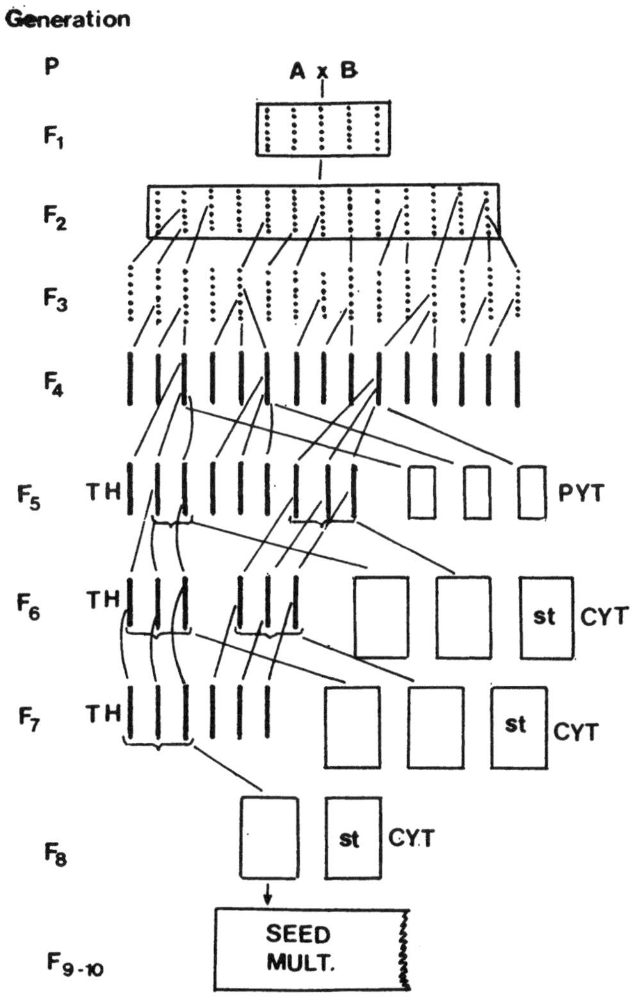

Selected parents are crossed.

F_1 plants grown with minimum competition, elimination of eventual self-pollinated plants and poor plants.

F_2 Seeds of each F_1 plant sown in a separate row, at distance. Best plants selected (SP).

F_3 Seeds of SPs sown in separate rows, selection of best plants repeated.

F_4 Seeds of best plants sown in separate rows, reduced spacing. Best rows (SR) and best plants (SP) selected.

F_5 Seeds of best plants sown in separate rows for homogeneity testing (TH), the remaining seeds from the best rows sown in preliminary yield trial (PYT).

F_6 Best lines from PYT included in comparative yield trial (CYT) using seeds from homogeneity test (TH).

F_7 Procedure continued in F_7 and F_8. Lines superior to the standard cultivar sent to official tests for registration, those approved promoted in commercial production.

Fig. 12.6 Pedigree method of selection after hybridization of self-pollinated plants

F_1 generation. F_0 seeds of various cross combinations are sown at a distance larger than that used in commercial production in order to facilitate plant growth and to obtain the maximum number of seeds (Fig. 12.6). The F_1 plants that are

not hybrids but were self-pollinated due to incomplete or late emasculation should be removed. Such self-pollinated plants, i.e., the female plant, can be recognized if the male parent possesses dominant traits (red color, awnless, etc.). The hybrid plants exhibit the dominant trait of the male while the self-pollinated plants exhibit the female's recessive trait (white colour, awned, etc.). If the female parent is the carrier of a dominant trait, self-pollinated plants cannot be recognized in the F_1 generation but can be in the F_2. Poorly developed F_1 plants should be discarded.

F_2 generation. Seed from each F_1 plant is threshed separately and sown in a separate row or a small plot for the F_2 generation. The spacing between the rows is still larger than normal.

The F_2 is the segregating generation and the rate of segregation is higher if parents differ in a larger number of genes:

♀ *AA bb CC DD ee FF gg* × ♂ *aa BB CC dd ee ff gg*.

In this cross, segregation in the F_2 will occur for the alleles *A* and *a*, *B* and *b*, *D* and *d*, *F* and *f*, and there will be no segregation in the *C*, *e* and *g* loci.

It has already been said in Chapter 11 that the F_2 generation offers the maximum gene recombinations which are not encountered in any other generation. The segregation of traits takes place in the progenies of each plant. The segregation ratios of different phenotypes and genotypes are closer to the theoretical ones if there are more plants in the F_2 generation. Fully uniform rows, i.e., those without segregating plants, contain self-pollinated plants which have not been recognized in the F_1 generation and they should be discarded from further work.

The first cycle of selection starts in the F_2 (Fig. 12.6). Those plants are selected (*SP*) whose phenotype indicates that they possess desirable recombinations of the parents' traits. As the F_2 is an early generation in which 50% of genotypes are still heterozygous, the selection should not be too severe. It is better to select a large number of plants which are expected to possess desirable gene recombinations.

In the case of plants which tiller, e.g., wheat, barley, oat, and if the F_2 populations consists of several thousand plants per cross combination, a single spike or tassel may be selected from a plant instead of the whole plant. The same procedure may be used in the subsequent generations of segregation.

Let us consider an example of a short and late cultivar crossed with a tall and early cultivar with the objective of developing a short and early cultivar.

The height of the F_1 plants is intermediate, indicating the presence of an additive gene effect. In the F_2 generation, the plants will segregate in height from those as tall as the tall parent to those as short as the short parent. Still, most plants will have intermediate stature. Regarding the length of the vegetative period, partial dominance for earliness is usually expressed in the F_1, but in the F_2, continuous variation is expressed. If major genes for height (Rht genes) or photoperiod (Ppd) are present, this may be observed by discontinuous/continuous variation in the F_2 generation.

Short and early plants should be selected first because these are the desirable recombinations of the parents' traits. Such plants are few in the F_2 generation and it is well to also select medium late and medium early plants because they will segregate in the F_3 generation and produce short and early plants in later generations. Taking into account the additive gene effect for stem height and several dominant genes for earliness, tall and late plants should be discarded because they cannot be expected to produce short and early progenies at a high frequency.

As both traits are inherited polygenically, transgressive segregation may be expected in the F_2. An extra effort should be made to find plants which are shorter than the short parent and earlier than the early parent.

The selection for these two characters is usually done visually. It is not difficult to visually estimate the height of plants in the field, especially if a range of tolerance is accepted (40–50 cm in soyabean, 60–70 cm in barley, 70–80 cm in wheat, etc.). On the other hand, it would be difficult to actually measure all of the plants when there are 1,000 to 5,000 F_2 plants per cross combination and there are hundreds of combinations to be assessed. Regarding earliness, for many plant species it is recommended to label early plants at the time of flowering because any differences cannot be perceived at the time of harvest.

The recombination of only two traits described above is taken as an illustration. However, gene recombinations are much wider when genetically diverse parents are crossed and attention should be paid in the F_2 generation to several characters. If two parents differ in the level of rust resistance, their progenies should be evaluated for that trait using an inoculation method or natural infection in the field. At the time of the final selection, a high grain productivity per plant should be taken into account together with a short stem and earliness.

If a population is selected for the resistance to low temperature, drought resistance, the contents of proteins, oils, or vitamins, then the selected F_2 plants should be analysed in the laboratory prior to the sowing of the F_3 generation.

F_3 generation. Plants selected (SP) in the F_2 generation are sown in the F_3 by the pedigree method, i.e., seeds of individual plants are sown in separate rows (Fig. 12.6). The distance between the rows remains the same as in the F_2.

F_2 plants homozygous for certain traits will produce uniform F_3 progenies for these traits while the other traits will segregate. For example, there will be short plants differing in earliness as well as plants of different height maturing at the same time.

The F_3 is the first generation after the hybridization of self-pollinated plants in which lines start to be formed. Thus, the selection in the F_3 is not based only on the assessment of individual plants but also of the progeny of one F_2 plant. Plants which possess the desirable recominations of traits are selected from prospective progenies as well as those progenies which are not satisfactory on the whole but contain some promising plants.

F_4 generation. Plants selected (SP) in the F_3 generation are sown in the F_4 by the pedigree method, i.e., plant-to-row. As the level of homozygosity reaches 87.5% in this generation, a large number of progenies display uniformity and lines are thus formed. Therefore, the selection in the F_4 generation is focused on progenies rather than on individual plants (Fig. 12.6).

Prospective progenies, i.e., lines, which are closest to the breeding objectives and visually uniform are selected for sowing in the F_5.

Since the selected progenies are not fully homozygous for certain traits, it is recommended to select about 5% of the plants before the harvest of selected rows (SR) and use their seeds in the F_5 for the test of homogeneity or homozygosity (TH) which should be sown by the pedigree method. The remaining seeds are sown in the F_5 in the preliminary yield trial (PYT).

Although the selection for yield is made along with the selection for other traits starting from the F_2 generation by selecting the most productive plants, objective data on the productivity of individual plants and lines cannot be obtained without a test for yield capacity (Fig. 12.6).

Prospective plants (SP) are also individually selected from those progenies which still segregate for certain traits.

To gather as many data as possible on the main agronomic traits of the selected progenies or lines, in addition to the visual evaluation, small amounts of seeds are taken from each selected row (SR) for testing for winter hardiness, quality and disease resistance.

F_5 generation. The seeds of progenies or lines selected in the F_4 are sown in the preliminary yield trial (PYT). The trial includes two or more replications, depending on the amount of available seed. The tested lines are compared against standard cultivar or cultivars. The trial should be as close as possible to the commerical production regarding the method of sowing, density, mineral fertilization and other cultural practices.

A test of homozygosity (TH) is established in a separate field. It includes individual plants from the lines in the preliminary yield trial (PYT) as well as rows from the selected F_4 plants which are not in the preliminary trial.

Phenological observations are made in both nurseries for earliness, resistance to diseases, resistance to lodging, etc. F_5 plants are quite uniform because their level of homozygosity is 93.8%. Nevertheless, the test of homozygosity (TH) shows clearly that not all F_5 rows from F_4 plants selected in the same row are completely identical and this cannot be observed in the preliminary yield trial (PYT).

Samples of the lines in the preliminary yield trial are analysed in the laboratory in the same year. The decision as to which lines will be included in the comparative yield trial in the F_6 and which will be discarded depends on the results of the laboratory analyses, observations in the field, and the yield performance in the PYT.

At that stage, there are several kilograms of seed from each line. This is a sufficient amount for detailed analyses of quality, tests for disease resistance, etc.

The progenies of individual F_4 plants which were not included in the preliminary yield trial in the F_5 are selected for the next PYT. Individual plants are selected for the test of homozygosity (TH).

F_6 generation. The best lines from the preliminary yield trial (PYT) are sown in a comparative yield trial (CYT) for yield performance. The seeds for the CYT are not taken from the preliminary yield trial, because of the risk of contamination between lines arising from the processes of harvesting and threshing, but are taken from the test of homozygosity (TH) (Fig. 12.6). If the test of homozygosity included 10 to 20 one-metre rows per line, there should be enough seeds produced for five replications in the comparative yield trial. The trial should include one or two stan-

dard cultivars and the cultural practices applied should be identical with those used in commercial production.

In addition to the comparative yield trial, the F_6 also includes a preliminary yield trial for the newly selected lines which should be conducted according to the methodology used in the F_5.

F_6 progenies of individually selected F_5 plants are sown in a separate plot. Here are grown TH rows of the lines in the comparative yield trial, TH rows of the lines in the preliminary yield trial and rows from F_5 plants individually selected from segregating rows.

The best lines are selected for further work on the basis of observations made in the field, results of laboratory and greenhouse analyses, and yield performance. The data on yield performance are more reliable than before because they are obtained in the comparative yield trial performed in the randomized block design in five replications. The lines with the highest yields and other desirable traits are included in a comparative yield trial in the F_7 generation.

The remaining material is screened for lines and individual plants for the next preliminary yield trial, a test of homozygosity, and for the selection of new lines.

F_7–F_{10} generations. The process of selection used in the F_6 is repeated in the F_7 and later generations with the aim of selecting as many good lines as possible.

The best lines are tested in comparative yield trials for about three years on average (F_6–F_8), possibly in several locations. Their average yields are assessed and the most superior of the best lines are submitted to official trials for varietal approval. In Yugoslavia, the Federal Commission for Variety Approval tests annual self-pollinated plant species for three years in a number of locations, depending on the production area of the crop in question; 15–20 locations for wheat, 11 for barley, nine for soyabean, etc. The trials are conducted after the randomized block design in five replications, with two standard cultivars which as a rule are the best commercial cultivars regarding yield or quality at the time of the testing. Those lines which outyield the standard or equal the yield of the standard but better it in other important traits are registered in the official catalogue as new cultivars and may be commercially released. Breeding institutes which own the new cultivars and seed-trading associations promote them to the market.

It has been a brief and general description of the pedigree method which, in fact, is a method of continuous individual selection after hybridization. The selection is based not only on the phenotype but also on the genotype which makes the method exceedingly efficient for the selection of superior lines from segregating generations. The pedigree method of selection from hybrid generations is the most frequently used method for the development of new cultivars of wheat, barley, soyabean, peas, tomato, pepper, and other self-pollinated species.

However, the pedigree method is not free of shortcomings. The main objection to it is the that it requires much work and manpower. The method calls for individual selection of plants, individual checking and threshing of plants, individual sowing, labelling of individual plants, etc. The work load reaches the limit when the number of cross combinations exceeds one thousand a year. Another important objection is that the method implies a continuous selection of a number of heterozygotes which would reach the homozygous state anyway in later generations, as in the bulk method. In addition to that, the effects of heterosis and dominance mask the genotypic value of selected plants for a long time, reducing the efficiency of selection. This shortcoming, however, is compensated by an early selection of homozygotes and a faster access to new lines for the testing of yield performance than in the case of the bulk method. Compared with the bulk method, the pedigree method reduces the competition among genotypes to a level at which it does not affect the success of selection.

To overcome these shortcomings, variants of the pedigree method have been introduced.

Pedigree method with discontinuous individual selection

This method, recommended by Akerman and MacKey (1948) in Sweden and Lupton and Whitehouse (1957) in England, is also called the method of F_2 progenies. It comprises the following procedure.

In the F_2 generation, desired plants are individually selected (SP). The progenies of each selected plant are sown in a separate row in the F_3. In the F_4, without any selection, all progenies are included in a preliminary trial for yield performance (PYT). In the F_5 and F_6, again without any selection, all progenies are tested for yield performance in comparative yield trials (CYT). The progenies with high yields and other favourable traits are selected for further work.

As the selection was done only in the F_2 generation, the progenies are fairly heterogeneous. The pedigree method should thus be applied in the F_7 and F_8 generations, i.e., individual selection within each set of progenies. The selected lines go through comparative yield trials in later generations and the best lines are registered as new cultivars.

The method of F_2 progenies is established on the idea that the inbreeding factor should be allowed to work for the breeder. The method is based on the additive component of genetic variance, not on the non-additive components (heterosis and dominance) which tend to disappear in later generations. The main point of the method is that differences between plants selected in the F_2 and their progenies in later generations are always larger than differences among lines obtained from a single source line which in itself has been selected in later generations. It ensues that the selection from a large number of initial progenies (F_2 progenies) warrants more success than a high differentiation within a limited number of initial lines, i.e., the pedigree method with continuous individual selection. The objections that the data on yield performance pertain to progenies which are not genetically homogeneous and that lines selected in later generations may be less productive, are based on the assumption that progenies with a high yield include a large number of superior lines than do progenies with a low yield.

All these explanations have a genetic background and the method of F_2 progenies could have been counted among the best methods of selection after hybridization of self-pollinated plants if it has not been burdened with practical problems.

Namely, at least one hundred plants should be selected in the F_2 so as not to make the genetic variability for selection too narrow. With entire sets of progenies discarded in the process, there usually remain in comparative trials about seventy progenies per cross combination. If a breeding programme includes about one hundred cross combinations a year, it means that there are 7,000 progenies in comparative yield trials per one cycle of crossing plus the same number of progenies from the previous and earlier years. The size of experimental material is thus too large to handle.

This is why the method of F_2 progenies or the pedigree method with discontinuous individual selection is only applied to a part of the cross combinations made, or at breeding stations which only make a limited number of cross combinations.

The method of single-seed descent

With the intention of accelerating and simplifying the methods of selection, modifications to the pedigree method have been proposed. One of them, the method of single-seed descent (SSD) deserves to be discussed in more detail.

The method had been originally proposed by Goulden (1941) and it was adapted for soyabean breeding by Brim and Cockerman (1961), Brim (1966), and Grafius (1965). The method implies that all progenies of each F_2 plant should be tested. As this is practically impossible, single seeds of each plant are randomly sampled in subsequent generations (Fig. 12.7). In the case of soyabeans, a pod with two or three seeds is sampled per plant. When a sufficient level of uniformity (homozygosity) is achieved, usually in the F_5 generation, the best plants are selected. In the F_6, seeds of selected plants are sown in separate rows. The selection is done on the level of the row (plot) because each row represents a uniform line. In the F_7, selected lines are tested for yield performance and the highest yielding lines as well as those which meet the breeding objectives regarding protein content, oil content, disease resistance and other traits are selected for further work. The procedure of individual selection is continued in further generations (Fig. 12.7).

It is possible to make additional time savings by this method. Since a single seed per selected plant is taken for the next generation, the experimental material may be grown and tested in a greenhouse in which light and temperature may be regulated. In that way, two to three generations may be tested each year.

The method of single-seed descent (SSD) offers a number of advantages but its shortcomings are numerous too. The main risk is that a portion of superior genotypes is omitted when a single seed per plant is sampled. By the same probability, inferior F_2 plants (those which do not possess desired gene recombinations), are selected. Furthermore, the number of plants in the F_4 is the same as in the F_2 and there occurs a genetic drift which inevitably leads to the elimination of many desired genes, especially if parents differ in a large number of genes (Sneep, 1977). In addition the sowing at a large distance between rows and in a greenhouse prevents testing for resistance to lodging and other traits which later on reduce the efficiency of selection in the field.

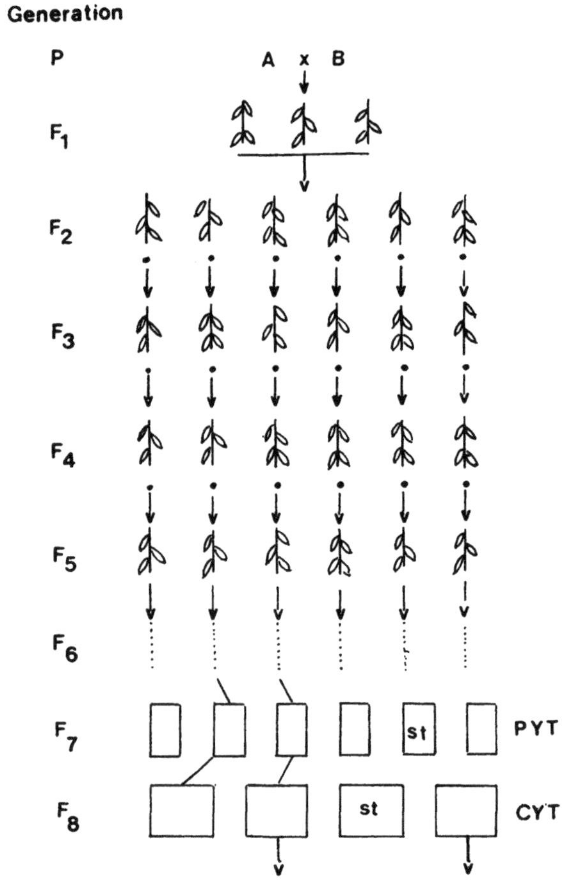

Generation	
P	$A \times B$ Crossing of varieties
F_1	Growing of F_1 plants
F_2	Growing of F_2 plants, random selection of a single seed from each plant
F_3	The same as in F_2
F_4	The same as in F_3
F_5	Selection of individual plants
F_6	Growing of selected plants in rows, selection of best rows (lines)
F_7	Testing the selected lines in a preliminary yield trial against a standard (PYT)
F_8	Yield performance testing of best lines in a comparative yield trial (CYT)

Fig. 12.7 Single-seed descent method applied to soyabean.
Source: Brim, 1966

The back-cross method

The back-cross method, $(A \times B) \times A$, has already been reviewed in Chapter 10 among the crossing methods. As this is also a method of selection (because the entire process of hybridization consists of the selection of plants to the type of the recurrent parent but with a trait derived from the donor parent) it will be described in detail in this chapter.

For example, the back-cross method is used as a selection method when a disease outbreak brings in question the further growing of a susceptible but otherwise good commercial cultivar. It is thus necessary to incorporate resistance to the disease in the cultivar, i.e., the objective is the improvement of an existing cultivar, not the development of quite a new cultivar.

The procedures of crossing and selection differ if the trait intended to be transferred from the donor is inherited dominantly or recessively. Let us consider an example of a trait controlled by a dominant gene.

A recurrent parent (*A*), susceptible to stem rust (rr), should be crossed to another cultivar (*B*), a donor of the gene for resistance (*RR*) to stem rust (*Puccinia graminis* f.sp. *tritici*).

The resultant F_1 plants are resistant but heterozygous (*Rr*) and they must be back-crossed to the recurrent parent (*A*). The progenies of the first back-cross (BC_1F_1) segregate to 50% resistant plants (*Rr*) and 50% susceptible plants (*rr*). To separate resistant from susceptible plants, they are tested for resistance to prevalent races of rust at the seedling stage in a greenhouse. The test is repeated in the greenhouse

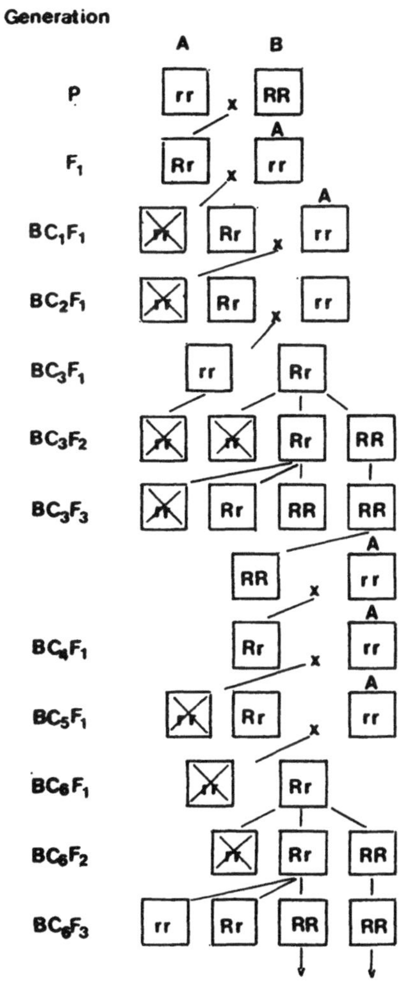

Generation	
P	Susceptible (*rr*) recurrent parent (*A*) crossed with donor (*B*) whose resistance (*RR*) is dominant.
	F_1 plants back-crossed to recurrent parent (*A*).
BC_1F_1 BC_2F_1	F_1 plants tested for resistance at the seedling stage, in greenhouse; resistant plants (*Rr*) back-crossed with recurrent parent (*A*) two times.
BC_3F_1 BC_3F_2	After third back-crossing, plants grown in field, tested for resistance, and only resistant plants with traits of recurrent parent are selected.
BC_3F_3	Selected BC_3F_3 progenies grown as lines and tested for yield performance. Yield-superior resistant plants back-crossed with cultivar *A*.
	After six back-crossings with recurrent parent and continuous selection for resistance, lines of cultivar *A* possessing resistance are obtained.

Fig. 12.8 Transfer of resistance to stem rust of wheat (controlled by a dominant gene) by the back-cross method

and in the field on mature plants. Only those plants whose resistance is confirmed should be crossed with the recurrent parent (BC_2F_1). The procedure is repeated for the third back-crossing, BC_3F_1 (Fig. 12.8).

As the breeding objective is to obtain plants to the type of the recurrent parent but resistant to rust, the F_2 generation (BC_3F_2) should be allowed to segregate in order to be able to select desirable plants. This generation is usually grown in the field at a larger spacing than normal, and inoculated to test its resistance level. Resistant plants to the type of the recurrent parent are selected and their seeds are grown in separate rows in the F_3 generation (Fig. 12.8).

The desirable resistant plants from BC_3F_3 are again crossed with the recurrent parent and those BC_4F_1 plants are selected which display resistance at the level of the donor. These plants are back-crossed with the recurrent parent once (BC_5F_1) or twice (BC_6F_1) (Fig. 12.8).

In principle, further back-crossing is not necessary because the recurrent parent has already been restituted by 96.9%. Lines are then selected by the pedigree method and tested for yield performance and resistance. The end product of this whole process of crossing and selection is the original commercial cultivar possessing a new trait which should secure its success in commercial production.

If the resistance to a disease or parasite is inherited recessively, the transfer of the resistance gene and the restitution of traits of the recurrent parent require a longer procedure of back-crossing and selection.

For example, a recurrent parent susceptible (*SS*) to leaf rust (*Puccinia recondita* f.sp. *tritici*) should be crossed with a resistant (*ss*) donor.

All F_1 plants are susceptible (*Ss*) because the resistance is inherited recessively. If the F_1 plants were back-crossed with the recurrent parent (*SS*), there would have been no resistant plants in the next generation. Thus, the F_1 plants are selfed to produce the F_2 generation.

The F_2 plants should be tested for resistance at the seedling stage and only resistant plants (*ss*) are back-crossed with the recurrent parent (*SS*). The progenies are again susceptible (*Ss*) but they are more to the type of the recurrent parent. To select resistant plants, the F_2 generation should be sown, tested for resistance, and selected for the traits to the type of the recurrent parent. The procedure is repeated until lines homozygous to the resistance to leaf rust and to the type of the recurrent parent are obtained.

The method described above of back-crossing may be modified depending on the plant species used, the genetic complexity of the trait intended to be transferred, and the desired degree of restitution of the recurrent parent. If the donor itself is a high-yielding cultivar with few negative traits, there is no need to go through six cycles of back-crossing. Three to four cycles may do just as well while the procedure is shorter and the success of breeding is accelerated.

A special advantage of the back-cross method is that it only requires a small number of plants till the BC_6F_2 so that two or three generations a year may be grown in a greenhouse or growth chambers, bringing additional time savings.

The back-cross method is highly efficient when improving a cultivar for one or two traits. The selection is still more successful if a desirable trait is controlled by one or by a pair of major genes, as is the case with genes of resistance to pathogens and, in some cases, to insects (e.g., Hessian fly in wheat).

The back-cross method is far less efficient if a desired trait is inherited polygenically or if it is linked with a negative trait.

In the 1920s and 1930s, when plant breeding was much less intensive than today, good cultivars used to stay in commercial production for a long period. It was reasonable at that time to go for the improvement of individual traits, primarily the resistance to pathogens which tended to vanish with the occurrence of new physiological races of the parasites.

Nowadays, when in many countries a large number of cultivars is developed annually and when the average commercial life of a cultivar is under five years, the method of full back-crossing is slow and conservative. If the improvement of a cultivar in a single trait is attempted, new cultivars will be developed in the meantime which will be superior in yield as well as other traits.

Nevertheless, the back-cross method has its uses in other aspects of plant breeding. It is still an irreplaceable method for incorporating male sterility in female lines of both self- and cross-pollinated species (see Chapter 16), for the development of series of monosomics, for the substitution of chromosomes (see Chapter 18), etc.

Methods of selection after hybridization in open pollinated plant species

Earlier in this chapter, methods of selection of open pollinated plants from random-mating populations, e.g., natural or local populations, have been mentioned. Regardless of the method, mass or individual, the selection is based on the female genotype while the contribution of the male remains unknown. Genotypes which recombine all or most of the desirable genes may occur only in a very large population and even then at a low frequency. As breeders work with small populations, such genotypes are usually not found and this explains the poor efficiency of mass or individual selection from random-mating populations.

Therefore, two approaches have been adopted in order to know more about the contribution of the male to its progenies: (1) to introduce planned hybridization to cross-pollinated plant species, and (2) to find and use the most adequate method of selection.

The method ear to row. A number of cultivars of open pollinated species, e.g., rye, has been developed by the method of individual selection from local populations. Different cultivars may be crossed in order to make new cultivars which possess either new gene recombinations, or a higher frequency of desirable gene recombinations; which had existed in the original cultivars or their source populations at a low frequency.

If two rye cultivars are crossed, their F_1 generation will not be as uniform as in wheat because the parental cultivars themselves are more or less heterogeneous and because there occur new gene recombinations which make the F_1 generation more heterozygous than any parent. It is therefore possible to start the process of selection already in the F_1 generation.

Phenotypically desirable plants are selected in the F_1 generation, harvested separately and sown ear to row. The rows are checked in the course of the season to remove unsatisfactory progenies before flowering to prevent them from pollinating the desirable progenies.

In the F_2 generation, the best plants are selected from the best rows, a half of their seeds is left in reserve and the other half is sown ear to row. Whole sets of progenies are selected in this cycle. The reserve seeds are then sown in order to avoid producing another generation of progenies coming from unknown male parents.

The selection continues in the F_4 and subsequent generations; similar progenies are grouped and multiplied together to produce new cultivars.

This method is based on the selection of the phenotype from a female which was pollinated by several males whose contributions cannot be estimated with precision. Since the crops grown for grain may be selected before flowering only for some traits, the method ear to row is successful only if the desirable trait is highly heritable and if a large sample is used. The latter requirement is imposed due to the risk of inbreeding depression which may bring yield reduction instead of yield increase.

Half-sib selection. With plants grown for vegetative bulk (lucerne, clover, grasses), the selection of female plants of the same genotype pollinated by more than one male may be more effective than with plants grown for grain. The progenies of the selected female plant are half-sibs which flower simultaneously. As vegetative traits are selected for, it is usually possible to remove undesirable plants before flowering and eliminate them as possible male parents to the next generation.

The half-sib selection is in fact based on general combining ability because the entire population serves as a tester for pollination of the female plants. The selection proceeds through several cycles, without preserving reserve seeds of selected plants. Success of selection is assured by the presence of superior genotypes, which in turn depends on sample size and the heritability of the selected trait.

Full-sib selection. If the selected female plant is crossed to a single male, it is known as full-sib selection. A prolonged full-sib selection tends to cause a rapid reduction of genetic variability in the population. The method is therefore applicable for the development of inbred lines but not for the development of cultivars from hybrid populations.

Polycross method. This method is a variant of half-sib selection and it is particularly suitable for plants which reproduce by seed and vegetatively (grasses, lucerne, etc.).

The work by the polycross method starts by selecting, on the basis of phenotype, about 50 best plants from a hybrid population. These plants are vegetatively reproduced in a greenhouse and transplanted in such an arrangement that each clone is surrounded by four other clones, to facilitate mutual intercrossing (polycross). The clones included in the polycross are also planted in a collection for multiplication. The best clones are threshed and individual progenies are tested in the next generation. The best progenies are selected for further work but their seeds are not collected. Seeds from the clones in the collection are used instead. The selected clones are then planted in an isolated plot to be mutually pollinated. If necessary, the cycle of selection may be repeated using reserve seeds, and the progenies of different clones are grouped.

The major shortcoming of the polycross method is that the selection of phenotypically selected clones is based on general combining ability. As each clone is surrounded by four other clones, data on specific combining ability are not available. Diallel crossing may be introduced after the polycross but it is seldom practiced because of time and difficulties involved in evaluation.

In the case where good combining ability is achieved on account of superdominance, the polycross methods leads to equilibrium in the gene frequency instead of the originally planned accumulation of desirable dominant genes. All this diminishes the success of selection but it should not be forgotten that more should not be expected with plants whose pollination cannot be controlled.

Recurrent selection

This method of selection was developed when a need was felt for increasing the success in breeding maize hybrids. In the U.S.A., random-mating maize populations have been used for a long time as source materials for the development of maize inbred lines. The initial success was high but later on the progress in the development of superior maize hybrids has not been large in spite of the large number of inbred lines produced. An idea was engendered that the most successful single- and double-cross hybrid combinations as well as synthetic cultivars might be used as source populations for the development of new inbred lines.

The starting premise was that the number of potentially available superior inbred lines depends on:

(1) the proportion of superior genotypes in the source population, and

(2) the success of selection in increasing the frequency of desirable genes in the process of inbreeding.

The higher the frequency of desirable genes, the larger the expected proportion of superior genotypes in a heterozygous population.

Therefore, a method which ensures a gradual increase in the frequency of desirable genes may be expected to render superior inbred lines and superior gene recombinations in hybrids. The idea came from Hayes and Garber (1919) and East and Jones (1920) in the U.S.A. The method itself was developed by Jenkins (1940) and Hull (1945). The latter author named it "*recurrent selection*".

The recurrent selection is realized in several cycles, each of which consists of:

(1) evaluation and selfing of selected plants,

(2) crossing of progenies of the selected selfed plants in all possible combinations,

(3) bulking equal amounts of seeds of each cross to develop new population, and

(4) development of inbred lines from the population.

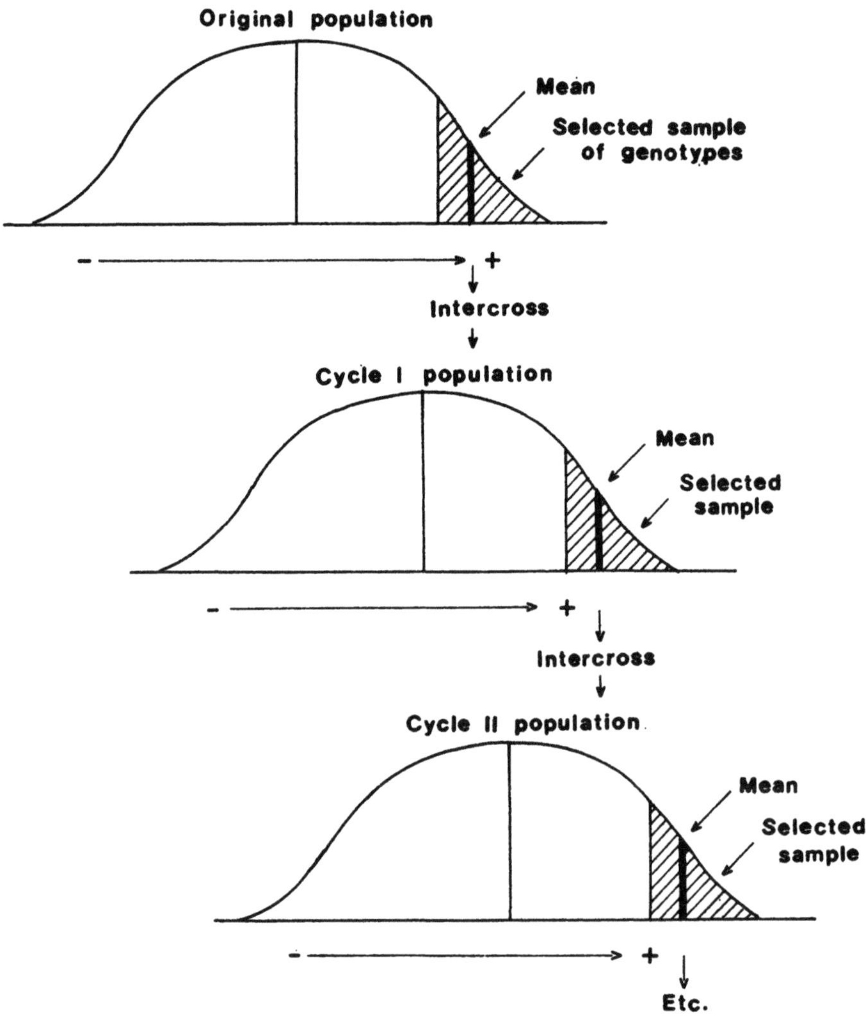

Fig. 12.9 Illustration of recurrent selection.
Source: Frey, 1983

Frey (1983) gives an illustration of recurrent selection where genotypes at the plus end of frequency distribution are selected and intercrossed through several cycles, i.e., as long as genetic variability exists in the population (Fig. 12.9).

A source material for recurrent selection may be random-mating populations, synthetic cultivars, single- or double-cross hybrids. For cross-pollinated species where extensive crossing and production of seed in each cycle is rather easy, recurrent selection is often practiced as a method of plant breeding, while for self-pollinated species, due to difficulties in obtaining hybrid seeds, it is rarely practiced. However, nowadays when male sterility is available in many species and there is the possible use of gametocides, then recurrent selection can be successfully applied on self-pollinated species too.

Four variants of the method of recurrent selection have been developed so far:

1) recurrent selection for phenotype,
2) recurrent selection for general combining ability,
3) recurrent selection for specific combining ability, and
4) reciprocal recurrent selection.

Recurrent selection for phenotype. In this method, a source material is screened and the best plants are selected and selfed. In the next generation, the progenies of each selected plant are sown in a separate row. The progenies (rows) are crossed in all possible combinations, the resulting hybrid seeds are bulked and prepared for sowing the next year as a synthetic population. This completes the first cycle of selection (Fig. 12.10).

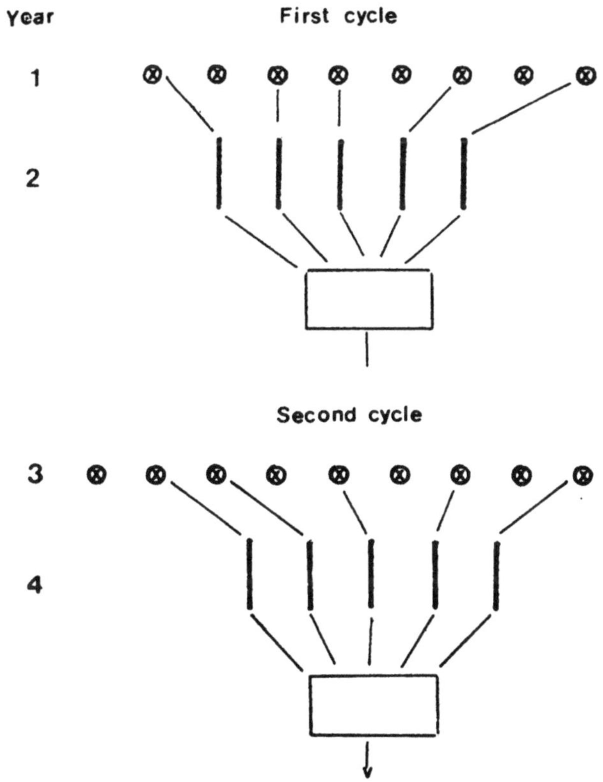

Fig. 12.10 Phenotypic recurrent selection. Where is possible to identify the desired selections before flowering intercrosses of selections may be made in the first year of each cycle, and the second year and progeny tests eliminated from each cycle.
Source: Briggs & Knowles, 1967

In the second cycle, the plants from the first cycle are selfed and the best plants are selected at the time of harvest. The seeds are sown in separate rows and the rows are crossed in all possible combinations. The obtained hybrid seeds are bulked and, if necessary, used for the third cycle of selection (Fig. 12.10).

Each cycle produces more and more uniform progenies and the degree of uniformity, i.e., still existing genetic variability, determines how many cycles of selection will be made. When it is found that there is no genetic variability any more, desirable inbred lines are maintained by selfing, ready to be tested for combining ability. The lines which produce hybrid vigour in the F_1 generation are then used for the production of hybrid seed.

One of the first successful uses of recurrent selection for phenotypes was described by Sprague et al. (1952). Starting from ten self-pollinated maize plants whose average content of oil was 4.97% in the S_1 generation, two cycles of recurrent selection for high oil percentage produced S_5 lines with the average oil content of 5.62% (S – selection generation).

Johnson (1956) applied the method of recurrent selection for a phenotype with long a vegetative period in *Melilotus*. After two cycles of selection, they succeeded in increasing the yield of green mass by about 50% in relation to the source cultivar. When the method of mass selection was applied, the yield increase of 19% was obtained after three cycles of selection as compared with the source population.

Recurrent selection for general combining ability. In this method, the plants selected in the first year are selfed and simultaneously crossed with a heterozygous tester (Fig. 12.11). If these two operations cannot be done on a single plant, a portion of seeds from the selfing is sown as S_1 in the subsequent year and these plants are crossed with the tester. In the course of the season, the plants are screened for desirable traits, most usually for yield performance, and the best hybrid progenies are selected. Their seeds are not used for sowing because the plants were in open pollination with progenies which were discarded, and so reserve seeds from selfed plants are used. Next year, the seeds of individual plants are sown in separate rows and these progenies are crossed in all combinations (Fig. 12.11). The resultant hybrid seeds are bulked to make a synthetic population which, after selfing and crossing with a tester, will serve for the next cycle of selection (Fig. 12.11).

The method of recurrent selection for general combining ability has been used fairly extensively for maize cultivars and populations in the U.S.A. It has been found that synthetic cultivars were superior in yield in relation to random populations while hybrids were superior in relation to synthetic cultivars (Lonnquist, 1961).

Recurrent selection for specific combining ability. This method differs from the method of recurrent selection for general combining ability only in relation to the tester for which a good and stable inbred line is used.

The most detailed study of the effect of recurrent selection on the production of superior maize inbred lines was conducted by Russell et al. (1973, 1977) in Iowa (USA).

Alph, a very heterogeneous maize cultivar, was used as the source population (C_0); B-14, a good and stable inbred line, was used as the tester. In the first year, plants selected from Alph were selfed and crossed with B-14. In the second year, the test-

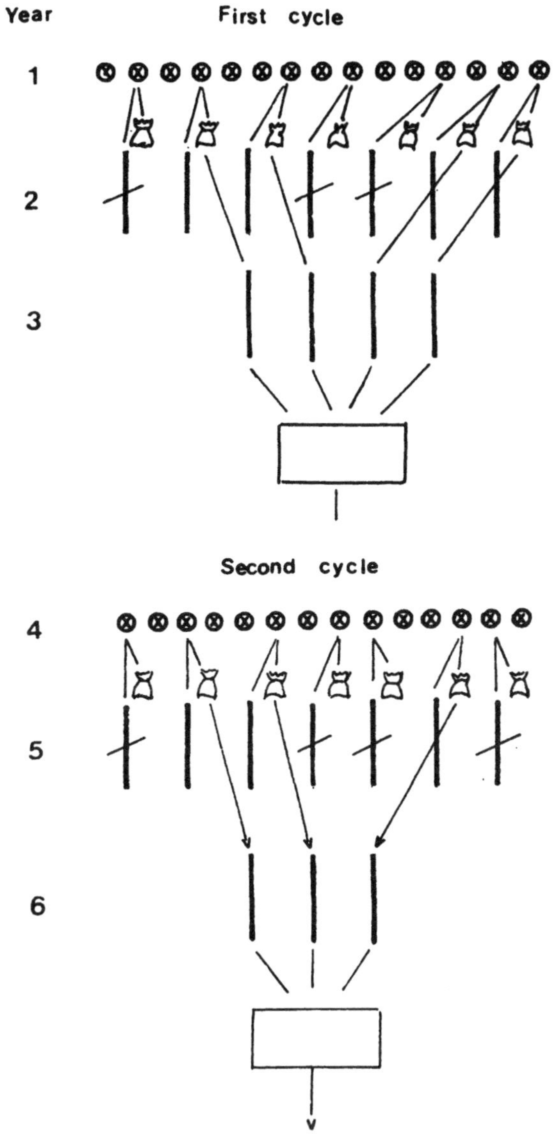

Fig. 12.11 Recurrent selection for general combining ability. Often each cycle is extended an extra year, with the first year being used for selfing and the second for increasing the selfed seed and making crosses to the tester. The third year of the diagram above may be extended into a fourth year to permit additional recombinations.
Source: Briggs & Knowles, 1967

crosses were examined for yield in comparative trials. In the third year, progenies of the S_1 plants from the first year were grown and crossed in all possible combinations to produce the C_1 population. Unlike the C_0 population which contained genotypes present in Alph, the C_1 population contained genotypes developed by crossing the best plants from Alph with B-14. In the fourth year, the C_1 population was used to commence the second cycle of selection which again included selfing the best plants, the crossing of their progenies with the tester line, B-14, the development of a new synthetic population, C_2, etc. The study included five cycles of recurrent selection. Each cycle lasted for four years, i.e., the whole programme lasted for twenty years (1949–1969).

A regression analysis of the cycles of recurrent selection for grain yield has shown that the yield increased by 309 kg/ha per cycle which may be considered quite a success.

This example shows that (by the continuous selection of superior plants, their crossing with the good tester, the development of the synthetic (hybrid) population, the selfing of the best plants from that population, the repeated crossing with the tester, etc.) it was possible to increase the proportion of superior genotypes in the populations of each cycle of selection as well as to increase the frequency of desirable genes in the inbred lines. The significant yield increase resulted from the selection for alleles with additive and partially dominant effects, i.e., the tester itself made the selection for both specific and general combining ability.

Reciprocal recurrent selection. This method implies the use of two populations, *A* and *B*. Each population is selected and selfed as in the case of recurrent selection for general combining ability. The population *A* is the tester for the population *B* and vice versa. The populations should be genetically diverse, i.e., they should be two cultivars, synthetic cultivars, or single-cross hybrids.

The method has been proposed by Comstock et al. (1949) who believed it to be better than the selection for general combining ability, or the selection of alleles with superdominant effect as well as of alleles with partially dominant effect.

References

Allard, R. W., 1960. Principles of Plant Breeding. John Wiley, New York, 485 pp.

Akerman, A., and Mac Key, J., 1948. The breeding of self-fertilized plants by crossing. In: Svalöf 1866–1946, Lund, p. 46–71.

Atkins, R. E., and Murphy, L. C., 1949. Evaluation of yield potentialities of oat crosses from bulk hybrid tests. Agron. J. 41: 41–45.

Borojević, S., and Potočanac, J., 1966. The development of the Yugoslav programme for creating high-yielding wheat varieties. Savremena poljoprivreda (Novi Sad), 11–12: 7–35.

Briggs, F. N., and Knowles, P. F., 1967. Introduction to Plant Breeding. Reinhold, New York, 426 pp.

Brim, C. A., and Cockerman, C. C., 1961. The inheritance of quantitative characters in soybeans. Crop Sci. 1: 187–190.

Brim, C. A., 1966. A modified pedigree method of selection in soybeans. Crop Sci. 6: 220.

Comstock, R. E., Robinson, H. F., and Harvey, P. H., 1949. A breeding procedure designed to make maximum use of both general and specific combining ability. Agron. J. 41: 360–367.

Frey, K. J., 1967. Mass selection for seed width in oat populations. Euphytica, 16, 341–345.

Frey, K. J., and Huany, T., 1969. Relation of seed weight to grain yields in oats *Avena sativa*. Euphytica 18: 417–424.

Frey, K. J., 1983. Plant population management and breeding. In: D. R. Wood (Editor), Crop Breeding. p. 55–87.

Goulden, C. H., 1941. Proc. 7th Inter. Genetics Congress, Edinbourgh, p. 132–133.
Gonzales, V., 1976. Evolution of production stability and other quantitative traits in a bulk population of oats. Ph.D. Thesis, Iowa State University.
Grafius, J. E., 1965. Short cuts in plant breeding. Crop Sci. 5: 377.
Hallauer, A. R., and Sears, J. H., 1969. Mass selection for yield in two varieties of maize. Crop Sci. 9: 47–50.
Harding, M., and Allard, R. W., 1969. Population studies in predominantly self-pollinated species. XII. Interactions between loci affecting fitness in a population of Phaseolus lunatus. Genetics, 61: 721–736.
Harland, H. V., Martini, M. L., and Stevens, H., 1940. A study of methods in barley breeding. U.S. Dept. of Agric. Tech. Bull. 720, 26 pp.
Hull, F. H., 1945. Recurrent selection for specific combining ability in corn. Jour. Amer. Soc. Agron. 37: 134–145.
Jennings, P. R., and Herrera, R. M., 1968. Studies on competition in rice. II. Competition in segregating populations. Evolution 22: 232–236.
Jenkins, M. I., 1940. The segregation of genes affecting yield of grain in maize. Jour. Amer. Soc. Agron. 32: 55–63.
Johnson, I. J., 1956. Further progress in recurrent selection for general combining ability in sweetclover. Agron. J. 48: 242–243.
Lonnquist, J. H., 1961. Progress from recurrent selection procedures for the improvement of corn populations. Nebraska Res. Bull. No. 197.
Lupton, F. G. H., and Whitehouse, R. N. H., 1957. Studies on the breding of self-pollinating cereals. 1. Selection methods in breeding for yield. Euphytica, 6: 169–185.
Romero, G. E., and Frey, K. J., 1966. Mass selection for plant height in oat populations. Crop Sci., 6: 883–887.
Russel, W. A., Eberhart, S. A., Vega, V. A., 1973. Recurrent selection for specific combining for yield in two maize populations. Crop Sci. 13: 257–261.
Sneep, J., 1977. Selection for yield in early generations of self-fertilizing crops. Euphytica 26: 27–30.
Sprague, F. G., Miller, P. A., and Brimhall, B., 1952. Additional studies of the relative effectiveness of two systems of selection for oil content of the corn kernel. Agron. J. 44: 329–331.
Suneson, C. A., 1949. Survival of four barley varieties in mixture. Agron. Jour. 41: 459–461.
Suneson, C. A., 1956. An evolutionary plant breeding method. Agron. J. 48: 188–191.
Wallace, B., 1981. Basic Population Genetics. Columbia Univ. Press, New York, 688 pp.

13. SELECTION FOR TRAITS AND GENETIC GAIN FROM SELECTION

The subject matter of the previous chapters was about how to develop genetic populations possessing a rich gene pool and also about which selection methods to use in order to efficiently find out and select superior genotypes, i.e., originators of new lines and cultivars.

Most quantitative traits which are usually set as a breeding target are controlled by a large number of genes, ten, twenty, or more. These genes may affect favourably or unfavourably the development and expression of a trait. As a rule, not all favourable traits are found in one genotype and all unfavourable traits in another. Each genotype contains a mixture of favourable and unfavourable traits. Therefore, hybridization should be undertaken to obtain genotypes with the highest possible percentage of recombinations of desirable genes.

Bailey (1977) provided an illustration of how to fix desirable genes in soyabean. Starting from the assumption that each parent possesses thirty gene loci with desirable alleles (and, by the same probability, thirty gene loci with undesirable alleles), the F_1 generation will be heterozygous for sixty gene loci. The probability of obtaining a homozygous line possessing sixty loci with desirable alleles practically does not exist since the frequency of lines possessing more than even 42 desirable alleles is only 2% (0.0196) at the heritability level of 25% or 7% (0.0660) at the heritability level of 50% and the intensity of selection of 10% (Table 13.1).

The percentage of lines possessing desirable genes may be increased to a certain extent by the method of back-crossing. Still, after four cycles of back-crossing with the recurrent parent, only 17% of lines will be superior to the better parent and only 0.4% of lines possess more than fifty desirable gene loci in the homozygous

Table 13.1 Probabilities of homozygous lines with more than k loci homozygous for the favourable allele, given sixty heterozygous loci in the F_1

k	No selection	Intensity of selection $h^2 = 0.25$	$h^2 = 0.50$
30	0.4487	0.8660	0.9534
33	0.1831	0.6293	0.8157
36	0.0462	0.3248	0.5443
39	0.0067	0.1057	0.2455
42	0.0005	0.0196	0.0660
45	–*	0.0019	0.0094
48	–	–	0.0006

* Probability less than 0.0001.
Source: Bailey, Jr., 1977

state. Having conducted a biometrical – genetic analysis, Bailey (1977) concluded that selection from populations of self-pollinated plants has its limits. If two parents are similar in the value of their traits, it is not probable that the selection will render a line possessing more than 80% of the total genetic potential of both parents. If a parent is quite superior to another one can only expect to obtain a line which possesses just two or three more desirable gene loci than the better parent, which will be fixed in the homozygous state.

This shows the complexity of breeding self-pollinated species for genotypes superior to the better parent. The main problems are the difficulties involved in working with a very large population, which by its very size limits the selection of homozygous genotypes possessing all desirable genes of both parents. This is why geneticists and breeders keep trying to develop methods of crossing and selection which would increase the frequency of desirable genes and the rate of their fixation in a genotype

Jensen (1970) proposed the so-called *diallel selective mating system* for the development of a dynamic genetic population to be used in breeding of self-pollinated plant species. The system implies a selection of parents (five to seven) according to breeding objective and their diallel crossing. The whole series is grown in bulk and in the F_2 there starts the mass selection for the recombination of the expected traits. The pedigree method is introduced in the F_6 to obtain new lines.

A portion of the F_1 plants is crossed in different combinations – $(1\times2)\times(3\times4)$, $(2\times3)\times(4\times5)$, an so on, i.e., a series of double crosses is formed, grown in bulk and mass selection starts from the F_2. Promising F_2 plants are selected and crossed (selective crossing). This is the beginning of the first selective series which is then used for mass and individual selection.

The advantages of this method are the large number of crossings, diallel in the first phase and selective in later phases starting from the F_2, the use of mass and individual selection and the possibility of including a new cross or selection at any stage of the programme. The method has not been sufficienty verified although Redden and Jensen (1974) report a successful increase of tillering in barley and wheat. The selection advances from the series of selfing after the F_2 were 8.7% and 18.5%, respectively; the advances from the series of selective crossing (a random mating of the selected plants) were 17.1% and 22.6%, respectively.

To ensure the success in hybridization of selected plants at different levels of the method, the authors recommend the use of male sterility and a special care in the choice of cultural practices to maximize the expression of the selected traits.

To improve and genetically enrich a source population, Eberhart et al. (1967) proposed the so-called *comprehensive breeding system* which is applicable to self- and cross-pollinated species alike. Several years should be spent in evaluating and selecting materials to be included in the source population. These materials should be genetically different sources of resistance to diseases and insects, and in addition there should be cultivars or lines possessing a number of favourable agronomic traits (resistance to lodging, desirable vegetation period, quality, etc.). More than a half of that germplasm should possess a high yield potential. Depending on the crop being dealt with, the source population should comprise locally adapted material as well as exotic material (material which is generally unsuitable for production but which does possess some extreme useful traits – dwarfness, branching, erect leaves, etc.).

Having based their programmes on these principles, CIMMYT (Centro International por Mejoramientes de Maiz y Trigo) in Mexico had, by 1975, developed fourty-eight maize populations which are extensively used in breeding programmes, especially in developing countries.

The same principles were adhered to during the development of one of the earliest maize populations, Iowa Stiff Stalk Synthetic (BSSS).

Selection for a single trait

One of the most thoroughly studied and most instructive examples in the literature on genetics and breeding is the 80-year programme of selection for oil and protein content in maize conducted in Illinois (USA).

The programme started in 1896 when 163 ears of the open pollinated population "Burr White" were analysed. After chemical analysis, 24 ears with a high protein content (IHP), 12 ears with a low protein content (ILP), 25 ears with a high oil content (IHO), and 12 ears with a low oil content (ILO) were selected. A number of researchers have worked on the programme, to mention some of them – Drs. Leng, Woodworth, Jugenheimer, Alexander, Lambert, Dudley, etc.

According to Dudley (1977), the forward selection phase was divided into four segments as follows:

Segment 1. Generations 0–9, mass selection based on chemical composition. The number of ears analysed and selected varied but approximately 20% of the ears analysed were selected. Each strain was grown in a separate isolated field.

Segment 2. Generations 10–25. 120 ears per strain were analysed and 24 were saved. Seed from each ear was planted ear-to-row. Alternate rows were detasseled and 20 ears were analyzed from each of the six highest yielding rows. Four ears were saved per row.

Segment 3. Generations 26–52 in IHP and ILP; generations 26–58 in IHO and ILO. Twelve selected ears were arbitrarily divided into two lots (*A* and *B*) of six ears. Seed within each lot was bulked and planted in the nursery. Silks in lot *A* were pollinated by a bulk sample of pollen from 15–20 plants in *B* while silks in *B* were pollinated with pollen from *A*. Thirty ears from each lot were analysed and the 12 most extreme of the 50 ears analysed were saved.

Segment 4. Generations 52–76 in IHP and ILP; 59–76 in IHO and ILO. The procedure was the same as in segment 3 but 80 lbs of nitrogen per acre were added to the soil.

The source population had contained 4.68% of oil and about 10% of protein on average.

One selection programme was aimed at high oil content (IHO), another at low oil content (ILO). Each generation in the IHO group was selected for high oil content, each generation in the ILO group for low oil content. The selection advance was slow and it tended to fluctuate due to environmental effects, but, after 76 genera-

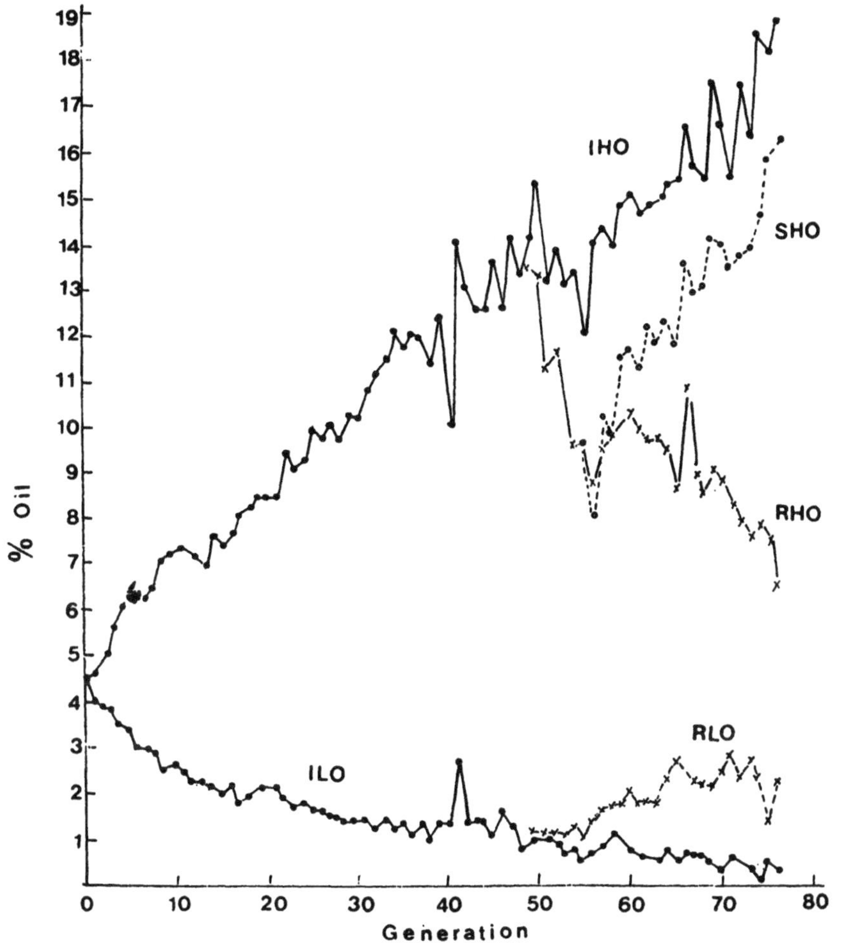

Fig. 13.1 Mean percent oil for IHO, ILO, RHO, RLO, and SHO plotted against generations. Source: Dudley, 1977

tions of selection, the average oil contents in the IHO and the ILO groups were 18.2 and 0.75%, respectively (Fig. 13.1).

The reaction of the population to the selection for protein was similar. After 76 generations of selection, the average contents of protein in the IHP group and the ILP group were 26% and less than 4%, respectively (Fig. 13.2).

A reverse selection was applied in the 48th generation, i.e., the IHO group of high oil content was selected for low oil content (RHO) and the ILO group of low oil content was selected for high oil content (RLO). The percentage of oil dropped rapidly in the former group while the increase in the latter group was slow (Fig. 13.1). After seven generations, the group RHO again started to be selected for high oil content (SHO), with evident progress (Fig. 13.1).

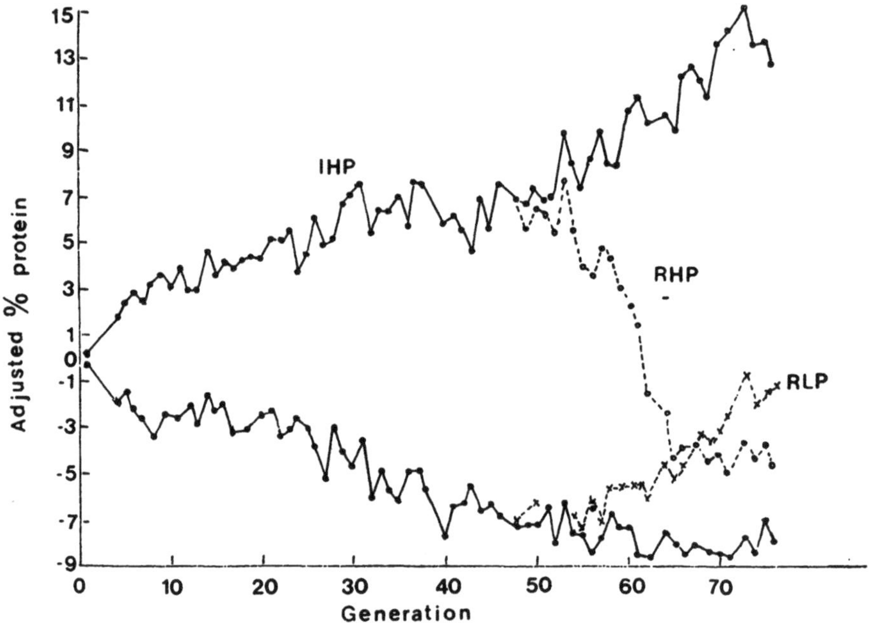

Fig. 13.2 Mean adjusted percent protein for IHP, ILP, RHP, and RLP plotted against generations. Source: Dudley, 1977

The reverse selection for protein – RHP and RLP – followed the pattern of the reverse selection for oil (Fig. 13.2).

As reported by Dudley (1977), neither the forward nor the reverse selection have reached the selection limit after 76 generations. Genetic variability was still present in the group ILO, which had the average oil content of 0.3%, although further selection cannot be effective because the limit value of zero percent oil has almost been reached. Regarding the other groups, even after the fourth segment of selection, the heritability percentages confirmed the presence of considerable genetic variability (in spite of some unexplained phenomena) which is an indication that effective selection may still be expected.

The gains in the groups IHO, ILO, IHP, and ILP, after 76 generations were 20, 6, 20, and $12 \times \sigma A$ (standard deviation), respectively. Expressed in the percentages in relation to the original means, these gains were 279, 92, 133, and 78%, respectively (Dudley, 1977).

Analysing the frequencies of desirable genes for both traits, Dudley (1977) stated that the frequencies of desirable genes for protein and oil in the source population were less than 0.37 and less than 0.25, respectively. The estimation of the number of genes showed that at least 54 loci differentiated the IHO from the ILO group and that at least 122 loci differentiated the IHP from the ILP group.

Dudley (1977) concluded that the fear to exhausting the genetic variability for grain yield is unfounded since the gene frequencies of 0.25 and 0.5 for only 50 to 200 loci are required to allow a gain of 20 A. "Thus, as far as quantitative traits are concerned, the most pressing problems may not be those of obtaining larger genetic variability but those of learning how to concentrate most efficiently the favourable alleles we now have and then proceed to do it" (p. 471).

Selection for multiple traits

Breeding programmes are very seldom aimed at selecting for a single trait but usually aim for multiple traits. Still, the most frequent objective is the selection for yield, i.e., for a complex of traits.

It should be kept in mind that the selection for one trait may have an effect on another trait, even if the genes controlling them are not linked. In that respect, the American programme of continuous selection for oil and protein content in maize described in the previous paragraphs has rendered very instructive data. The selection for high oil content resulted in small but well-filled grains and short ears, while the selection for low oil content resulted in long ears with large but shrivelled grains. The selection for high protein content produced short ears with well-filled flint grains while the selection for low protein content produced long ears with large and dent grains. Regarding the yield of grain, the groups with high percentages of oil and protein had about fifty percent lower yields than the source population.

This is an illustrative example of selection for extreme plus or minus values. It also indicates the presence of a strong interdependence among the traits which cannot be always observed in a period of ten years, i.e., during the process of the development of a cultivar.

Although the yields of grain, protein, oil, or sugar are frequently the final breeding objective, the yield is not the only trait selected for but rather the result of the selection for a number of other traits. A high yield simply cannot be obtained without the selection for these other traits too.

In the chapter on selection methods it was mentioned that the best plants, the superior genotypes, the best lines, etc. should be selected. Here are examples for some traits and some crops to illustrate what may be the outcome of such selection.

Truncation selection

Each year, the breeder faces hundreds of F_1 cross combinations, several hundred thousand F_2 plants, thousands of F_3–F_5 plants and progenies and hundreds of lines in the F_6 and later generations. The situation is similar with genetic populations and the development of inbred lines of cross-pollinated plants. The questions are asked: how to select the best genotypes and how to obtain superior cultivars? How to find the highest yielding genotypes by selecting plants in the F_2 generation of self-pollinated plants or in the S_1 generation of cross-pollinated plants? As it is not feasible to measure thousands of plants, it is inevitable that the breeder must select for a few traits which characterize the desired ideotype or model variety, and these traits must be yield components or must have the highest effect on the yield itself.

In the case of breeding self-pollinated plants by the pedigree method, the main traits that should be selected for in the F_2 and later generations of segregation are: vegetation period, i.e., earliness or lateness, plant height for resistance to lodging, and the resistance to diseases and pests. These traits may be accurately assessed before flowering or harvest and the breeder may decide what is more appropriate to apply, negative or positive selection. Negative selection implies the physical removal or labelling of plants and progenies which are too late (if the breeding objective is earliness), or are too tall and thus lodge or which are susceptible to major diseases, etc. Positive selection implies that desirable plants and progenies are labelled in the course of the growing season and are selected for earliness, desired plant height, resistance to diseases and pests, and other traits at harvest, when they may be visually checked.

The selection of cross-pollinated plant species such as lucerne, clover, grasses, etc. is similar but the progenies will be more different from the desirable phenotype than in the case of self-pollinated plants because the contribution of the male parent is either unknown or less known.

Direct selection for yield is not efficient in early generations of segregation (F_2–F_4), unless in a certain sense by the method of F_2 progenies. Namely, lines in later generations may be much different from those selected in previous generations because of the segregation of traits in the progenies, larger spacing than in commercial production, or because the evaluation is done on the basis of the individual plant and not on the basis of the plot. Some information on yield is nevertheless obtained because it may be assumed that a plant with a favourable vegetation period, a short stem resistant to lodging, resistance to major pathogens, etc. will also possess a high genetic yield potential. Besides, each breeder has his own favourable plant type, his own concept of the ideal cultivar, and so he adds his personal criteria of "the best plant" or "the best line" to the conventional criteria.

The described system of selection is also called *elimination selection* (ex selection) *or independent culling level.*

Tandem selection

There is a similar system called *tandem selection* which comprises several cycles of selection. In the first cycle, a trait (e.g., short stalk) is selected for until a desired level is achieved. In the second cycle, another trait is selected for (e.g., resistance to rusts) but only in plants with a short stalk because tall plants would have been eliminated due to the susceptibility to lodging even though they had been resistant to rusts.

Since there exist genetic correlations among individual traits, the selection for one trait may unintentionally affect another trait. This may be advantageous if the traits are positively correlated. If, however, they are negatively correlated, the success of selection for one trait may be limited by the reaction of the other (example of maize). Therefore, genetic correlations must receive due attention because they are more difficult to overcome by breeding than are negative correlations occurring among traits in the course of ontogenetic development under the effect of environmental factors.

The success of selection increases in proportion to the genetic advance for the selected trait. This is why the parameters determining genetic advance as well as the method of calculating genetic advance deserve to be discussed in detail.

Genetic advance under selection

Genetic advance or gain under selection is the difference between the mean genetic value of selected lines and the mean genetic value of the source population or parents. According to Allard (1960), genetic advance under selection (G_s) depends on the following parameters:

$$G_s = (k)(\sigma_A)(H)$$

where k is a selection differential taking into account the mean phenotypic value of the q selected and n original lines, the phenotypic standard deviation, and selection intensity, i.e., the percentage of plants selected for the foundation of next generation (q/n). Since k is expressed as standard deviation units, it is fairly stable and it is most dependable on selection intensity. For example, the values of k for 1, 2, 5, 10, 20, and 30% of selected lines are 2.64, 2.42, 2.06, 1.76, 1.40, and 1.16, respectively (Fig. 13.3).

According to Allard σ_A is the phenotypic standard deviation of the mean yields of the n original lines.

H is the heritability coefficient estimated as the ratio obtained by dividing genotypic by the phenotypic variance. Thus:

$$H = \frac{\sigma^2 a}{\sigma^2 A}$$

Since a selected phenotype is in fact the result of the interaction of a genotype and environmental factors, experiments described in Chapter 6 have to be carried out in order to distinguish among these effects and to establish the genotypic value of the trait in question and the genetic advance under selection.

The example of Hanson et al. (1956) (cited by Allard, 1960) and their work on Korean lespedeza (*Lespedeza stipulacea*) (a legume grown in the U.S.A. for hay making) will be used to illustrate a method of calculating genetic advance under selection. Three lespedeza cultivars were crossed; 284 plants were randomly sampled from the F_2, their progenies were studied in the F_3, and all seeds of a single F_3 progeny were tested in the F_4. The experiment was conducted in two replications, two locations and it lasted for two years.

The results were processed by the analysis of variance to separate the effect of the environment from the effect of genotype, as well as to separate the effect of the environment into the effect of location and the effect of year. The three hybrid populations of lespedeza were measured for the total yield. The data in Table 13.2. show that population no. 1 had a low genotypic variance (2.480) compared against the error variance (16.578). The interaction genotype vs. year contributed to the latter value; however, there were no interaction between genotype and location.

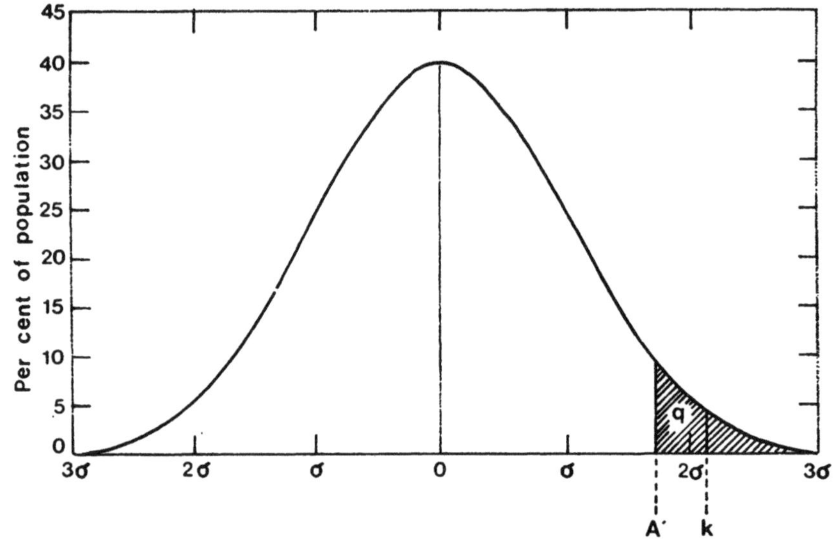

Fig. 13.3 Theoretical distribution of mean yields of lines compared in replicated-yield trials. When the top 5 per cent of the lines are selected, that is, those with yields larger than A' (= 1.65σ) the mean of the q selected families is expected to be k (in standard measure) and $k\sigma$ when expressed in terms of the actual units in which the measurements were taken.
Source: Allard, 1960

Table 13.2 Estimates of components of variance for total yield in Korean lespedesa

Population	σ^2_a	σ^2_{al}	σ^2_{ay}	σ^2_{aly}	σ^2_e
1.	2.480	0.000*	2.057	3.099	16.578
2.	3.354	0.000*	858	2.070	17.432
3.	2.084	1.777	1.745	1.461	15.071

* Negative estimates for which the most reasonable value i zero.
Source: Allard, 1960

The genotypic variance (σ^2_a) for population no. 1 was 2.480, for population no. 2 3.354. The phenotypic variance (σ^2_A) is calculated from all variances:

$$\sigma^2_A = \sigma^2_a + \frac{\sigma^2_{al}}{L} + \frac{\sigma^2_{ay}}{Y} + \frac{\sigma^2_{aly}}{LY} + \frac{\sigma^2_e}{RLY}$$

The dividends are the variances of individual factors given in Table 13.2; among the divisors, L stands for the number of locations, Y for the number of years, and R for the number of replications. If the values obtained for population no. 1 are put in the formula:

$$\sigma^2_A = 2{,}480 + (0000/2) + (2{,}057/2\times 2) + (3{,}099/2\times 2\times 2) = 6{,}355.48.$$

Heritability may now be calculated:

$$H = 2{,}480/6{,}355.48 = 0.3902 = 39.0\%.$$

The phenotypic standard deviation is the square root of the phenotypic variance:

$$\sigma_A = \sqrt{6{,}355.48}.$$

Finally the selection differential k is calculated. As only 5% of the total hybrid population were dealt with, the value of k is 2.06.

Therefore, the genetic advance under selection for population no. 1 is:

$$G_s = (2.06)(\sqrt{6{,}355.48})(0.3902) = 64 \text{ grams/plot}.$$

Thus the genetic advance under selection for population no. 1 was 64 grams of total yield per plot or 7.7% (64/829). The genetic advances for the other two populations were calculated in the same manner (Table 13.3).

The populations differed in the values of heritability, population no. 2 having the highest one. The genetic advance in relation to the average of the population increased in proportion with the increase in heritability. Attention should also be paid to the mean value of the total yield. Although population no. 1 was inferior to population no. 2 in heritability and genetic advance, its mean value for the yield was higher, indicating that new lines should be selected from it. In spite of its lower total yield, population no. 2 should also be used for selection, on account of its high heritability since lines with a high yield may be expected to be derived from it (Table 13.3).

Table 13.3 Heritability and estimated genetic advance under selection for total yield in three populations of Korean lespedesa

Population	Heritability	Genetic advance		Mean in g/plot
		g/plot	% of the mean	
1.	39.0	64	7.7	829
2.	52.8	86	11.6	741
3.	34.2	55	7.3	751

Source: Allard, 1960

References

Allard, R. W., 1960. Principles of Plant Breeding. John Wiley & Sons. New York, 485 pp.

Bailey, T. B., 1977. Selection limits in self-fertilizing populations following the cross of homozygous lines. Proc. Inter. Confer. Quant. Genetics. Ames, USA, p. 399–412.

Dudley, J. W., 1977. 76 generations of selection for oil and protein percentage in maize. Proc. Inter. Conf. Quant. Genetics, Ames, USA, pp. 459–473.

Eberhart, S. A., Harrison, M. N., and Ogada, F., 1967. A comprehensive breeding system. Der Züchter 37: 169–174.

Hanson, C. H., Robinson, H. F., and Comstock, R. E., 1956. Biometrical studies of yield in segregating populations of Korean lespedesa. Agron. J. 48: 268–272.

Jensen, N. F., 1970. A diallel selective mating system for cereal breeding. Crop Sci. 10: 629–635.

Redden, R. J., and Jensen, N. F., 1974. Mass selection and mating systems in cereals. Crop Sci. 14: 345–350.

14. PLANT BREEDING FOR RESISTANCE TO PARASITES

Breeding for individual traits, such as winter-hardiness, drought resistance, high contents of oil or proteins, quality, etc., will not be discussed in this book because the methods used are so specific for each trait and crop that the reader should refer to the literature sources on the breeding of specific crops.

However, breeding for resistance to parasites does deserve a chapter in this book for the following reasons.

(1) The resistance relates to two living organisms, a plant and a parasite, each of which has its own development and inheritance, and which interact mutually and separately with the environment. All this makes breeding for resistance a complex but inevitable task.

(2) Diseases and pests develop on the leaf surface in most cases, reducing photosynthetic activity, and hence, the yield.

(3) Harmful organisms (fungi, bacteria, viruses, nematodes, insects) live on and within plants and cannot be completely eliminated. Rather, they should be controlled at an economically tolerable level.

(4) It is essential to breed for all kinds of resistance to pathogenic organisms in order to reduce the chemical means of control to a minimum, so as to lower production costs, increase the nutritional value of agricultural products and to improve the environment.

Fungi, bacteria and viruses are the most frequent pathogens of plants. In the course of evolution, a certain coexistence has developed between parasites and host plants which allows both of them to survive. It means that natural selection favours neither exceedingly aggressive and virulent parasites nor extremely resistant plants; the former would destroy the hosts which, ultimately, would be their own end and the latter would not give the chance to parasites of finding a host and so no survival chance.

This status of natural co-adaptation and balance may be interrupted by large changes in the environment. The intensive development of agriculture in this century has profoundly disturbed this status. For example, in underdeveloped agricultural conditions, the inoculum of powdery mildew in wheat (*Erysiphe graminis* f.sp. *tritici*) can hardly provoke an epidemic of the desease because a different variety of wheat grows in the neighbouring plot, or growing next to it are oats, rye, other grasses or an altogether different species from another family upon which this particular disease cannot develop. Trying to achieve still higher yields and production per unit area, modern agriculture has introduced monoculture which involves growing pure varieties and the application of high doses of mineral fertilizers, especially nitrogen. This has created very favourable conditions for the development of pathogenic organisms. When a pathogen occurs in the most favourable spot in a field, it spreads rapidly over the entire field and, if the same cultivar is grown in the neighbouring fields, it causes an epidemic. The consequences are reductions in yield and quality which in turn bring about large economic losses. Of course, it does not mean that the intensive agriculture should be abandoned; however, efficient measures of protection against diseases and pests should be undertaken.

There are four main methods of controlling pathogenic organisms:

(1) development of cultivars resistant to certain pathogens,
(2) manipulation of agricultural practices (avoidance of monoculture, crop rotation, inter-cropping, etc.),
(3) biological control of parasites,
(4) chemical protection of crops.

The parasite – host relationship

Growers are interested in having a healthy crop in which large economic damages cannot be inflicted by parasites. Whether a crop is healthy or diseased depends on the relationship established between the parasite and the host plant. This relationship is so specific and complex that Loegering (1966, 1984) considers the pathogen – host association as the third entity. He coined the term "agricorpus", to distinguish between the pathogen as one entity, the host as another entity and the pathogen – host as the third entity.

Complete resistance is an ability of the host to prevent any multiplication of the parasite. The resistant plant is in fact considered to be hypersensitive, using the products of metabolism (toxins) to prevent the parasite from invading the plant and thriving on it.

In addition to complete resistance, there is also incomplete or partial resistance characterized by individual disease lesions. Slow rusting also belongs to the same category (Parlevliet, 1985).

Tolerance is yet another kind of genetic resistance. Some cultivars are, in fact, susceptible to a pathogen which develops on them. However, these cultivars tolerate the attack without suffering a significant yield reduction. Simons et al. (1961) and K. Borojević (1976) state that the degree of tolerance of a cultivar may be calculated from the ratio between yields in an infected and an uninfected plot, the tolerance being higher as the ratio approaches unity.

It is not a simple task to develop a cultivar resistant to several pathogenic species belonging to the same genus (e.g., leaf rust, stem rust, and yellow rust) and it is even more difficult to achieve resistance to pathogens from different genera (*Puccinia*, *Erysiphe*, *Septoria*, *Fusarium*, *Helminthosporium*). This is why the discovery and development of cultivars possessing tolerance to at least some pathogens is an important part of plant breeding, deserving more attention than it has had in the past.

Races of pathogens and genes for resistance in hosts

A study of Stakman and associates conducted in the U.S.A. at the beginning of this century showed the presence of a specific relationship between a pathogen and its host. They found that different wheat cultivars reacted differently to two rust isolates (*Puccinia graminis* f.sp. *tritici*) (Stakman and Piemeisel, 1917). It was a proof that cultivars possessed different genetic constitutions regarding resistance to pathogens as well as the fact that the pathogenic population itself possessed different genetic constitutions regarding its virulence, i.e., the capacity of infecting wheat plants. This study opened the door to a series of studies which showed the presence of high genetic variability within populations of various pathogenic species, i.e., the existence of many biotypes or physiologic races of pathogens. Numerous genes for resistance or susceptibility have been identified in host plants by testing varieties with pathogenic cultures. To make the results comparable, an international system of identification of resistance genes in host plants and physiological races of pathogens has been established. The international pathogenicity surveys for wheat rusts have been very useful in breeding for specific resistance (Stubbs et al., 1974; Bošković, 1980; Bošković and Browder, 1976; etc.).

The analysis of parasitic populations was started by testing different varieties, so-called differentials. They had to be different genetically with regard to the degree of resistance in order to allow identification of pathogenic races and resistance genes in host plants. For example, there are 12 differential varieties for leaf rust in wheat, 14 for powdery mildew in wheat, 9 for downy mildew in potatoes, etc. New findings in genetics led to the introduction of isogenic lines as differentials in the analysis of parasitic populations. Each isogenic line possesses different individual gene(s) for resistance to certain parasitic species but an almost identical genetic background for the other traits. For example, the isogenic lines of the wheat cultivar Thatcher, which possess different *Lr* genes, are used as differentials for leaf rust.

The identification of pathogenic races is based on the types of reaction or infection.

Resistance to rusts in wheat is assessed on the basis of infection type and severity (Table 14.1). The infection types ranging on the scale from 0 to 4 are mostly used for the screening of seedlings in a greenhouse. The reaction type is described in more detail by adding symbols + or — to the above figures. The reaction types observed on mature plants in the field are usually designated *R* (resistant), *MR* (medium resistant), *MS* (medium susceptible) and *S* (susceptible). Equally important for a field test is the need to establish the severity of attack which is rated on the scale from 0 to 100 percent. For example, the designations 10*R*, 25*MR*, 60*S* mean that 10, 25, and 60 percentages of leaf area have the infection types – resistant, medium resistant, and susceptible, respectively (Fig. 14.1). Reaction type and severity are used to calculate the coefficient of infection which determines the degree of resistance of a host plant.

The resistance or susceptibility to other pathogenic organisms, for which it is difficult to distinguish the type of infection, is evaluated by determining the degree of infection, on the scale from 0 to 5 or from 0 to 9, 0 standing for infection-free plants, 1 for resistant plants, and 5 or 9 for highly susceptible plants.

Table 14.1 Infection types produced by physiologic races of *Puccinia graminis tritici* on differential varieties of *Triticum* spp.

Infection type*	Varietal response
	Resistant (*R*)
0	IMMUNE – No uredia developed
0;	ALMOST IMMUNE – Hypersensitive flecks sometimes present
1	VERY RESISTANT – Uredia minute; surrounded by distinct necrotic areas
2	MODERATELY RESISTANT – Uredia small to medium; usually in green islands surrounded by a decidedly chlorotic or necrotic border
	Susceptible (*S*)
3	MODERATELY SUSCEPTIBLE – Uredia medium in size; coalescence infrequent; no necrosis, but chlorotic areas may be present, especially under unfavorable growing conditions
4	VERY SUSCEPTIBLE – Uredia large, and often coalescing; no necrosis, but chlorosis may be present under unfavorable growing conditions
	Mesothetic (*X*)
X	HETEROGENEOUS – Uredia variable, sometimes including all infection types and intergradations between them on the same leaf; no mechanical separation possible; on reinoculation small uredia may produce large ones, and vice versa

* Plus and minus signs are used to indicate variation within a given type: ++ and = indicate the upper and lower limits, respectively, of each type. The symbol = indicates a variation between + and − for the type. The symbol c indicates chlorosis.
Source: Stakman et al. 1962

A large number of physiological races of pathogens have been identified on the basis of the infection type on standard differential varieties, for example:

Puccinia graminis f.sp. *tritici*	over 300 races
Puccinia graminis f.sp. *avenae*	over 16 races
Puccinia recondita f.sp. *tritici*	over 180 races
Helminthosporium gramінearum	over 30 races
Ustilago tritici	over 20 races
Tilletia tritici	over 10 races
Erysiphe graminis f.sp. *hordei*	over 90 races
Phytophthora infestans	over 16 races.

The number of physiological races that may be identified in a pathogenic population depends on the number of differential varieties of the host species and the number of interactions between the pathogen and the host (*R*, *MR*, *MS*, *S*, etc.). According to Van der Plank (1968), this may be calculated by the formula:

$$X = K^n,$$

where X is the number of physiological races, K is the level of interaction between the pathogen and the host, and n is the number of differentials possessing different genes for resistance.

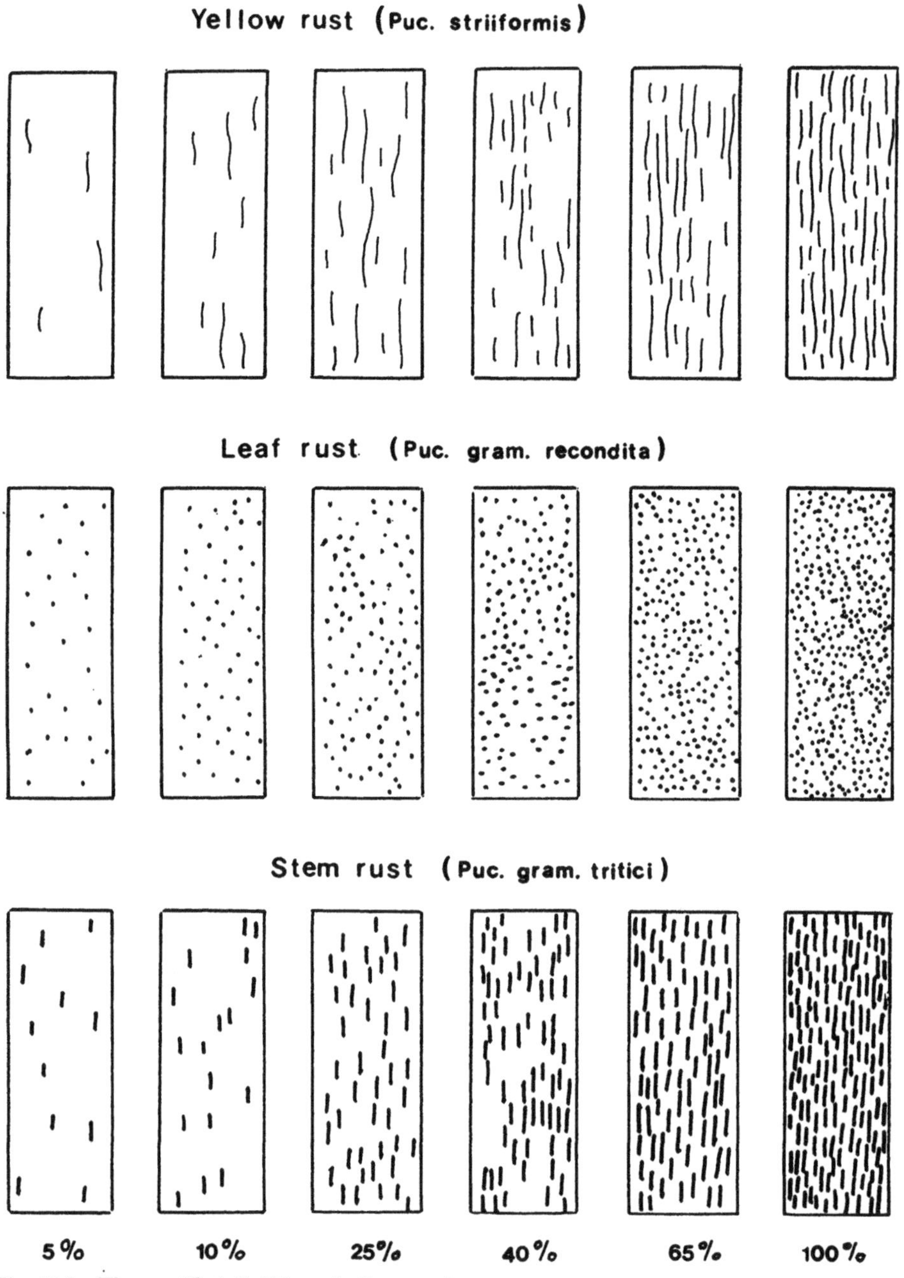

Fig. 14.1 The modified Cobb's scale for cereal rusts

On the other hand, genes for resistance have been identified in host plants, e.g., the genes $Sr_1 \ldots Sr_{32}$ for the resistance to stem rust in wheat, the genes $Lr_1 \ldots Lr_{35}$ for the resistance to leaf rust in wheat (Browder, 1980), the genes $Pg_1 \ldots Pg_{13}$ for the resistance to stem rust in oats, the genes $R_1 \ldots R_4$ for the resistance to blight in potatoes, etc. McIntosh (1983, 1988) supplies detailed data about this.

Flor (1942, 1955, 1971) was the first to offer a genetic explanation of plant resistance to pathogens by his gene-for-gene theory, "for every gene for resistance in flax there was a corresponding gene for pathogenicity in the flax rust fungus". Namely, if a gene for resistance exists in a host plant, a certain pathogenic race is not capable of causing infection. In that case, the cultivar is resistant to that race of the pathogen and the pathogen is avirulent. If the pathogen race can infect the cultivar, then the

Table 14.2 Race deciphering key of oat stem rust (*Puccinia graminis* Pers. f.sp. *avenae* E. et H.). Source: Mac Key, 1974

Patho-type	Ratio in race complexity	Combination of genes of virulence	Conventional race no.	Host type with gene of resistance			
				A	B	D	E
1	1	- - - -	1	□	□	□	□
2	4	a - - -	11	▤	□	□	□
3		- b - -	1A	□	▤	□	□
4		- - d -	3	□	□	▤	□
5		- - - e	2	□	□	□	▤
6	6	a b - -	11A	▤	▤	□	□
7		a - d -	4	▤	□	▤	□
8		a - - e	8	▤	□	□	▤
9		- b d -	3A	□	▤	▤	□
10		- b - e	2A	□	▤	□	▤
11		- - d e	7	□	□	▤	▤
12	4	a b d -	4A	▤	▤	▤	□
13		a b - e	8A	▤	▤	□	▤
14		a - d e	6	▤	□	▤	▤
15		- b d e	7A	□	▤	▤	▤
16	1	a b d e	5A	▤	▤	▤	▤

□ = resistant ▤ = mesothetic or susceptible reaction

pathogen possesses a gene for virulence and the host plant lacks genes for resistance, i.e., it is susceptible.

A simple and instructive example which proves the gene-for-gene theory is the identification of the races of stem rust in oats presented in Table 14.2. Sixteen races or pathotypes of *Puccinia graminis avenae* have been identified on four oat cultivars – Richland (the carrier of gene *A*), Rodney (*B*), White Tortor (*D*), and L-703 (*E*). It may be seen that each cultivar is resistant to some disease races and susceptible to others, and that races exist which do not possess genes for virulence. There are also races that possess only one gene for virulence, or two and three genes which may be different (Table 14.2).

The pathogenic races are not uniformly distributed in time and space (Table 14.3). Some races of *Puccinia graminis avenae* (e.g., 7, 12, 6, 13) were more prevalent than the other races (e.g., 3*A*, 4*A*); however, the percentage of prevalence may vary from one year to another. As regards the distribution of the races in space, a large number of races was determined in Sweden and in the U.S.A. while some races were found only in Sweden (3, 4, 3*A*), and some only in the U.S.A. (2, 5, 8, 10) (Table 14.3).

Regarding the genes for virulence, there was not a single race without a gene for virulence, a small number of the races (3 and 2 and 5) possessed only one gene (*d* or *e*), the majority of the races possessed two or three genes, and only two races, 6*A* and 13*A*, possessed all four genes for virulence (*abde*) (Table 14.3).

This example was chosen because of its simplicity in the presentation of the genetic mechanism of the host – pathogen interaction. Understandably, there may be a larger number of pathogenic races. The number of genes for virulence or resistance may also be larger and their effects may be different.

Numerous studies show that the identification of physiological races of pathogens and genes for resistance in hosts is an endless job. It is more advantageous for plant breeding if research effort is directed to gaining better knowledge of the parasite – host – environment specificity (Loegering, 1984; Browder, 1985) and if emphasis is placed on sources of resistance and their usefulness rather than on description of pathogenicity of pathogen populations (Bošković, 1988).

Race-specific resistance

Knowing that it is very difficult to achieve resistance to a large number of physiological races of a pathogen and practically impossible to achieve resistance to all races, and having realized that the frequency and distribution of races varies in time and space, breeders direct their efforts towards the incorporation of resistance to the prevalent races of certain pathogens. Particularly intensive is the breeding for resistance to the pathogens rusts, powdery mildews and smuts in cereals, potato and tomato blights; leaf stripe in maize, etc. Cultivars have been developed which possess resistance to several prevalent races of pathogens, i.e., to the races which existed in actual time and space. This resistance type is called "race-specific resistance".

Cultivars possessing race-specific resistance have contributed considerably to the control of pathogens in many agroecological areas. However, it frequently happens that a cultivar loses its resistance and new epidemics, similar to those which pestered the crop at the time of the development of the cultivar, break out anew. This loss of resistance may be attributed to at least two reasons.

The first one is that a resistant cultivar greatly increases selection pressure against the prevalent races, thereby decreasing their frequencies in the pathogenic population, in some cases even to below the level of significance. In consequence, those races which have had low frequencies and to which the cultivar is not resistant can greatly rise in their frequencies. Thus, new races can become prevalent and the cultivar turns susceptible.

Table 14.3 Race spectra of oat stem rust in Sweden and the United States in 1956–1959

		Relative prevalence (%) of oat stem rust race number/pathotype															
Year	*n*	1	11	1A	3	2,5	11A	4	8, 10	3A	2A, 5A	7, 12	4A	8A, 10A	6, 13	7A, 12A	6A, 13A
		----	*a---*	*-b--*	*--d-*	*---e*	*ab--*	*a-d-*	*a--e*	*bd-*	*-b-e*	*--de*	*abd-*	*ab-e*	*a-de*	*-bde*	*abde*
Sweden																	
1956	32	–	–	–	31	–	–	28	–	6	–	13	–	–	13	3	6
1957	61	–	–	–	7	–	–	7	–	3	–	11	3	–	13	25	31
1958	96	–	–	–	24	–	–	14	–	5	–	20	3	–	15	6	14
1959	133	–	–	–	12	–	–	9	–	7	–	16	5	–	10	21	20
1956–59	322	–	–	–	16	–	–	12	–	6	–	16	4	–	12	15	19
U.S.A.																	
1956	476	–	–	–	–	16	–	–	15	–	–	66	–	–	1	2	–
1957	522	–	–	–	–	12	–	–	21	–	0	59	–	–	2	6	0
1958	286	–	–	–	–	14	–	–	26	–	–	54	–	–	1	5	–
1959	230	–	–	–	–	7	–	–	11	–	–	59	1	–	10	10	2
1956–59	1514	–	–	–	–	13	–	–	19	–	0	60	0	–	3	5	0

Source: Mac Key et al. 1963

The second reason is that a cultivar, when it enters commercial production, becomes susceptible because of the occurrence of new physiological races of a pathogen which did not exist at the time of breeding the cultivar. These new races develop by mutations which, considering the exceptional multiplication capacity of microorganisms (e.g., *Puccinia recondita* may produce 10^{11} spores per day per hectare), and in spite of a low frequency of mutations (10^{-8}), may produce about one thousand mutants per locus/day/hectare (Parlevliet and Zadoks, 1977). New races of microorganisms may also develop as a consequence of gene recombinations.

Race-specific resistance does not provide lasting protection. Pathogenic populations are not efficiently controlled and new epidemics are a constant threat. After much work expended in the U.S.A. and Canada on the identification of the prevalent physiological races of *Puccinia graminis* f.sp. *tritici* and the development of wheat cultivars resistant to them, the outbreaks of race 56 in 1935 and of race 15*B* in 1953 and 1954 caused epidemics which brought yield reductions of over 50 percent in some regions. However, it was observed during these epidemics that those cultivars which had no genes for resistance to the prevalent races exhibited a medium resistance over a long period of time.

These intriguing phenomena called for adequate explanations. So, Van der Plank (1963, 1968) came up with a new theory of vertical and horizontal resistance.

Vertical and horizontal resistance

Van der Plank (1963) defines under vertical resistance all cases when a cultivar is resistant to some pathogenic races and is susceptible to other races of the same pathogen. If, however, a cultivar is equally resistant to all races of a pathogen, then it is a case of horizontal resistance.

Van der Plank illustrated differences between vertical and horizontal resistance in the example of resistance in potatoes to blight (*Phytophthora infestans*) (Fig. 14.2). Two potato cultivars, Kennebec and Maritta, possess the gene R_1 which ensures vertical resistance to races (0), (2), (3), (4), (2.3), (2,4), (3.4), and (2,3,4). However,

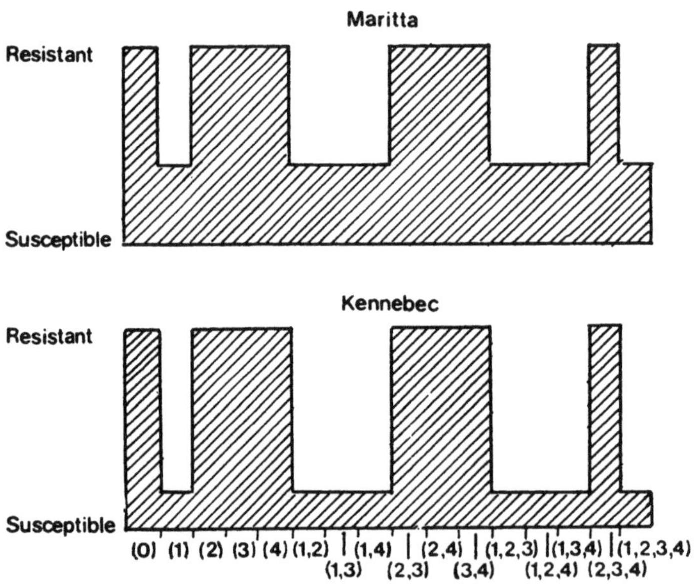

Fig. 14.2 Diagram of the resistance to blight of the foliage of two potato varieties, Kennebec and Marita, both with the resistance gene *R*1. The resistance is shown shaded to 16 races of blight. To races (0), (2), (3), (4), (2, 3), (3, 4), and (2, 3, 4) the resistance of both varieties is vertical. To races (1), (1, 2), (1, 3), (1, 4), (1, 2, 3), (1, 2, 4), (1, 3, 4), and (1, 2, 3, 4) resistance is horizontal, small in Kennebec and greater in Maritta.
Source: Van der Plank, 1963

the cultivars differ regarding their reactions to the other races, (1), (1,2), etc., the resistance to which is not controlled by the gene R_1. The infection is faster on Kennebec than on Maritta – when the leaves of the former cultivar turn brown, the leaves of the latter cultivar are still green. In other words, Maritta exhibits a higher horizontal resistance (also called field resistance) than does Kennebec (Fig. 14.2). The cultivars which possess the other genes for resistance – R_2, R_3, R_4, or their combinations – do not differ on average in reaction from the cultivars which possess only the gene R_1.

Certainly, there are potato cultivars which do not possess the *R* genes at all, but they are not automatically susceptible and they do differ in their reaction to the pathogen. For example, the cultivar Katahdin is highly susceptible to leaf blight and it displays a low horizontal resistance in field conditions. Conversely, the cultivar Capella resists the pathogen more effectively; the spreading of the infection is slow, necrotic spots on leaves are few, and the leaves and stem remain green for a long time even when conditions favour the development of the pathogen. It is evident that the cultivar Capella possesses a high horizontal resistance (Van der Plank, 1963).

Using this and other examples, Van der Plank (1963, 1968) explained the genetic basis of vertical and horizontal resistance. Plant resistance controlled by a single or several major genes (monogenic and oligogenic resistance) falls in the category of vertical resistance. This resistance type comes to expression when the pathogen does not possess corresponding genes for virulence. If the pathogen does possess genes for virulence and the plant lacks genes for resistance, then infection takes place. This is the gene-for-gene relationship which results in the specific resistance to certain races of the pathogen. The expression "vertical resistance" may thus be considered synonymous to race-specific resistance.

Conversely, the resistance controlled by a number of minor genes (polygenes) which exhibit small but cumulative effects is called horizontal resistance. According to Van der Plank, this type of resistance is not based on the gene-for-gene relationship and it is therefore non-race-specific.

These two resistance types are not entirely separate. Some cultivars possess both vertical and horizontal resistance and others possess either one of them.

Owing to their different genetic backgrounds, vertical and horizontal resistance affect differently the epidemiology of pathogens, i.e., the health condition of crops. The importance of horizontal resistance increases if a pathogen spreads from one crop possessing horizontal resistance to another crop with the same resistance type, i.e., when two cultivars have the same genetic resistance. Vertical resistance gains importance when the cultivar from which the infection spreads is genetically different, i.e., it has different vertical resistance from the cultivar to which the infection spreads. The epidemic is either stopped or decelerated and the result is less damage. As Van der Plank put it, vertical resistance postpones the outbreak of infection by reducing the initial inoculum while horizontal resistance decelerates infection after it has already taken place. He supports this by example of the pattern of the spreading of potato blight to cultivars that possessed vertical resistance and those that possessed horizontal resistance. On average, it took 15.5 days for the infection to develop from degree 1 to degree 3 in six cultivars possessing *R* genes and 26.7 days in six cultivars lacking *R* genes (according to Kirste, 1958, cit. Van der Plank, 1968).

It has already been said that vertical resistance is not longlasting because of the shifts in the prevalence of physiological races and the occurrence of new mutations in the pathogen. It should be added that despite the fact that a large number of races has been identified in many pathogens, genes for resistance which are "easy to handle" have mostly been incorporated in plant materials. As a rule these are major genes, a single gene or a pair of genes, which extend easily observable effects. There are presently in use only twenty or so genes for resistance to *Puccinia graminis* f.sp. *tritici* (*Sr* genes) although over three hundred races of the pathogen have been identified. The majority of these races is probably controlled by a number of polygenes which do not exhibit easily identifiable effects. In other words, the resistance to the majority of these races has been left to natural selection and to breeders who select plants exhibiting a low degree of infection and thus a slow and delayed development of the pathogen, i.e., plants possessing a field resistance which is a synonym for horizontal resistance. Thus, in spite of the lack of coordinated work on the development of horizontal resistance, there is a number of commercial cultivars possessing horizontal resistance to *Puccinia graminis* f.sp. *tritici* or other pathogens. Understandably, there are cultivars which possess both types of resistance.

Van der Plank's (1963, 1968) theory of vertical and horizontal resistance has made an important impact on the understanding of plant resistance to pathogens and on the direction of plant breeding for disease resistance in the last two decades. Robinson (1976) contributed to the further development of Van der Plank's theory by providing definitions for a number of concepts and by introducing new terms and example. However, the genetic explanation of resistance has not yet been generally accepted and it is still a subject of discussion.

Parlevliet and Zadoks (1977) claim that the division of resistance into horizontal resistance (to which stability, polygenic inheritance, and non-specific effect on pathogenic races are attributed) and vertical resistance (defined by instability, monogenic inheritance, and specific effect on pathogenic races) risks generating a false concept in that there exist two kinds of resistance as well as two kinds of genes for resistance. They maintain that such division is permissive for research purposes only, but unacceptable to describe the actual situation in nature.

Furthermore, if it is assumed that the majority of genes for resistance are major genes which operate on a gene-for-gene basis, there is no reason to doubt that minor genes too operate on the same basis. It is relatively simple to test the effect of major genes in the host on pathogenic races but it is very difficult to test genes with additive effects on a gene-for-gene basis. Parlevliet and Zadoks (1977) designed a model for testing genes with small effects. Later on, Parlevliet (1985) offered a simple theoretical explanation of non-race specificity or horizontal resistance. A hypothetical example is demonstrated in Table 14.4.

The four races of the pathogen can be identified only in model *B* with large gene effects. When the gene effects as well as interaction effects become smaller, the experimental error (not shown in the Table) becomes of such a range that no races can be identified and one speaks of non-race-specific resistance. Parlevliet (1985)

Table 14.4 Hypothetical Example Demonstrating Disease Severity When the Gene Effects[a] between Host and Pathogen Are Additive (Model A) or When They Interact According to a Gene-for-Gene Relationship (Model B)

Model[b]	Host cultivar		Gene effects large				Gene effects small			
			Pathogen race							
			–	*a*-	*b*-	*ab*	–	*a*-	*b*-	*ab*
A	–	–	40	60[c]	60	80	50	55[c]	55	60
	R	–	20	40	40	60	45	50	50	55
	–	*S*	20	40	40	60	45	50	50	55
	R	*S*	0	20	20	40	40	45	45	50
B	–	–	80	80	80	80	50	50	50	50
	R	–	40	80	40[d]	80	45	50	45[d]	50
	–	*S*	40	40	80	80	45	45	50	50
	R	*S*	40	40	40	80	40	40	40	50

[a] Only the resistance (*R*, *S*) and pathogenicity (*a*, *b*) alleles are shown.

[b] In Model A the resistance alleles reduce the disease severity with 20% (large) or 5% (small), and the pathogenicity alleles add 20% or 5% to the disease severity. In Model B the resistance alleles reduce the disease severity with 40% (large) or 5% (small), the pathogenicity alleles *a* and *b* cancel the effects of *R* and *S*, respectively.

[c] Variation in absence of interactions.

[d] Variation due to differential interaction.

Source: Parlevliet, 1985

concludes: "Apparently, non-race-specific resistance can arise from two genetic systems: when the host and the pathogen genes have small effects and operate on a gene-for-gene basis (model *B*) and when the host and pathogen genes, whether small or large in effect, do not operate on a gene-for-gene basis (model *A*). Genes with large effects operating in a non-specific way (model *A*) are not believed to exist in hostspecialized pathogen systems. In cereal-cereal rust systems resistance genes with large effects most likely operate on a gene-for-gene basis and are race-specific; resistance genes with small effects, too small to study them individually (polygenes), give a non-race-specific pattern whether they operate on a gene-for-gene basis or not" (p. 505).

According to Van der Plank (1978), the difference between these two resistance types is not in the fact that vertical resistance is monogenic and horizontal resistance is polygenic, because the latter one may be monogenic too, but in the fact that vertical resistance can be interpreted on the basis of a gene-for-gene system and horizontal resistance cannot. Van der Plank (1978) offered a hypothesis that vertical resistance is determined by the process of polymerization and horizontal resistance by the process of catalysis. He abandoned this theory later on and refers only to protein polymorphism in connection with resistance (Van der Plank, 1984).

Martin and Ellingboe (1976) came out with a hypothesis that horizontal resistance is in fact a masked vertical resistance. Namely, if the host's gene for vertical resistance is subdued by the corresponding gene for virulence in the pathogen, it is still strong enough to offer protection in the form of horizontal resistance. Van der Plank (1978) rose against this hypothesis by giving examples of horizontal resistance to *Botrytis cinerea* in grapes and horizontal resistance to *Erwinia caratovora* in potatoes although the resistant cultivars of both species do not possess genes for vertical resistance.

In his most recent book, Van der Plank (1984) offers some new explanations and views on pathogens – host specificity and resistance/susceptibility to diseases (Table 14.5).

Table 14.5 The Diagonal Check for Specificity in a Gene-for-Gene Relation[a]

	Plant				
Pathogen	*R1R1*[b]	*R2R2*	*R3R3*	*R4R4*	*R5R5*
v1v1	Susceptible[c]	Resistant	Resistant	Resistant	Resistant
v2v2	Resistant	Susceptible	Resistant	Resistant	Resistant
v3v3	Resistant	Resistant	Susceptible	Resistant	Resistant
v4v4	Resistant	Resistant	Resistant	Susceptible	Resistant
v5v5	Resistant	Resistant	Resistant	Resistant	Susceptible

[a] Plant reaction when resistance genes *R1*, *R2*, *R3*, *R4*, and *R5* at five loci interact with virulence genes *v1*, *v2*, *v3*, *v4*, and *v5* at five loci in the pathogen.

[b] Resistance is assumed tobe dominant, and *RR* can be replaced by *R* Virulence is assumed to be recessive: This is usual but not invariable or necessary for the illustration, recessive resistance and dominant virulence are known.

[c] The plant is susceptible.

Source: Van der Plank, 1984.

"The resistance genes R_1, R_2 ... are identified and numbered individually, they are specific genes. But in relation to the *v* genes, i.e., in relation to whether disease will or will not occur, resistance is unspecific. This is the inescapable logic of the diagonal (or more precisely the offdiagonal) arrangement (Table 14.5). The diagonal check explains the paradox. A resistance gene in the host is identified by the virulence allele, *not the avirulence allele*, in the corresponding locus in the pathogen. That is, it is identified in a state of susceptibility, not resistance". (p. 7) "There is race-specific susceptibility but not race-specific resistance" (p. 9).

This new concept of Van der Plank provoked new discussions. Browder and Eversmeyer (1986) believe that "A specific relationship within a parasite: host system may result in resistance, but another specific relationship may result in susceptibility. Specificity is a characteristic of parasite: host associations rather than of single organisms within the associations" (p. 379). Van der Plank (1986) responded by adhering to his thesis that specificity is in susceptibility and that there is no such thing as specific resistance to disease.

Van der Plank (1984, 1986) also criticizes the gene-for-gene hypothesis, maintaining that it cannot offer a viable explanation of specific avirulence. It should be mentioned here that the gene-for-gene hypothesis is exposed to criticism by other scientists too. Day et al. (1983) pointed out that the hypothesis may satisfactorily explain the rules governing the interactions between one kind of host resistance and the pathogens forced to respond to it, but it ignores most other forms of resistance, such as host's tolerance to infection, durability of resistance over time, genotype-environment interactions, etc. According to these authors, "Any variation in the host population which reduces the probability of infection will be favored by natural infection, and conversely, any variant in the parasite which enables it to overcome the defences of the host and reproduce successfully will also be favored. Thus there will be a constant flux of reciprocal adaptive responses in host and pathogen as each is influenced by the presence of the other. Evolution is an *ad hoc* process; it improvises on the variation available. It would be therefore very surprising if, under these circumstances, a single mechanism underlay all plant host – parasite interactions" (p. 424).

All these hypotheses and discussions on plant resistance to parasites put forward in recent years, made a step forward to a better understanding of host – parasite relationships and better genetic interpretation of the resistance/susceptibility phenomenon. How much they contributed to plant breeding for resistance to parasites is another question.

Breeding for disease resistance

In the last fifty years, the work on the incorporation of disease resistance in plants has mostly been directed towards vertical or race-specific resistance. This primarily stands for self-pollinated plants such as wheat, barley, oats, rice, flax, tomato, and vegetatively propagated plants, e.g., potato, and much less for open pollinated plants such as maize, rye, or lucerne.

The incorporation of resistance to the prevalent races of a certain parasite into a new cultivar gives this cultivar a decisive advantage over the previously grown susceptible cultivar. The new cultivar becomes most popular in a matter of a few years and every farmer wishes to grow it.

The spring wheat cultivar Ceres may be taken as an example of such a cultivar which gained wide popularity on account of its resistance. The cultivar possessed race-specific resistance to the prevalent races of *Puccinia graminis tritici.* It was introduced into commercial production in North Dakota in 1926, soon became the dominant cultivar and stayed there till 1935 when the occurrence of race 56 wiped it out from production (after Waldron, 1935 – according to Van der Plank, 1963). The same disaster befell the leading wheat cultivars in the Fifties because of their susceptibility to race 15 of *Puccinia graminis* f.sp. *tritici.*

As phrased by Robinson (1976), vertical resistance lasts long enough to support a boom of a new cultivar, but not long enough to prevent its downfall when the boom reaches its maximum. This is the so-called "boom-and-bust" cycle.

It has already been said that race-specific resistance does not ensure a lasting protection against parasites because each new resistant cultivar increases the selection pressure on the prevalent races of the pathogen to which it is resistant. Consequently, the frequency of the prevalent races goes down, other races become prevalent, to which the cultivar is not resistant. Also, new races occur by mutation or gene recombination, which have not existed at the time of the development of the cultivar.

Vertical resistance reduces the initial inoculum by postponing the outbreak of infection. As the infection spreads from plant to plant, from field to field, vertical resistance is more lasting if all plants of the cultivar do not belong to the same genotype regarding their resistance or if another cultivar with different resistance genes is grown in the neighbouring field. Vertical resistance cannot prevent auto-infection, i.e., infection from one leaf to another leaf of the same plant, but it can prevent alo-infection, i.e., infection from plant to plant and from field to field provided that the plants and cultivars possess different vertical resistances (Van der Plank, 1963; Robinson, 1976).

In other words, the value and duration of vertical resistance increases with any increases in the genetic diversity within the cultivar and the number of cultivars possessing different vertical resistances.

Genetic diversity within cultivars of self-pollinated plant species used to be maintained in varietal populations or varietal mixtures. Such varieties are not suitable for intensive agricultural production for a number of reasons. To find a replacement for them, a concept of the multi-line variety has been developed. The concept was put in practice for oats by Jensen (1952) and for wheat by Borlaug (1959) (see Chapter 21).

A pathogenic race capable of infecting a certain plant is not capable of causing an epidemic if the neighbouring plants are resistant. This is a way of maintaining a stable field resistance of a crop without large negative effects on yield. As the array of races continuously changes in prevalence and composition, corresponding changes must be introduced in lines constituting a multi-line variety. This has turned out to be a tedious chore to breeders and seed processors alike, which explains why multi-line varieties never became really popular.

The growing of several cultivars possessing different vertical resistance has been a fairly successful method of controlling certain pathogenic species and preventing large-scale epidemics.

If a cultivar grown in a field is attacked by old pathogenic races because of their prevalence, or by new races because of their virulence, the epidemics will spread in that field but not to other fields in which other cultivars, resistant to these races, are grown. If several cultivars are grown in a region in alternate fields, the epidemics will be curtailed and economically important yield reductions will not occur.

Unlike vertical resistance, horizontal or non-race-specific resistance cannot reduce the initial inoculum but it may decelerate an epidemic after its outbreak which in itself may frequently be enough to prevent extensive damage in yield or quality. If the cultivar grown in the adjacent field also possesses a horizontal resistance, the infection will continue to spread slowly and the epidemic will be prevented. Therefore, the value of horizontal resistance increases together with the uniformity of cultivars.

The degree of horizontal resistance can be measured only after a cultivar has lost its vertical resistance because whatever remains after the loss of vertical resistance is genuine horizontal resistance. However, if vertical resistance is lost, breeders discard all susceptible materials without paying attention to whether the materials do or do not possess horizontal resistance. Van der Plank (1963) mentions the example of the potato cultivar Vertifolia which first lost horizontal resistance to blight in consequence to the incorporation of resistance genes R_3 and R_4 and then it lost the vertical resistance (the so-called *vertifolia effect*).

Horizontal resistance does not warrant a complete protection but it does warrant a lasting protection. It is thus advantageous to cross cultivars possessing different degrees of horizontal resistance, and then to try to introduce a certain degree of vertical resistance.

As much as Van der Plank's theory of vertical and horizontal resistance to diseases introduced a new understanding of the host-parasite relationship and disease control, it aroused much perplexity about which way to go in the development of genetic resistance to diseases. Vertical or race-specific resistance has been evaluated as unstable and short-lived, horizontal or non-race-specific resistance as stable and long-lasting. The interest in the former dwindles but it is not clear how to approach and handle the latter.

If horizontal resistance is the remainder after the loss of vertical resistance, it is not sufficient to protect a cultivar. Because, if it had been sufficient, the epidemics would not have broken out in the first place. If good field resistance in some cultivars is considered as horizontal resistance, then it has already been quite extensively used by breeders, e.g., slow rusting, durable resistance, etc. It deserves to be mentioned here that vertical resistance is by no means complete because it provides resistance only to prevalent races of a pathogen but not to non-prevalent races which also exist.

There is no doubt that the epidemiology of the physiological races of pathogens should continue to be studied. Such investigations keep track of changes taking place in pathogenic populations which is an important piece of information in the breeding for disease resistance. The screening of cultivars for reaction to pathogenic populations should be intensified in order to select the most resistant ones. Such cultivars are prospective sources of resistance genes, no matter if oligogenes or polygenes. These genes should be transferred to cultivars superior in agronomic traits by hybridization (complete or incomplete backcross method) and by selection of resistant genotypes in segregating generations (see Chapter 12). The duration and effectiveness of resistance depend on genetic changes in pathogenic populations and the dynamics of breeding programmes aimed at the development of new cultivars. If certain cultivars retain resistance for five to ten years, it should be considered a long enough period because cultivars are replaced rapidly in intensive agricultural production not only because of the loss of disease resistance but primarily because of the changes in agricultural practices and the growers' demands for better and better cultivars. If a cultivar stays resistant for twenty years on account of horizontal resistance built into it, it will be replaced earlier anyway because higher-yielding cultivars will be developed. in the meantime.

Nowadays, the best sources of resistance to be used in breeding programmes can be found in international nurseries. Most resistance genes have been transferred from wild relatives of cultivated crops which saves breeders the trouble of having to go the long way of interspecific hybridization. Of course, a certain number of geneticists and breeders must work on the incorporation of new genes for resistance from wild species into agricultural crops.

It is frequently pointed out that local varieties and populations which are nowadays becoming extinct, used to be more tolerant to parasites, because of their genetic diversity and heterogeneity, than are the new high-yielding cultivars which are genetically uniform and homogeneous. This is also an excuse for a more intensive use of pesticides. It remains to check whether such statements are really true or not.

Van der Plank (1984) holds the opinion that high yield potential is incompatible with high resistance. High yield means bigger sink capacity and bigger sink means lost resistance. He also maintains that the resistance involved in sink-induced loss is horizontal. That loss is associated with high-sugar resistance. Therefore, a question is posed whether the breeding for high yields is a danger in itself or not.

It should be said at this point that the generally accepted picture of the resistance to parasites in high-yielding cultivars does not reflect any inferiority in relation to older cultivars. High-yielding cultivars are grown in dense populations, with high doses of mineral fertilizers (especially nitrogen), frequently on irrigation, and these factors create very suitable conditions for attacks by diseases and pests. If old local varieties had been grown in such conditions their being susceptible to lodging (e.g., cereals and forage crops) would mean that the catastrophy would have been complete.

We cannot be completely satisfied with the present level of breeding of high-yielding varieties or hybrids for resistance to parasites, but, considering their number, there are many which possess quite satisfactory resistance or tolerance. The real risk is when a susceptible cultivar gains popularity and covers more than 50% of the acreage in a large region. Such a situation should be counteracted by growing several cultivars in a region or even on a large farm. The cultivars should differ in resistance to major pathogens but should have equal genetic yield potentials.

Resistance to bacteria

Besides a large number of pathogenic fungi, there is also a large number of pathogenic bacteria, especially those from the genera *Pseudomonas*, *Xanthomonas* and *Erwinia*, which cause diseases in many plant species.

In recent years, several tomato cultivars resistant to bacterial wilt (*Pseudomonas solanacearum*) have been developed, especially in the U.S.A. (Henderson and Jenkins, 1971, – according to Russell, 1978). Some cultivars which exhibited resistance in the U.S.A. were susceptible in India. This indicates the existence of different races of the pathogen as well as the race-specific resistance of these cultivars.

The resistance to bacterial wilt is considered to be polygenic and unstable. Resistance genes from *Lycopersicon pimpinellifolium* secure a somewhat higher resistance than the genes from *L. esculentum* but they are linked with small fruits and poor quality which makes them less attractive for breeding programmes.

An apple and pear disease caused by the bacterium *Erwinia amylovora* may be so harmful that infected branches, and sometimes even whole trees, must be cut down. Differences in susceptibility have been observed among apple and pear varieties but resistant varieties have not been developed yet (Eden-Green and Billing, 1974, – according to Russell, 1978).

Plant resistance to viruses

Viruses cause diseases in many plant species. Viruses are transmitted from plant to plant by contact, vegetative propagation and grafting, pollen and seeds, or fungi and insects as vectors.

The types of plant resistance to viruses are immunity, resistance to virus infection, resistance to virus multiplication, resistance to virus distribution in the plant, tolerance, and resistance to virus vectors.

Viruses are particularly detrimental to plant species which propagate by tubers (potato) and cuttings (a number of horticultural plants) because, in addition to the primary infection of these parts, there occurs the secondary infection on growing plants.

As it is exceedingly difficult to achieve resistance to many viruses, growers resort to virus-free materials obtained by tissue culture.

Some viruses inflict considerable damage to vegetables and flowers grown in greenhouses. Several tomato cultivars have already been developed which are resistant to tomato mosaic-virus (TMV) on the basis of resistance genes transferred from wild species of *Lycopersicon*. Unfortunately, this resistance is linked with some undesirable traits such as chlorosis and low productivity. Nevertheless, several commercial cultivars resistant to mosaic-virus have been released (Walter, 1967, – according to Russell, 1978).

Barley, wheat, oats, and other cereals are attacked by barley yellow dwarf virus (BYDV) which causes leaf yellows, retarded growth, and underdevelopment of secondary tillers. It may reduce yields by more than 50%. Barley yellow dwarf virus is transmitted by aphids (*Aphidae*) which could be controlled by insecticides. However, as wheat and barley are grown in very large areas treatment by chemicals would be costly, so it is more economic and also safer to develop resistant cultivars.

Schaller et al. (1963) tested over 7,000 barley varieties and found only about 100 tolerant varieties, most of those coming from Ethiopia. These varieties are nowadays used in breeding programmes for resistance to BYDV. However, there is only one gene (Yd_2) which is incorporated in all cultivars and there is a risk of losing that resistance (Schaller, 1977).

A limited number of wheat cultivars exhibit resistance to BYDV. A few years ago, many wheat samples were tested for reaction to BYDV at the National Institute of Agricultural Research in New Delhi, India. The samples included about 20 entries from Novi Sad and one of them, the line NS-879/4, was found resistant. The Novi Sad wheat breeders had not been aware of that trait in their line because they had not tested their materials for resistance to the virus. The line NS-879/4 was developed from the cross of the variety Brkulja-4 and the line NS-725. The former variety had been obtained by crossing two Italian varieties, Mara×Funo, and the line NS-725 by crossing the Italian variety R-37 and the German variety Gohlds-121. It still remains unknown as to which of the parents possesses the resistance to BYDV, but in any case a new source of resistance to the virus has been obtained and presently it is used in many breeding programmes.

Breeding for resistance to insects

A large number of insects parasitize cultivated plants causing considerable reductions in yield and quality. Many plants to not possess any resistance to certain types of insects, especially those whose larvae or imagoes feed on leaves, fruits, or seeds, e.g., gypsy moth, locust, grain weevil, etc. In cases when it is estimated that a certain insect population has become so large that it will cause economic losses, plant protection by chemicals is recommended.

Only those insects against which either passive or active resistance may be developed will be discussed further. Those are insect species whose imagoes or larvae feed on plant juice or which oviposit in certain plant parts.

The mechanisms of plant resistance to insects are classified according to Painter (1951) in three main categories:

(1) *Non-preference.* Due to some traits of a plant species, either morphological or physiological, insect avoid it, do not feed on it, and do not oviposit on it.

(2) *Tolerance.* Some plant species are capable of tolerating an attack by insects without much reduction in yield or quality.

(3) *Antibiosis.* A capacity of a plant species to either kill insects feeding on it, hamper their development, or reduce their reproductive power and vitality. Therefore, antibiosis is considered an active or genetic resistance.

Genetic resistance to insects

Gallun and Kush (1980), who conducted extensive studies on the genetics of resistance to insects in plants, stated that there are horizontal and vertical resistances to insects, analogous to the classification for plant diseases given by Van der Plank (1963).

The horizontal resistance of a plant species implies that it is resistant to all biotypes (races) of an insect species. Therefore, it is a non-specific resistance controlled by a large number of minor genes or polygenes, as indicated by continuous variability from non-resistance to resistance in segregating generations of the crosses between resistant and susceptible varieties.

The vertical resistance implies a differential reaction of host varieties to various biotypes (races) of an insect. It is thus a race-specific resistance which is usually controlled by major genes or oligogenes. This is confirmed by discontinuous variability to resistant and susceptible plants in segregating generations.

Regarding the expression of resistance genes in the hosts, it may be recessive, dominant, or incompletely dominant in the intra-allelic relationship, and additive, epistatic, or complementary in the inter-allelic relationship.

The insect capable of attacking the host plant possesses virulent genes and the one incapable of attacking the host plant possesses avirulent genes.

So far, genetic resistance has been studied most extensively for Hessian fly (*Mayetiola destructor*) on wheat, corn moth (*Ostrinia nubilalis*) on maize, and some insect species which attack rice and cotton (Gallun, 1980).

Testing various populations of Hessian fly on different wheat cultivars, Gallun and his associates at Purdue University at Lafayette (USA) determined the existence of five loci for virulence, labelled *t*, *s*, *m*, *k*, *a*, and eight biotypes of the insect (analogous to fungus races) whose genotypes are given in Table 14.6.

Table 14.6 Genotypes of eight Hessian fly biotypes

	wheat varieties[1]				
Biotype	Turkey	Seneca	Monon	Knox 62	Abc
GP	*tt*	*S–*	*M–*	*K–*	*A–*
A	*tt*	*ss*	*M–*	*K–*	*A–*
B	*tt*	*ss*	*mm*	*K–*	*A–*
C	*tt*	*ss*	*M–*	*kk*	*A–*
D	*tt*	*ss*	*mm*	*kk*	*A–*
E	*tt*	*S–*	*mm*	*K–*	*A–*
F	*tt*	*S–*	*M–*	*kk*	*A–*
G	*tt*	*S–*	*mm*	*kk*	*A–*

[1] Reccessive and dominant alleles represent virulence in insect and susceptibility in wheat and vice versa.
Source: Gallun, 1977

The reaction of the wheat cultivars to the Hessian fly biotypes is shown in Table 14.7. The cultivar Turkey was susceptible to all biotypes because it lacked resistance genes (*tt*). The cultivar Seneca was resistant to biotype *GP* which has at least one dominant gene for avirulence in the locus which corresponds to the locus with a dominant resistance gene in wheat. Biotype *D* was homozygous in the loci *t*, *s*, *m*, *k* and it was virulent on the cultivars Turkey, Seneca, Monon, and Knox 62 but avirulent on the cultivar *Abc* which possesses the corresponding resistance gene (Tables 14.6. and 14.7).

Further investigations showed that some major genes (H_3, H_6, H_9 and H_{10}) which convey resistance of wheat to Hessian fly may interact differently and consequently different degrees of resistance may occur (Stebbins et al., 1982). Maas et al. (1987) found that the wheat cultivar Marquillo possesses a gene for resistance to biotype *D* of Hessian fly, distinct from all other known genes for resistance, which is partially dominant and was expressed most consistently at 16 °C.

The resistance of wheat and several other grasses to cereal leaf beetle depends very much on the size and density of leaf hairs (pubescence). Larval density and larval feeding damage were lower on pubescent-leaved cultivars than on wheat lines without a pubescent leaf surface. The yield loss of susceptible lines ranged from 17.6%

Table 14.7 Response of five wheat varieties to eight Hessian fly biotypes

Biotype and genes for virulence	wheat varieties and genes for resistance				
	Turkey (nema)	Seneca (H_7 i H_8)	Monon (H_3)	Knox 62 (H_6)	Abc (H_5)
GP (no)	*S*[1]	*R*[1]	*R*	*R*	*R*
A (*s*)	*S*	*S*	*R*	*R*	*R*
B (*s*, *m*)	*S*	*S*	*S*	*R*	*R*
C (*s*, *k*)	*S*	*S*	*R*	*S*	*R*
D (*s*, *m*, *k*)	*S*	*S*	*S*	*S*	*R*
F (*k*)	*S*	*R*	*R*	*S*	*R*
GG (*m*, *k*)	*S*	*R*	*S*	*S*	*R*

[1] *S*– (susceptible, *R* – (resistant)

Source: Gallun, 1980

to 25.6% (Webster et al., 1982). On the basis of the length and density of leaf hairs, Wellso (1986) expressed mathematically the degree of resistance of wheat cultivars to cereal leaf beetle. Elden et al. (1986) found a negative correlation between clones of lucerne with pubescent stems and the degree of damage by larvae of potato leafhopper (*Empoasca fabae*, Harris).

The leaf pubescence of common wheat (the cultivars Vel, Downy) provides a potentially valuable resistance to *Rhopalosiphum padi* (L.) which may contribute to the control of the virus disease BYDV, as compared with the glabrous-leaved cultivars (Arthur, Abe) which were susceptible.

Some wild species of potato contain granular trichomes on their foliage which convey resistance to potato leafhopper (Tingey and Laubengayer, 1986). Such trichomes were also found with some wild species of *Lycopersicon* and play an important role in the resistance to both tobacco hornworm (*Manduca sexta*) and Colorado potato beetle (*Leptinotarsa decimlineata*) (Kennedy et al., 1987).

Satisfactory levels of resistance to Hessian fly (*Mayetiola destructor*), leaf aphid (*Schizaphis graminum*), cereal leaf beetle (*Oulema melanopus*) and wheat stem sawfly (*Cephus cinctus*) have been successfully built up in wheat cultivars. Several barley cultivars resistant to these insects have also been developed.

Some maize varieties and lines possess resistance to corn moth (*Ostrinia nubilalis*), corn earworm (*Heliother zea*), corn rootworm (*Diabrotica virgifera*), etc. Also, attempts have been made to develop genetic resistance to aphids (Aphidae) in lucerne (*Acyrthosiphon pisum* and *Theriophis maculata*) and other plant species.

Integrated plant protection against diseases and pests

Many cultivated plants had been alienated from the nature in the process of their domestication. In the period of modern breeding, numerous cultivars and even species have been made to suit the demands and needs of the man. It was a step towards further alienation which made these plants more useful but less capable to resist parasites. Nowadays, a large number of cultivars would not be able to support themselves without protection supplied by the man.

Each plant species is attacked by a large number of pathogenic organisms and it is not possible to achieve genetic resistance to all of them, sometimes not even to the major ones. Genetic resistance should be strived for wherever it is possible because it is the safest and the most efficient measure of parasite control. Of course, other avenues should not be neglected, such as the development of passive resistance (when a cultivar avoids a major attack because its development cycle does not coincide with that of the parasite) or careful agricultural practices (such as crop rotation, precise sowing dates, mixed cropping) etc. Measures of biological control of parasites should be undertaken whenever possible and profitable. The parasites which remain unhindered by these measures should be controlled by chemicals.

This approach may be called "the integrated protection of crops". It does not imply by any means the application of pesticides against all diseases and pests which befall a crop in the course of growing season but on the contrary, it involves the combining of genetic resistance with all other measures including the minimum application of pesticides.

The growers of fruit, grapes and some vegetable crops go as far as to chemically treat their crops 15–20 times in the course of growing season. Programmes of development of cultivars resistant to certain diseases are rather neglected because pesticides are "so efficient". If active and passive resistance to several pathogens have successfully been developed in maize, wheat, etc., there should be no biological obstacles to develop resistance to some pathogens in fruit trees, vegetables or ornamentals. There will always be several parasites which will have to be controlled by pesticides. But, there is a large difference between 15 treatments and 3 to 5 treatments. The chemicals leave harmful residues which accumulate in plants and animals and then in humans. The genetic consequences of these residues are not known but, unfortunately, it is certain that they will not be beneficial.

It is therefore an imperative to all plant geneticists, pathologists and breeders to keep on developing resistance in all plant species in order to reduce the use of chemicals and to protect the environment.

References

Borlaug, E. N., 1959. The use of multiline or composite varieties to control airborne epidemic diseases of self-pollinated crop plants. Proc. 1st Inter. Wheat Genet. Symp. Winnipeg, p. 12–27.

Borojević, Katarina, 1976. Tolerance to *Puccinia recondita* f. sp. *tritici* in the population of *Triticum aestivum* ssp. *vulgare* after mutagenic treatments. Eucarpia Meeting, Novi Sad 7–10 June 1976. Genetika 8 (3): 233–239.

Borojević, S., and Potočanac, J., 1966. (The development of the Yugoslav programme for creating high-yielding wheat varieties). Savremena poljoprivreda, XIX, 7–36.

Bošković, M., and Browder, L., 1976. A comparison of pathogenicity of *Puccinia recondita* f. sp. *tritici* in Europe, the United States and Canada. Plant Dis. Report 60: 278–280.

Bošković, M. M., 1980. The use of some new differentials in international pathogenicity surveys of *Puccinia recondita* f. sp. *tritici*. Proc. of the Fifth Europ. and Mediter. Cereal Rusts Conf., Bari, Italy, p. 185–190.

Bošković, M., 1988. International pathogenicity survey of wheat leaf rust pathogen and sources of resistance. Cereal Rusts Bul. (In press).
Browder, L. E., 1980. A compendium of information about named genes for low reaction to *Puccinia recondita* in wheat. Crop Sci. 20: 775–779.
Browder, L. E., 1985. Parasite: host, environment specificity in cereal rusts. Ann. Rev. Phytopatol. 23: 201–222.
Browder, L. E., and Eversmeyer, M. G., 1986. Parasite: host specificity and resistance/susceptibility, two concepts, two perspectives. Phytopathology 76, 4: 379–381.
Browning, J. A., and Frey, K. J., 1969. Multiline cultivars as a means of disease control. Ann. Rev. Phytopath. 7: 335–382.
Browning, J. A., 1974. Relevance of knowledge about natural ecosystems to development of pest programs in agro-ecosystems. Proc. Am. phytopath. Soc. 1: 190–199.
Day, P. R., 1974. Genetics of host-parasite interaction. W. H. Freeman, San Francisco, 238 pp.
Day, P. R., Barrett, J. A., and Wolfe, M. S., 1983. The evaluation of host-parasite interaction. In: T. Kruge, C. P. Meredith and A. Hollander (Editors), Genetic Engineering of Plants, Plenum Press, New York, p. 419–430.
Dotlačil, L., and Apltauerová, M., 1978. Pollen sterility induced by Ethrel and its utilisation in hybridisation of wheat. Euphytica 27: 353–360.
Elden, T. C., Elgin, Jr. J. H., and Soper, J. F., 1986. Inheritance of pubescence in selected clones from two alfalfa populations and relationship to potato leaf-hopper resistance. Crop Sci. 26: 1143–1146.
Flor, H. H., 1942. Inheritance of pathogenicity in *Melampsora lini*. Phytopathology 32: 653–669.
Flor, H. H., 1955. Host-parasite interactions in flax rust, its genetics and other implications. Phytopathology, 45: 680–685.
Flor, H. H., 1971. Current status of the gene-for-gene concept. Annual Rev. of Phytopathology 9: 275–296.
Gallun, R. L., 1977. Genetic basis of Hessian fly epidemics. In: P. R. Day (Editor), The Genetic Basis of Epidemics in Agriculture, Ann. N.Y. Acad. Sci. 287: 1–400.
Gallun, R. L., and Khush, G. S., 1980. Genetic factors affecting expression and stability of resistance. In: F. G. Maxwell and P. R. Jennings (Editors), Breeding Plants Resistant to Insects, John Wiley, New York, p. 64–86.
Hooker, A. L., 1983. Breeding to control pests. In: D. R. Wood (Editor), Crop Breeding. Amer. Soc. Agr. Inc. Madison, Wis. pp. 199–229.
Jensen, N. E., 1952. Intravarietal diversification in oat breeding. Agron. J. 44: 30–34.
Kennedy, G. G., Nienhuis, J., and Helentjaris, T., 1987. Mechanisms of anthropod resistance in tomatoes. Tomato Biotechnology, p. 145–154.
Loegering, W. Q., 1984. Genetics of the pathogen-host association. In: W. R. Bushenell and A. P. Roeffs (Editors), The Cereal Rusts. Academic Press, New York, p. 165–192.
Maas, F. B., Patterson, F. L., Foster, J. E., and Hatchett, J. H., 1987. Expression and inheritance of resistance of "Marquillo" wheat to Hessian fly biotype D. Crop Sci. 27: 49–52.
Mac Key, J., 1974. Systematic approach to race-specific disease resistance. In: "Induced Mutations for Disease Resistance in Crop Plants", Proc. Res. Coord. Meet. FAO/IAEA, Novi Sad, 1973, IAEA, Vienna STi/Pub/388. p. 9–22.
Mac Key, J., 1977. Strategies of race-specific phytoparasitism and its control by plant breeding. Genetika (Beograd), 9: 237–255.
Mac Key, J., 1978. Genetics of race-specific phytoparasitism on plants. 14th. Inter. Congr. Genet. Moscow, USSR, August 21–30.
Mac Key, J., 1980. Use of monogenetically resistant lines in deciphering races, revealing strategies of pathogens and planning efficient breeding programmes. 19th Eucarpia Congress, Leningrad, September 16–20.
Martin, T. J., and Ellingboe, A. H., 1976. Differences between compatible parasite/host genotypes involving the Pm4 locus of wheat and the corresponding genes in *Erysiphe graminis* f.sp. *tritici*. Phytopathology, 66: 1435–1438.
Maxwell, F. G., and Jennings, P. R., 1980. Breeding Plants Resistant to Insects. John Wiley & Sons, New York, 683 pp.

McIntosh, R. A., 1983. A catalogue of gene symbols for wheat (1983. edition). Proc. 6th Inter. Wheat Genet. Symp. kyoto, Japan, pp. 1197–1254.
McIntosh, R. A. 1988. Catalogue of gene symbols in wheat. Proc. 7th Inter. wheat genet. Symp. Cambridge, UK. p. 1225–1323.
Parlevliet, J. E., and Zadoks, J. C., 1977. The integrated concept of disease resistance; a new view including horizontal and vertical resistance in plants. Euphytica, 26: 5–21.
Parlevliet, J. E., 1985. Resistance of the non-race-specific type. The Cereal Rusts II: 501–525.
Painter, R. H., 1951. Insect Resistance in Crop Plants. Macmillan, New York, 520 pp.
Roberts, J. J., and Foster, J. E., 1983. Effect of leaf pubescence in wheat on the bird clerry oat aphid (Homoptera: Aphidae). J. Econ. Entomol. 76: 1320–1322.
Robinson, R. A., 1976. Plant Pathosystems. Springer-Verlag, Berlin. 184 pp.
Russel, G. E., 1978. Plant Breeding for Pest and Disease Resistance. Butterworth, London, 485 pp.
Schaller, C. W., Rasmusson, D. C., and Qualset, C. O., 1963. Sources of resistance to the yellow-dwarf virus in barley. Crop Sci., 3: 342.
Schaller, C. W., 1977. Utilizing genetic diversity in the improvement of barley cultivars. California Agriculture, 31: 18.
Simons, M. D., and Browning, J. A., 1961. Seed weight as measure of response of oats to crown rust infection. Proc. Iowa Acad. Sci. 68: 114.
Stakman, E. C., and Levine, N. W., 1922. The determination of biologic forms of *Puccinia graminis* on *Triticum* spp. Minn. Agric. Exp. Stat. Tech. Bull. (St. Paul), No. 8.
Stakman, E. C., and Piemeisel, R. J., 1917. A new strain of *Puccinia graminis*. Phytopathology 7: 13.
Stakman, E. C., Stewart, D. M., and Loegering, W. Q., 1962. Identification of physiological races of *Puccinia graminis* f.sp. *tritici*. USDA, E. 617 (Revised) 55 pp.
Stebbins, N. B., Patterson, F. L., and Gallun, R. L., 1982. Inter-relationship among wheat genes H_3, H_6, H_9 find H_{10} for Hessian fly resistance. Crop Sci. 22: 1029–1032.
Stubbs, R. W., Fuchs, E., Vecht, H., and Basset, W., 1974. The International Survey of Factors of Virulence of *Puccinia striiformis* Westend in 1969, 1970 and 1971. Wageningen, Sep. 1974.
Tingey, W. M., and Laubengayer, J. E., 1986. Grandular trichomes of a resistant hybrid potato alter feeding behaviour of the potato leafhopper (Homoptera: Cicadellidae). J. Econ. Entomol. 79: 1230–1234.
Van der Plank, J. E., 1963. Plant Diseases: Epidemics and Control. Academic Press, New York, 349 pp.
Van der Plank, J. E., 1968. Disease Resistance in Plants. Academic press, New York/London, 206 pp.
Van der Plank, J. E., 1975. Principles of Plant Infection. Academic Press. New York, 216 pp.
Van der Plank, J. E., 1978. Genetic and Molecular Basis of Plant Pathogenesis. Springer-Verlag, Berlin, 167 pp.
Van der Plank, 1984. Disease Resistance in Plants (2nd edition). Academic Press. New York, 194 pp.
Van der Plank, J. E., 1986. A reply to L. E. Browder and M. G. Eversmeyer. Phytopathology, 76, 4: 382.
Watson, I. A., and Singh, D., 1952. The future for rust resistant wheat in Australia. J. Austral. Inst. Agr. Sci. 18: 190–197.
Webster, J. A., Smith, Jr. D. H., and Hoxie, R. P., 1982. Effect of cereal leaf beetle on the yields of resistant and susceptible winter wheat. Crop. Sci. 22: 836–840.
Wellso, S. G., 1986. Cereal leaf beetle (Coleoptera: Chrysomelidae) and winter wheat: host plant resistance relationships. The Great Lakes Entomologist, p. 191–197.
Zadoks, J. C., 1972. Modern concepts of disease resistance in cereals. In: The Way Ahead in Plant Breeding. Proc. 6th. Congr. Eucarpia, 89–98.

15. BREEDING FOR YIELD

It was stated in the chapters "Breeding methods" and "Breeding for individual traits" that yield is the main and ultimate goal of breeding. The yield (in a broad sense the yield of grain, fruit, biomass, dry matter, and in a narrow sense the yield of proteins, oil, sugar, fiber) is so important for the advancement of the production of food, feed and raw materials for industry that it is the major target in plant breeding and encompasses all other breeding objectives. It is therefore obvious why plant breeding is concerned primarily with breeding for yield.

The yield is the final result of a plant's life cycle. It ensues that the yield is not a simple character but a complex character, a set of characters, or a supercharacter. It includes a number of quantitative components which are polygenic in their genetic base. Furthermore, the yield bears testimony of all the events that take place in the course of a plant's ontogeny and it is therefore under the heaviest pressure of the environment. Thus, the heritability of yield is relatively low and it is not possible to estimate accurately the success of breeding for yield and the magnitude of genetic gain. That is why particular attention and great efforts must be devoted to breeding for yield.

Since new cultivars are usually made by hybridization or by a combination of hybridization and vegetative propagation, it takes at least 10 years of crossing, selection and testing to develop a new cultivar of an annual plant species. The process is much longer with perennial species. Two questions are asked repeatedly: how to shorten that period and how to readily recognize genetically superior genotypes? To answer them, we must solve two problems beforehand:

(1) in which segregating generation should the selection for yield start, and
(2) which selection criteria should be given preference when breeding for yield?

As selection criteria are more important than generations of segregation in respect to the success of breeding, main selection criteria will be reviewed first. Those are:

(1) selection for yield *per se*,

(2) selection for yield via yield components,

(3) selection indices,

(4) selection for yield via a plant model,

(5) morpho-physiological and biochemical criteria.

Selection for yield per se

Selection for yield may be done by measuring the yield of a selected plant, progeny or line. The success of selection will depend on the following factors:

(a) the generation in which selection is started,
(b) the method of growing segregating generations,
(c) the method of yield testing in space and time.

Selection for yield may start in the F_2 generation. One may visually select the best plants according to the desired plant type or measure the yield of individual plants and select the best.

It is the experience of a large number of breeders that the success of visual selection for yield in the F_2 and the F_3 is relatively low because the obtained progeny is frequently different from the selected one. Selection by measuring the yield of individual plants is slightly more successful. A large environmental variability is present in both cases because the tested plants are spaces wider apart than in commercial production. The success is also small because dominant and epistatic gene effects are strong in early generations and these disappear in later generations of self-pollinated plants.

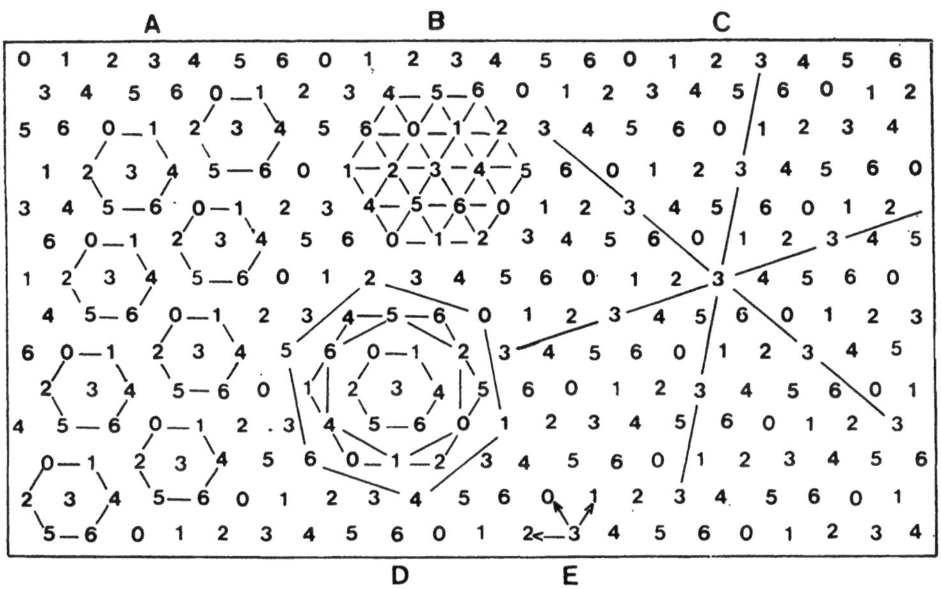

Fig. 15.1 Some characteristics of the hexagonal method of sowing. A. The progenies no. 3 are comparatively tested in the hexagonal fields. B. Each plant no. 3 undergoes six comparisons. C. A certain plant may be viewed from different directions. D. Superiority is ascertained by comparison with the six neighbouring plants. If the superiority extends to the outer rings of six, it is an indication of an outstandingly high yielding capacity. E. Three directions for observation of replications with different random distribution of genotypes in the replication.
Source: Fasoulas, 1976

The large environmental variability is primarily conditioned by soil heterogeneity (differences in physical structure, mineral and organic substances, etc.). It should be remembered that these differences in a field are not distributed at random but follow a certain pattern. That could be counteracted by randomly distributing the plants of the F_2 generation, but this is seldom practiced.

Soil variability should be made random in relation to plants. and Fasoulas (1973, 1976, 1981, 1988) proposed the *hexagonal method* or *honeycomb design* of sowing in an attempt to solve the problem. According to that method, each F_2 plant (as

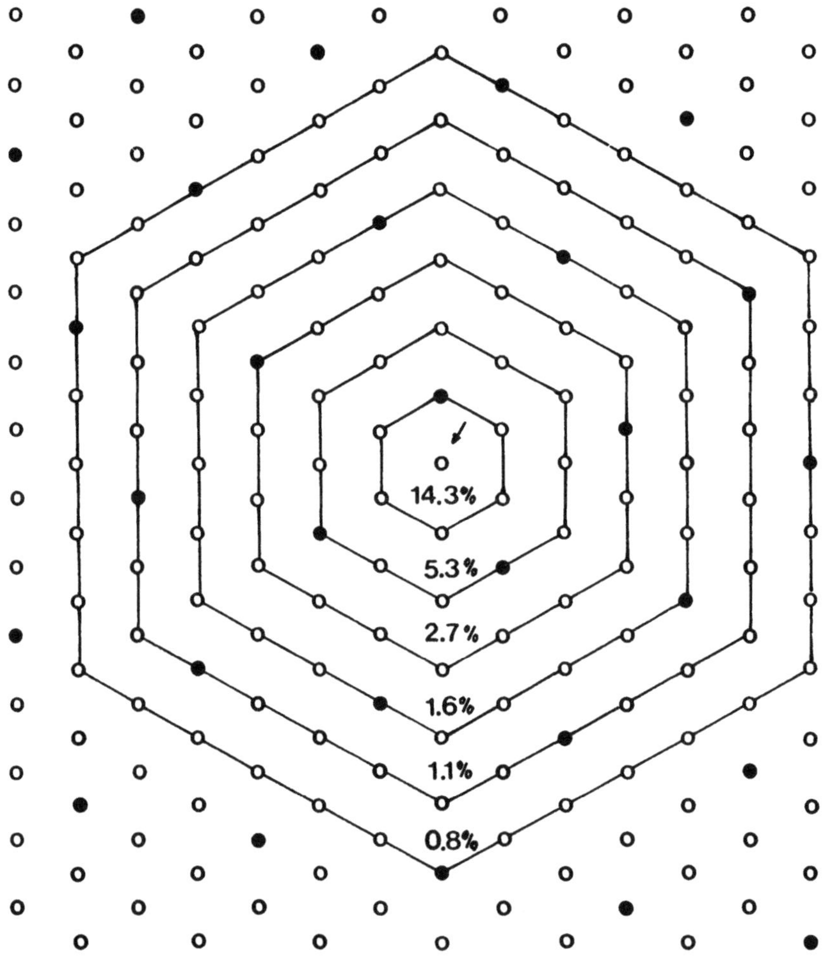

Fig. 15.2 When a plant in the center of hexagon outyields the surroundings six plants, it is selected by 14.3% intensity of selection (1/7). If it outyields the 18 plants included within the second concentric hexagon, it is selected by 5.3% selection pressure (1/9) and so forth. Thus, the desired selection pressure is determined by the size of the hexagon. Since each plant at the moment of evaluation is placed in the center of the hexagon, the hexagon is not fixed but moving. Solid circles represent check plants.
Source: Fasoulas, 1988

well as plants in other segregating generations) is comparatively tested against the six neighbouring plants and it is selected only if it outyields all of them (Fig. 15.1). The progenies of the selected plants are tested in the F_3 and other generations according to the same method, provided that, of the progeny from one F_2 plant, several individuals are randomly distributed in the replications. When a plant in a hexagon outyields the six neighbouring plants, it is selected by 14.3% intensity of selection (Fig. 15.2).

Furthermore, Fasoulas considers that different F_2 genotypes may come to their full expression only if free of competition, i.e., if they do not compete with quite different genotypes (allo-competition). The competition for nutrients, water and light may be avoided by sowing at distances, 50–85 cm for small grains and 120 cm for maize and sugar beet.

From the genetic point of view, the advantages of the hexagonal method are large. It solves the problem of soil variability which prevents the tested genotypes from performing to their best and which makes the evaluation of yielding potentials faulty. If different genotypes in the F_2 are grown in allo-competition, the real value of some genotypes may be obscured and they may be lost for ever. This explains why breeding still depends on chance. Fasoulas (1976) states that the maximum exploitation of soil fertility and the maximum yield per unit area are obtained only in conditions of iso-competition, i.e., the growing of a single genotype. In conditions of allo-competition, more competitive genotypes (higher stem, more intensive tillering, etc.) may bring higher yields although they do not actually have a higher genetic potential than the genotypes they excel. Therefore, superior genotypes in the F_2 and F_3 should be evaluated free of competition. Fasoulas (1973, 1988) also pointed out that intensive agricultural practices better differentiate the yielding potential of genotypes than do stress conditions.

Fasoulas (1973, 1976) made a significant contribution to plant breeding by insisting that it should be based on an unbiased evaluation of genotypes rather than on subjective views and chance. In his latest publication "The Honeycomb Methodology of Plant Breeding", Fasoulas (1988) elaborates in detail on the theoretical and practical aspects of the honeycomb methodology. In answering criticism that correlations must inevitably be different between the yield of a genotype bred free of competition (a large spacing between the plants) and the yield of its progenies grown in a dense stand, Fasoulas emphasizes that farmers nowadays grow cultivars which are in most cases monogenotypic pure lines, hybrids or clones, i.e., under the conditions of iso-competition. Since allo-competition is negatively correlated with yielding ability, selection should be practiced in the complete absence of allo-competition.

In connection with this, three advantages of large spacing between plants are secured: "(1) the confounding effects of allo-competition are controlled, (2) phenotypical differentiation which improves heritability and reduces environmental effects is maximized and (3) inbreeding which brings gene fixation indispensable for progress is enhanced" (Fasoulas, 1988, p. 39). Wide plant spacing and optimum growing conditions allow the best phenotypical expression and hence the best chances of identifying superior genotypes, while density and other conditions which suppress the yield of individual plants have the opposite effect. Therefore, in the honeycomb methodology, the units of evaluation and selection at all stages of the breeding programme are individual plants instead of field plots.

The recommended large spacing between plants in the segregating generations precludes the growing of a large number of plants per cross combination or a large number of cross combinations per year, and actually Fasoulas worked earlier on 20–40 cross combinations per year. Many breeders consider that number too low for obtaining a large number of superior lines. The shortcoming is compensated by the effectivness of the hexagonal method, i.e., the fact that genotypes selected will be truly superior. However, the selection from a large number of cross combinations may compensate even better for the loss of some superior genotypes in early generations.

The honeycomb methodology requires very large fields for growing breeding materials and measuring the yields of an exceedingly large number of plants. This is a kind of job that not every research institution can afford. Fasoulas (1988) emphasizes that all procedures of sowing, harvesting and measuring can be mechanized and computerized and these should not make problems. In any case, the honeycomb methodology is not used much in plant breeding in spite of its sound theoretical background. Fasoulas' (1988) considerations about the superiority of homozygosity over heterozygosity or inbred vigour will be discussed in connection with heterosis in Chapter 16.

Many researchers have tried to overcome the unreliability of single-plant selection for yield in the F_2 and F_3 generations by sowing selected plants in 2–3 replications, even if a single row represents one replication, and comparing them against controls at frequent intervals. In some instances the values obtained for yield were in good agreement with the values obtained in subsequent generations (Briggs and Shebeski, 1968; O'Brien et al., 1978) but there were also cases when the method was found unfit for selecting superior genotypes in early generations (Knott and Kumar, 1975; Fischer and Kertesz, 1976). Comparing F_3 yield tests with the single seed descent technique (SSD), Knott (1979) concluded that F_3 selection based on a two-replicate test with SSD procedure did not justify the extra work involved.

Shebeski and Evans (1973) in Canada designed a method of selection for yield from the F_2 to the testing of homozygous lines. The method has the following sequences.

(1) 20–30 combinations of crossing, selected on the basis of carefully matched parental pairs, performed annually.
(2) 150 F_1 seeds from each cross combination are sown in three replications (one row, one meter long) together with controls and parents. Only 2–3 cross combinations are selected for further work on the basis of a visual check for plant type, height, earliness, disease resistance, etc.
(3) 5,000 seeds from F_1 plants of each selected cross combination are sown in the stand 30×30 cm to avoid allocompetition and to obtain the maximum number of grains per plant.
(4) 1,000 best plants from each F_2 cross combination are selected and in the F_3 from each plant three rows are sown, five metres long, 300 seeds per row. The

yield of each F_3 plot is evaluated, compared against the control, and only 10–20 progenies (plots) with the highest yields are selected for further work.

(5) In the F_4 all progenies are sown replicated in bulk in three rows, five metres long, 300 seeds per row. The 50 best plants are selected in each F_4 plot on the basis of their yield.

(6) The procedure from F_4 is repeated in F_5. The 30 best lines per cross combination are selected on the basis of the yield per plot.

(7) In the F_6, the selected lines are comparatively tested in two locations, three replications, in plots with three rows, five metres long, 300 seeds per row. The lines with the highest yields are selected for testing in the F_7.

(8) in the F_7, the best lines are tested for yield in comparative trials in 4–5 locations, four replications, in plots with four rows, five metres long, with a row-to-row distance of 30 cm.

The best lines selected on the basis of their yielding performances are considered to be new cultivars and they may be registered for commercial production.

The advantages of Shebeski's method are that selection for yield starts already in the F_3, allo-competition is avoided and the problem of soil variability is overcome by sowing the progenies of a single plant in three rows (replications) starting from the F_3. It is thus easy to single out superior genotypes. Furthermore, new lines can be tested in several locations already by the F_6. It should be mentioned that Winnipeg coordinated the entire programme with CIMMYT in Mexico so that their experimental material was grown in Mexico in winter and in Canada in summer. In that way two generations a year may be produced which further accelerates the process of breeding.

The major shortcoming of the method is the small number of cross combinations per year which lowers the probability of deriving combinations with the best combining abilities, (which is the same problem as with Fasoulas's method). There still remains an open question: what are the common points of selection for yield in a thin stand, without competition, and commercial yield obtained in a thick stand in which competition, although only iso-competition, is present? The problem of stand density is imposed for crops which are normally grown in thick stands, e.g., wheat, barley, oats, forage grasses. It is less important for crops grown in thin stands, e.g., soyabean, pea, tomato, pepper, etc.

Harvest index (HI)

Harvest index, used as a selection criterion for yield increase may be counted among the methods of selection for yield *per se*. Harvest index (HI) is a relative ratio between the yield of grain (fruit) and the yield of total biomass (grain + straw + root) at the time of harvest. As it is difficult to measure the weight of roots, the total biomass usually implies only the above-ground plant parts and about 20% may eventually be added to compensate for the weight of roots.

It has been found that short cultivars (wheat, barley, sorghum, etc.) have a higher harvest index (more than 40%) and thus a higher grain yield than tall cultivars (less than 40%). An idea arose to measure the harvest index in early segregating generations and use it as a selection criterion for yield improvement.

Investigations by Fischer and Kertesz (1976), Bhatt (1977) and Bhatt and Derera (1978) showed that wheat plants with an increased harvest index in the F_2 and F_3 generations produced progenies with a high harvest index, i.e., that HI may be used as a selection criterion for yield in early generations. Conversely, Allan (1983), who studied wheat genotypes with differing plant heights – dwarfs (Rht1 and Rht2 genes), semi-dwarfs (Rht1 or Rht2) and tall genotypes (Rht0) – concluded that the use of HI as a selection criterion for grain yield appeared more feasible among inherently taller than shorter backgrounds. Similar conclusions had been drawn by McVetty and Evans (1980) who maintained that potentially high-yielding tall cultivars could be selected from F_2 space-planted populations using productivity and, in contrast, for semi-dwarf cultivars using the harvest index.

Our studies (Kraljević-Balalić and Borojević, 1988) showed that in conditions of space planting there was no correlation between HI and yield per plant. Thus, it appears that HI cannot be used as a reliable criterion when selecting for grain yield, especially when the weight of roots is taken in account.

Since wheat and other cereals and grasses are in fact "communal plants", as aptly named by Donald (1981), they "have features in accord with the success of the crop community rather than of the plant itself..., and may give low individual plant yields, but, when grown in a pure stand at a density sufficient to induce interplant competition and full exploitation of the environment, they are capable of high crop yields" (p. 235).

Therefore, the harvest index calculated for space-planted plants is not free of the shortcomings attributed to selection for yield in early segregating generations. If, however, HI is calculated on the plot basis, it may be a reliable criterion in selection for yield.

The improvement of the harvest index as an avenue for further increasing the genetic yield potential will be dealt with in more detail at the end of this chapter.

There are many other methods of selection for yield *per se* starting from the F_2 generation. It would not be an exaggeration to say that every breeder has a variant of his own.

Selection for yield via yield components

The main yield components of grain-producing plants are: plant number per unit area (X), grain number per plant (Y), and absolute grain mass (Z). Grafius (1964) made a geometrical diagram of yield components and their effect on yield. As a rule, the realized yield is the balance of yield components. When the contribution of one component increases, another one decreases and vice versa. Maximum yield is the realization of the most favourable balance among yield components. Such an approach gave rise to the idea that yield components must be genetically correlated and that when breeding for yield components, one actually breeds for yield itself.

It would be particularly advantageous if superior genotypes regarding grain yield could be produced by selection for grain mass in the F_2 and F_3. Studies by Bhatt (1972), Edwards et al. (1976), and other researchers indicated that breeding for increased grain mass does increase grain yield. That selection criterion does not warrant higher yields in later generations on account of the negative correlations between grain mass and grain number as well as grain mass and plant number/$m2$. The selection for the two yield components of grain mass and grain number per plant or spike, proved more successful (McNeal et al., 1978).

In spite of its advantages, the method of selection for yield via yield components is not widely used. According to Frey (1971), the reasons against it are: (a) the relationship between yield and yield components is not always linear, (b) the relationship changes under the effect of environmental conditions, and (c) it is sometimes more difficult to evaluate the yield components than the yield itself.

The unreliability of selection for yield components arises from their inherent compensating capacities. In that respect, Adams (1967) drew up some very valuable conclusions. Studying the yield components of beans – pod number per plant (X), grain number per pod (Y), and grain mass (Z), Adams found that components were regularly negatively correlated but also that these correlations were not of a genetic nature. Negative correlations were not found when the plants were sown at a spacing of 45 cm and they did not have to compete. Conversely, all three components were negatively correlated when the plants were sown at a 7.5 cm spacing (Table 15.1).

Table 15.1 Phenotypic correlations between components of yield in F_3 rows of navy beans grown at two levels of inter-plant competition. East Lansing, 1960

Plant spacing in plots	Number of plots	r_{xy}	r_{xz}	r_{yz}
45 cm	308	−0.09	0.06	0.17**
7.5 cm	257	−0.25**	−0.22**	−0.27**

* $P < 0.05$
** $P < 0.01$
Source: Adams, 1967

Negative correlations resulted from the relationships between the yield components in the course of plant ontogeny. The fact that the correlations were negative bore no effect on yield level since negative correlations among the components were found in low-yielding and high-yielding lines alike. The difference was that the former lines had higher negative correlation between $XY = (-0.51)$ and the latter ones had a higher negative correlation between $YZ = (-0.45)$.

It cannot be said on the basis of the above results that there are no genetic correlations among certain traits, especially the final yield components. As breeding is always aimed at multiple traits, it is essential to know their correlated response in order to manipulate them more easily. Experimental data and important theoretical explanations of this were provided by Grafius (1978). It will be discussed in more detail when summarizing the methods and approaches in breeding for yield.

Selection indices

To make selection for yield more reliable, attempts have been made to find those selection criteria which would reduce the subjective evaluation of traits and take more into account the effect of multiple characters on yield as the main goal. The resultant theory of selection indices has found wide application in livestock breeding but it has not advanced much in plant breeding.

Plant selection indices were worked out by Smith (1936) on the basis of Fisher's (1936) concept of discriminating function. Livestock breeding indices were calculated by Hazel (1943). To make a selection index for a desired character, the following information should be available:

(a) the relative economic importance of the trait,
(b) the genotypic and phenotypic variances of the trait,
(c) the phenotypic and genotypic covariances among traits.

Relative economic importance of the trait is established on the basis of the value of the trait (price, etc.) and it is expressed in points. To obtain data for variances and covariances, it is necessary to conduct experiments at least in two replications, two locations and two years, and analyse the tested material for the desired trait.

According to Lin (1978), the basic formulae for the calculation of selection index (I) and compound genotype or net gain (H) are:

$$I = \sum_{i=1}^{m} b_1 \times x_1 = X'b,$$

$$H = \sum_{i=1}^{n} a_1 g_1 = g'a,$$

where

$X' = (X_1 X_2, \ldots, X_m)$ – m order vector for known phenotypic values,

$g' = (g_1 g_2, \ldots, g_n)$ – n order vector for unknown genotypic values,

$a' = (a_1 a_2, \ldots\ a_n)$ – n order vector for known relative economic values,

$b' = (b_1 b_2, \ldots, b_n)$ – a vector from m index coefficients to be calculated.

To obtain these values, one must solve a series of simultaneous equations and other formulae; for more details see the original publications cited in the next.

Smith-Hasel's index takes into account several traits but it considers the compound genotype as a single trait and it is therefore impossible to have control over index components. That index is called the *estimated selection index* or the *indirect selection index.*

The breeder frequently wishes to improve one trait keeping another one constant, or he wishes to improve one trait on account of another. In that case one should use the *restricted selection index* worked out by Kempthorne and Nordskog (1959), Tallis (1962), etc.

Livestock breeding indices include additional parameters such as *family index* (Lush, 1935, 1947), *optimum index* (Henderson, 1963), *performance index* (Young and Tallis, 1961), etc.

When reviewing plant selection indices, we should mention the *base index* worked out by Brim et al., (1959) and Williams (1962), based on the rating of each trait according to its relative economic value. The index is easy to calculate.

As it is difficult to estimate the relative economic value of traits selected for, the results of breeding may diverge largely from the values estimated on the basis of selection indicies. Pesek and Baker (1969a) thus introduced the *desired gain index*, determined by the breeder himself and used this to calculate the index coefficient for each trait.

The following examples are given in order to illustrate the usefulness of selection indices.

Matzinger et al. (1977) studied a synthetic population of tobacco (*Nicotiana tabacum*) which comprised eight cultivars in diallel crosses, randomly pollinated during six generations. The study had the following independent breeding objectives: (1) to reduce plant height, (2) to increase leaf number and (3) to carry out index selection for increased leaf number on shorter plants. The purpose of index selection was to breed against the negative correlation between plant height and leaf number. Experiments were conducted in four replications, two locations and two years, in five breeding cycles. In the case of tobacco, the number of leaves represented in fact the yield.

The selection for reduced stem brought about the decrease of 6.52 cm per cycle, i.e., less than expected (Table 15.2). The number of leaves was reduced by −0.46

Table 15.2 Predicted and observed response to selection for decreased plant height and increased number of leaves in Nicotiana tabacum

Characters	Decreased plant height		Increased number of leaves	
	Predicted	Observed	Predicted	Observed
Height, cm	−10.19+	−6.52**	6.63+	4.74**
Leaves, number	−.95+	−.46*	1.35+	1.28**
Yield, grammes	−4.26+	−4.16*	3.23+	1.36
Flowering, days	−2.16+	−.70+	1.89+	2.27**
Alkaloids, %	0.169+	0.086**	−0.094	0.007
Sugars, %	−.660	−.044	0.271	−0.121
Leaf length, cm	−.023	−.012	0.216	−0.045
Leaf width, cm	−.693+	−.993*	−1.205+	−0.841
Internode, cm	−.290+	−.213*	−0.253+	−1.74**

+ Variance or covariance used in predicting response was twice its standard error.
*, ** Significantly different from zero at 5% and 1% levels, respectively.
Source: Matzinger et al., 1977

leaves/plant/cycle, or a half of the expected reduction (−0.95). As the result of interdependence, there occurred reductions in other traits which were not selected for, including yield, percent of alkaloids, leaf width and internode length (Table 15.2).

Selection for the increased number of leaves per plant brought about the linear increase of 1.28 leaves/plant/cycle, i.e., the expected value was almost reached (Table 15.2). The result of the correlations between the traits was an increase in stem height by 4.74 cm/cycle, prolonged vegetative period, etc. (Table 15.2).

It had been planned to achieve an average reduction of stem height by 4.26 cm and increase the leaves by 0.212 per plant. The actual reduction in height was only 2.00 cm, but the increase in the number of leaves was larger than expected (0.541) (Table 15.3).

Table 15.3 Average predicted and observed response of H and component characters in index

	Δ_H	Δ_1	Δ_2
Predicted	5.94	−4.26	0.212
Observed	5.83**	−2.00*	0.541*

*, ** Significantly different from zero at 5% and 1% levels, respectively.
Source: Matzinger et al., 1977

The index selection reached only 31% of the reduction in stem height achieved by the direct selection for plant height and 42% of the increase in leaf number achieved by the direct selection for leaf number. Nevertheless, it may be considered a success since the index selection was carried out contrary to the genetic correlation between the traits selected.

Brim et al. (1959) studied the effect for six traits in two soyabean populations. They were the following traits: oil content, protein percent, grain yield, resistance to lodging, vegetative period and grain mass. Each trait was considered in relation to three ratios of the price of oil versus protein. The highest selection gain was obtained by including all six parameters in the index, then the next highest gain was obtained by five parameters e.g., leaving out the grain yield, etc. Considerable differences were obtained between both, the populations and the price ratios. The authors concluded that index selection has many shortcomings, especially in the estimates of the economic values of parameters, and because of sample error magnitude. However, they also stated that the shortcomings of index selection cannot be mended by applying direct selection or other methods.

As index selection is not invariably better than selection for a trait *per se*, several comparative studies have been undertaken to evaluate the gain achieved by using different types of selection indices.

Eagle and Frey (1974) studied the efficiency of five selection methods (a) selection for increased grain yield alone, (b) selection for increased straw yield alone, (c) selection index I which maximized the expected genetic gain from breeding for increased economic profit, (d) basic selection indices under (a) and (b) but expressed via their economic values, and (e) the method of independent culling levels. Their experimental materials included 1,200 randomly selected F_9 oat lines derived from a population obtained by mixing F_2 seed of 75 combinations of crossing.

There was a high positive correlation (0.78) between grain yield and straw yield indicating that breeding for higher grain yield will also bring about an increase in straw yield, and vice versa. Grain yield was assigned double economic value but since the heritability for straw was higher than that for grain, the relative significance of straw was much higher than that of grain yield.

The basic index brought the highest grain yield, although the authors admitted that the experiments conducted in several replications, locations and years, combined with extensive analyses and calculations, do not justify the efforts invested. They recommend the method of independent culling levels for all traits which may be objectively evaluated in the course of vegetative growth before harvest, and the remaining materials may then be selected on the basis of indices.

In another study on the same materials, Rosielle and Frey (1975) used the restricted selection index for increased grain yield in oats with restricted tasselling and plant height, as compared with the unrestricted selection for each individual trait, i.e., increased grain yield, earlier tasselling, reduced stem height and increased harvest index. Their results showed that grain yield increase could be achieved only if earliness and stem height were slightly restricted. Larger restrictions in these two traits brought about corresponding yield decreases. If other restrictions had been included in the consideration, e.g., resistance to lodging, disease resistance, quality, etc., the yield increase would have been further limited. In such cases it is better to start selection by eliminating highly heritable traits first and than to select for yield and other traits with low heritability. It also turned out that the indirect selection for yield by harvest index was as successful as the restricted selection index.

Pesek and Baker (1969), using an open-pollinated plant species, simulated selection for traits X and Y from the F_6 to F_7 by the methods of tandem selection and index selection. The latter method was more successful in all combinations of different parameters. As regards the former method, if two traits are negatively correlated, it is important to select for the economically more important trait in the first breeding cycle because it may happen that no genetic gain is achieved for the second trait. Conversely, index selection is more efficient when the absolute values of negative correlation are large. In principle, there exists a general trend for index selection to increase in efficiency with the decrease in heritability of the traits selected for.

The application of selection indices is more justifiable with open-pollinated plants because the selected genetic population keeps changing and the obtained genotype (progeny) may differ considerably from the selected one. It is frequently the case that only the female's contribution is known while the male's remains unknown.

Elgin et al. (1979) used four methods of selection – tandem selection, modified method of elimination, estimated selection index and base index – to improve six lucerne traits: the rate of regeneration after cutting and the resistance to five pathogens. The selection conducted through five cycles, showed with regard to the total gain that the method of base index was the most efficient and the method of tandem selection the least.

Most papers on the use of selection indices deal with maize. Moll et al. (1975)

selected the cultivar "Jarvis" for grain yield and ear height using five selection criteria: (A) selection for yield alone, (B) selection for ear height alone, (C) and (D) restricted selection index based on the estimate that the desired change in height exceeds 5.1 cm or falls short of 5.1 cm, as compared with the source population, and (E) selection index for maximum yield increase taking into account the positive correlation with ear height.

The selection for yield alone (A) did bring a yield increase as compared with the control. The methods of selection for reduced ear height (B and C) brought the anticipated reductions in all sub-populations as compared with the source population. The index selection for the increase in ear height to 5.1 cm caused a slight yield reduction. The selection for maximum yield (E) increased both the yield and ear height (Fig. 15.3).

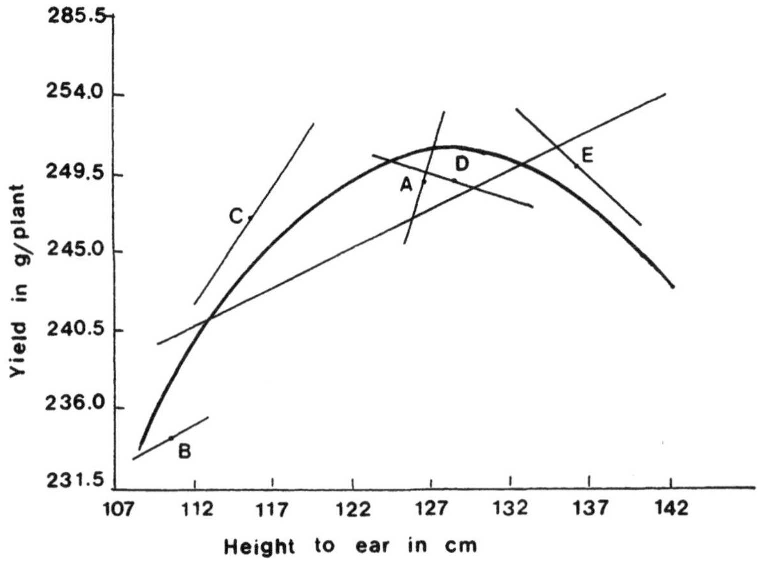

Fig. 15.3 Linear regressions for yield and ear height of selected populations of maize for each selection criterion, and general linear and square function.
Source: Moll et al., 1975

That study showed that the genetic correlation between yield and ear height is not linear and that there exists an optimum height which gives the maximum yield. The optimum ear height for the materials studied ranged between 116 and 129 cm. It is undoubtedly related to the number and position of leaves below and above the ear, which regulate the rate of shading in dense stands. Still, selection for reduced ear height alone brought a significantly lower ear height than the other selection methods.

Kaufman and Dudley (1979) compared two maize populations for the effect of selection for grain yield, protein content, and grain mass by the method of direct selection for individual traits, estimated selection index and desired gain index.

In general, index-based selection brought higher selection gains than did direct selection for individual traits.

To demonstrate the procedure of determining selection indices and calculating expected selection gain, a statistical analysis is presented for selection indices for a maize population studied by Mišević (1981) at the Maize Institute in Zemun Polje, Yugoslavia.

The objective was to increase oil content and grain yield in the synthetic maize population U1 by crossing within the population according to model I of Compstock and Robinsson (1952). The progenies of four female plants crossed with the same male plant were a half-sib family, and the progenies of individual female plants crossed with one male plant were a fullsib family. The experimental material was grown in a randomised block design in three replications and in two locations. Fifty half-sib families and 200 full-sib families were studied in each locality.

The experimental data were processed by the analysis of variance and covariance since the following parameters were necessary to apply selection indices:

- additive genetic variance of a trait – g_{ii}, g_{jj},
- phenotypic variance of a trait – f_{ii}, f_{jj},
- genotypic covariance between two traits – g_{ij},
- phenotypic covariance between two traits – f_{ij},
- desired gain in a trait – q,
- relative value of a trait – a,
- selection intensity – k.

It was planned to increase grain yield and oil content by using the desired gain index and the estimated selection index.

Desired gain index

It was calculated on the basis of genetic covariance between oil content and grain yield that the grain yield of the population U1 may be increased by 30%. The desired oil content was 8.5%.

The analysis gave the following figures:

g_{11} (grain yield) = 361.16,
g_{22} (oil content) = 0.3847,
g_{12} = 1.2911,
f_{11} = 1,060.01,
f_{22} = 0.7255,
f_{12} = 4.967,
q_1 = 31.02 (the difference between the measured and the desired grain yield),
q_2 = 2.12 (the difference between the measured and the desired oil content),
k = 1.4 (the coefficient for selection intensity of 20%).

The desired gain index for simultaneous selection for two or more traits is calculated after the formula of Brim et al. (1959), Pesek and Baker (1969):

$$I = b_1x_1 + b_2x_2 + \dots b_nx_n$$

where x = the measured level of the trait, b = coefficient of desired gain.

As there were two traits in our example, the coefficients b_1 and b_2 were calculated by the equations:

$$b_1 = q_1 g_{22} + q_2 g_{12},$$

$$b_2 = q_1 g_{12} + q_2 g_{11}.$$

Including the values of additive variance and covariance and desired gain for grain yield (trait 1) and oil content (trait 2), in the questions, the coefficients were calculated as follows:

$$b_1 = 31.02 \times 0.3847 + 2.12 \times 1.2911 = 14.67,$$

$$b_2 = 31.02 \times 1.2911 + 2.12 \times 361.16 = 805.21.$$

To simplify further calculation, one of the coefficients was converted to unity by dividing both coefficients by the value of coefficient b_1 (14.67):

$$b_1 = 14.67/14.67 = 1,$$

$$b_2 = 805.21/14.67 = 54.89.$$

Therefore, the progenies of the population U1 were selected for grain yield and oil content by the index:

$$I = \text{yield} + 54.89 \times \text{oil percent}.$$

If we are to choose the best of three progenies which have the following grain yields and oil contents:

P_1 yield = 125 $\quad I_1 = 125 + 54.89 \times 8.20 = 575.10,$
oil % = 8.20

P_2 yield = 135 $\quad I_2 = 135 + 54.89 \times 7.20 = 530.21,$
oil % = 7.20

P_3 yield = 118 $\quad I_3 = 118 + 54.89 \times 7.90 = 551.63,$
oil % = 7.90

it may be concluded that P_1 is the best progeny for it has the highest index for the combined trait of grain yield and oil content.

Having measured and calculated the necessary values, we may estimate the gain which will be obtained by the selection of half-sibs.

The improvement or gain in grain yield ($\Delta\, HS_1$) at the selection intensity of 20% may be:

$$\Delta\, HS_1 = \frac{b_1 g_{11} + b_2 g_{12}}{b_1^2 f_{11} + 2 b_1 b_2 f_{12} + b_2^2 f_{22}} \times 1/2k,$$

$$\Delta\, HS_1 = \frac{1 \times 361.16 + 54.89 \times 1.2911}{\sqrt{1^2 \times 1{,}060.61 + 2 \times 1 \times 54.89 \times 4.967 + 54.89^2 \times 0.7255}} \times 0.5 \times 1.4,$$

$$\Delta\, HS_1 = \frac{432.03}{\sqrt{3{,}791.72}} \times 0.7,$$

$$\Delta\, HB_1 = \frac{432.03}{61.58} \times 0.7 = 4.91,$$

It ensues that the expected yield gain is 4.91 grams per plant or 258 kg/ha, since there were 52.500 plants/ha.

The estimated gain in oil content ($\Delta\ HS_2$) is calculated by the formula:

$$\Delta\ HS_2 = \frac{b_1 g_{12} + b_2 g_{22}}{\sqrt{_1^2 b g_{11} + 2b_1 b_2 f_{12} + b_2^2 f_{22}}} \times 0.5k,$$

$$\Delta\ HS_2 = \frac{1.2911 + 54.89 + 0.3847}{61.58} \times 05 \times 1.4.$$

$$\Delta\ HS_2 = \frac{22.41}{61.58} \times 0.7 = 0.25$$

It means that an increase in oil content of 0.25% may be expected.

Estimated selection index

Relative economic importance of a trait may also be used as a selection criterion. In the maize population U1, the target was maximum yield increase. Thus, the relative economic value of yield was designated as 1 and the relative economic value of oil content was designated as 0, i.e., $a_1 = 1$, $a_2 = 0$.

In that case, selection index (I) was expressed in the same form as desired gain index, and coefficients b_1 and b_2 are calculated by the following equations:

$$b_1 f_{11} + b_2 f_{12} = a_1 g_{11} + a_2 g_{12},$$

$$b_1 f_{12} + b_2 f_{22} = a_1 g_{12} + a_2 g_{22}.$$

When the calculated values are included into the equations, then

$$b_1\ 1{,}060.61 + b_2\ 4.967 = 1 \times 361.16 + 0 \times 1.2911,$$

$$b_1\ 4.967 + b_2\ 0.7255 = 1 \times 1.2911 + 0 \times 0.3847.$$

The result for b_1 in the first equation is:

$$b_1 = \frac{361.16 - 4.967 b_2}{1{,}060.61}.$$

Substituting that expression in the second equation, it reads:

$$4.967\ \frac{361.16 - 4.967 b_2}{1{,}060.61} + 0.7255 b_2 = 1.2911,$$

$$1{,}793.88 - 24.67 b_2 + 768.93 b_2 = 1{,}368.79,$$

$$b_2 = \frac{-425.09}{745.26} = -0.55.$$

Substituting the value for b_2 in the equation for b_1, it reads:

$$b_1 = \frac{361.16 - 4.967 \times (-0.55)}{1,060.61},$$

$b_1 = 0.34.$

To reduce coefficient b_1 to 1, both coefficients should be divided by 0.34:

$b_1 = 0.34/0.34 = 1,$

$b_2 = -0.55/0.34 = -1.62.$

The index for the selection of best families in the population, established on the basis of relative economic value, will have the following form:

$I = \text{yield} - 1.62 \times \text{oil}\ \%.$

If, for example, we are to choose the best of the three families used in the calculation of desired gain index, then

P_1 yield $= 125$ oil % $= 8.20$	$I_1 = 125 - 1.62 \times 8.20 = 111.72,$
P_2 yield $= 135$ oil % $= 7.20$	$I_2 = 135 - 1.62 \times 7.20 = 123.34,$
P_3 yield $= 118$ oil % $= 7.90$	$I_3 = 118 - 1.62 \times 7.90 = 105.00.$

According to this selection criterion, family P_2 is the best. The result differs from that obtained by desired gain index where family P_1 was the best.

The next step is to estimate the rate of yield and oil content increase (ΔHS_1 and ΔHS_2, respectively) by half-sib selection on the basis of the constructed index of relative economic value. The calculation is done by the formulae used for the desired gain index. Since in this case the intensity of selection is 5%, the value of $k = 2.06$.

$$\Delta HS_1 = \frac{361.16 - 1.62 \times 1.2911}{\sqrt{1,061.61 + 2 \times (-1.62) \times 4.967 + 2.62 \times 0.255}} \times 0.5 \times 2.06,$$

$$\Delta HS_1 = \frac{359.07}{32.35} \times 1.03,$$

$\Delta HS_1 = 11.43.$

The increase is thus 11.43 g/plant or 601 kg/ha for a stand of 52.500 plants/ha. The estimated increase in oil content is:

$$\Delta HS_2 = \frac{1.2911 - 1.62 \times 0.3847}{32.34} \times 1.03,$$

$\Delta HS_2 = 0.02\%$ of oil.

Thus, the application of the estimated selection index on the basis of the relative economic values of the selected traits brings about increases of yield by 601 kg/ha and increases of oil content by 0.02% in each cycle of selection of the maize population U1.

* * *

The attempts made so far with the use of selection indices in plant breeding have not brought the expected results. On one hand, the number of such studies has been too small; on the other, the selection criteria which are the most effective for different populations have not been found, especially for open-pollinated plants. It is therefore necessary to continue studies on the value of selection indices used in plant breeding.

Selection for yield via plant model

After World War II plant modeling was introduced as a guidance of what should be achieved by plant breeding. That subject was discussed in Chapter 8 which includes an illustration of our concept of developing high-yielding wheat cultivars (Borojević, 1971, 1978).

That model is used not only as a guide for the choice of parents, but also for incorporating genes for certain traits, and for securing optimum growing conditions in order to allow the full expression of genetic yield potential. The model also serves as a guide in selection for yield from the F_2 until new cultivars are obtained.

There are difficulties involved in the selection for yield *per se*, as well as in selection based on yield components or selection based on multiple selection criteria. In view of this, the method of selection for yield via a plant model is an attempt to obtain a reliable procedure for the selection of superior genotypes by combining visual and metrical techniques of selection. The programme of development of high-yielding wheat cultivars carried out over the last 25 years in Novi Sad, Yugoslavia, may be used as an illustration of how the method works.

The pedigree method is used for all hybrid materials. To allow all genotypes to fully express their potentials, selected plants or spikes from the F_2 generation are planted in separate rows, two metres long, with 20 cm distance between the rows and 5 cm between plants in the row. The stand differs from that recommended by Fasoulas (1973) to eliminate allo-competition, but our estimate is that the competition among plants is not great since there are no large differences in plant height (70–100 cm) in the segregating generations.

The selection process starts in the F_2 by selecting plants (SP) or a spike per plant (SH) whose phenotype meets the requirements of the model. The decision whether to select an entire plant (SP) or a spike per plant (SH) depends upon the traits of the parents included in the cross and upon the number of plants per cross combination. If in the F_2 generation some plants are found which closely resemble the model, the entire plant is selected (i.e., all spikes from a plant) in order to obtain the maximum amount of seeds and to get a clear view of its progeny in the F_3. If there is a large number of plants in the F_2, only one spike is selected from a promising plant in order to cover a large number of genotypes.

Each F_2 cross combination includes 2,000–5,000 plants. An advantage is given to shorter plants of 70–80 cm, but those of 80–90 cm are selected too if the stem is strong enough to secure a high resistance to lodging in dense stands with high nitrogen doses. Stem height used to be measured but now it is estimated visually since there is already a profound experience regarding the desired height.

Much attention is paid to the spike's characteristics. We select visually the spikes with a large number of spikelets and a large number of grains per spikelet. It has been found that spikes of medium density are most productive in the agroecological conditions of continental Yugoslavia. Long and sparse spikes are selected only for dwarf plants (below 70 cm).

The next selection criterion is the length of vegetative period. According to the model, the optimum time of maturation is between those of the cultivars San Pastore and Bezostaia-1, i.e., from June 25 to July 5. Earlier or later plants are also taken into consideration if they have a satisfactory level of disease resistance, the desired stem height and a high spike productivity.

A special selection criterion in the F_2 generation is the resistance to disease. All materials are evaluated for resistance to powdery mildew (*Erysiphe graminis*), leaf rust (*Puccinia recondita* f.sp. *tritici*), stem rust (*P. graminis* f.sp. *tritici*), *Septoria* and *Fusarium* species. Resistance to leaf rust is tested in a greenhouse and in the field by inoculation with a mixture of the rust races present in the region. The other pathogens are evaluated on the basis of natural infection in the field. Resistant and medium resistant plants are labelled with special tags and selected if they satisfy other criteria, primarily stem height and spike fertility.

The selected plants and spikes are harvested individually. Each selected plant or spike is observed visually for seed size and colour and only those with a large number of well-developed seeds are planted. Depending on the cross combination, 10–50% of spikes or seeds may be eliminated after the comparison against the control varieties. In some instances, if the seeds are poor, an entire cross combination may be eliminated despite the fact that the selected plant had positive phenotypic traits.

Seeds of each selected F_2 plant are sown in five two-metre rows, 20 cm distance between the rows and 5 cm between plants in the rows. The selected spikes (SH) are sown in the same manner but as a spike per row. Since plant characters still segregate considerably in the F_3, the method of sowing as in the F_2 is retained in order to avoid competition between plants and for us to be able to recognize desired genotypes easily.

Since the formation of lines begins in the F_3 generation, not only plants and spikes but also entire rows are selected especially in cases when parents are morphologically similar or when one back-crossing has been performed, i.e., when the row or small plot is rather uniform.

From a selected row (SR), 10–20 spikes (SH) are taken first and the remaining plants are harvested together. It is of no use to measure the yield of the selected row (SR) because the progenies were not grown in replications. We only check the seeds and keep for sowing those SRs which have the largest number or best-developed seeds. The amount of seed thus obtained suffices for sowing in a plot of 2.5 m2, in 2–4 replications, in the preliminary yield trial (PYT) for testing new lines. The 10–20 selected spikes are harvested individually and sown spike per row (1 m× 20 cm) for the test of homozygosity (TH) (see pedigree method in Chapter 12).

Best plants (SP) or spikes (SH) are selected from segregating F_3 rows according to the criteria set up by the model. All seeds are visually checked prior to sowing and only those SPs and SHs are retained which have a large number of well-developed seeds.

Since F_4 plants are 82.5% homozygous, SHs and SPs are not sown in two-metre rows but in one and 3 to 5 one-metre rows, respectively. The distance between rows is 20 cm, the distance from seed to seed is 1–2 cm. The density thus approaches the conditions of commercial production, although is in fact about only half of it.

According to this selection procedure, the selected rows (SR), i.e., F_4 lines in the preliminary yield trials (PYT) represent the first test for yield. To make the obtained values truly representative of the genetic yield potential of the selected lines, sowing methods and density as well as agricultural practices should resemble as closely as possible those in commercial production. Since the available amounts of seed are still small, it is preferred to increase the number of replications (2–4) and to reduce plot size to 2.5 m^2 (10 rows × 2.5 m, row-to-row distance 10 cm). The sowing density is 500–600 germinable seeds per 1 m^2. Agricultural practices are quite intensive, as in commercial production, where the nitrogen dose ranges between 130 and 150 kg/ha and weeds are controlled by herbicides.

In the preliminary yield trials, new lines (L) are comparatively tested against 2–3 standard cultivars. Those are the leading cultivars in the country at the moment of testing, either for their yield performances or quality. The standards used in recent years in the entire country were Superzlatna (high yield) and Partizanka (high quality), to which the best new cultivars are added.

The lines in the preliminary yield trials (PYT) are tested for winter-hardiness in cold chambers and for quality by micromethods (gluten content, protein content, and sedimentation (SDS) test). Taking into account the results of these analyses and observations in the field, lines which outyield or equal standard cultivars are taken for comparative yield trials (CYT) in the next year. However, the actual seed for sowing is not taken from the preliminary yield trial (PYT) but from the homozygosity test (TH), i.e., from individual rows of selected spikes or plants of that particular entry. If in the homozygosity test (TH) segregation does not occur within the rows and if all rows are uniform (which is rarely the case in F_4), we take the seed of all SPs or SHs for sowing in a comparative yield trial (CYT) the next year. If differences are found among individual rows, then the line selected on the basis of its yield in PYT is divided into 2–3 sublines which are sent back to the preliminary yield trial because there is not enough seed to include the sublines in the comparative yield trial (CYT).

The comparative yield trial (CYT) is performed in a 5 m^2 plot, in five replications, in rows 10–15 cm apart with 500–600 seeds/m^2, i.e., as in commercial production. Agricultural practices are very intensive, including 150 kg/ha of nitrogen as well as corresponding amounts of phosphorus and potassium.

The earliest lines put into the comparative yield trial (CYT) are F_4s but usually they are F_5s and later generations. The line stays in the comparative yield trial for three years at the most. If by that time it does not satisfy in yield and quality, it is discarded from further testing. Best lines are sent to the Federal Commission for Varietal Approval and registered as new cultivars if they pass a three-year test in 15–20 locations.

The described procedure of selection for yield via a plant model comprises in fact selection for ideotype characteristics and selection for final yield components up to the F_4 and direct selection for yield *per se* in the F_4 and subsequent generations. The difference from the methods of Fasoulas and Shebeski is that the F_3 or F_4 materials for the preliminary yield trial are not selected by measuring the yield, but are selected on the basis of plant type and other parameters prescribed by the model. After that, lines are selected on the basis of grain yield (Fig. 15.4).

Fig. 15.4 Comparative yield trials (CYT) with new wheat lines by random block design in 5 replicates, basic plot being 5 m², Novi Sad (Yugoslavia)

Another distinction of our method is that the F_2 and F_3 generations, in which genotypic segregation is most intensive, are sown wider apart to avoid allo-competition, although the spacing is smaller (2 m × 20 cm × 4–5 cm) than practiced by many wheat breeders. In generations after F_4, when the material is fairly homozygous and allo-competition is less effective, the method of sowing simulates the conditions of commercial production as closely as possible. Selection in the F_4 and subsequent generations is based upon the row, i.e., on the progeny because the general impression about a row is invariably closer to the reality when the row hybrid materials in thin stands, with plants 10 cm apart between rows. In such conditions, there developed large and productive spikes, especially in early generations, as a result of strong non-additive gene effect (dominance and epistasis) which were lost in later generations.

Regarding genotypic competition, there are at least three possible reactions of materials selected in a thin stand and subsequently tested for yield in a thick stand. Firstly, the best genotypes from the thin stand remain superior in the thick stand, or secondly, the best genotypes from the thin stand lose their superiority in the thick stand, or thirdly, some genotypes remain indifferent to stand density. If there is the probability that the three possibilities will occur in equal ratios, than at least one third of the selected material will have to be discarded on account of selection being made in the thin stand alone. The majority of breeders believe therefore that crops commonly grown in thick stands should not be selected in thin stands but in conditions closely resembling those encountered in commercial production.

The fact that there are several forms of competition among plants convinced us that the yield measured in a thin stand, such as the one we use in the F_2 and F_3, does not give a realistic picture of the tested genotypes. Instead, we give preference to selection based on the plant model (SP) and especially to an increased number of spikes (SH) in the F_2, assuming that the majority of superior genotypes will be selected.

The described procedure of selecting for yield enables us to handle a large number of cross combinations. Between 500 and 1,000 crosses of wheat are made annually. The importance of the number of cross combinations for success in breeding was discussed in Chapter 11. It will only be mentioned here that in spite of the obvious advantages of the methods of Fasoulas and Shebeski regarding the selection of superior genotypes in the F_2 and F_3, it would be difficult to prove that the work on 30 cross combinations or the testing of lines from only 2–3 cross combinations would secure equal success in breeding new cultivars as the work on 100, 200 or more than 500 cross combinations.

Differences in yield testing in preliminary trials (PYT) and comparative trials (CYT) are not only based on plot size and the number of replications, but also in that the PYT includes a line for one year only. If the line performs well, it is included in the CYT and if it does not, it is discarded. The line which goes forward to the CYT has the chance to be tested for yield for 1–3 years, while the discarded line is never given a second chance. However, it may happen that the intensity of selection varies from one year to another on account of differences between the limiting environmental factors. The number of discarded lines is much larger in years of intensive attacks by powdery mildew, rust or *Fusarium*, than in years with less severe attacks. In the latter years, more lines will proceed from the PYT to the CYT, but again they may be eliminated from further testing if they yield below the standard cultivars. These risks are successfully counteracted only by testing a large number of lines (500–1,000) in the preliminary yield trial and 300 to 500 lines in the comparative yield trials.

Morpho-physiological and biochemical criteria of selection for yield

In view of difficulties encountered when selecting for yield *per se*, via yield components or on the basis of selection indices, attempts have been made to find other criteria which could be used for selecting superior genotypes in early generations. Since 90% of organic matter is formed by the process of photosynthesis, an idea was promoted to use leaf area, leaf form and duration of photosynthetic activity in leaves as indicators for selecting productive genotypes.

Leaf area duration

Among the above parameters of photosynthetic apparatus, leaf area duration (LAD) has been found to have the highest positive correlation with grain yield (Watson et al., 1963; Fischer and Kohn, 1966; Lupton et al., 1967; Loomis and Williams, 1969; Bingham, 1971; Borojević et al., 1980). Nevertheless, the actual measurement of LAD, limited even to the flag leaf, is totally impractical requiring more work than measuring yield itself. However, it stands to reason to visually select for the longest LAD since such plants are characterized by a long activity of the flag leaf and the top internode, lasting after the spike has turned yellow. This parameter is closely related to the resistance to foliar diseases, although two varieties may have the same level of disease resistance one will "stay green" longer than the other (Duvick, 1983).

Leaf architecture

Another morphological criterion for selection, obtained on the basis of photosynthetic activity studies, is the position of leaves or the leaf angle. It was found when grasses were grown in a thick stand that erect leaves were superior to horizontal ones (Duncan, 1969). In recent years, attention started to be paid to genotypes with erect leaves, especially in the case of semi-dwarf or dwarf cultivars.

In a paper discussing the effect of selection for ideotype characters in barley, Rasmusson (1984) gives data showing that at the beginning of the breeding programme, the erect leaf line was inferior in yield to the control cultivar. However, in the 5th cycle of breeding, improvements had been obtained and some erect leaf lines reached the yield level of the control. As Rasmusson (1984) states, it is impossible to know whether the leaf angle trait does influence the yield or not.

We have arrived at a similar conclusion regarding leaf architecture in wheat. Among several high-yielding cultivars, two with erect leaves were registered. However, they gave lower yields even in a very dense stand than did the best cultivars with curved leaf architecture (Borojević and Kraljević-Balalić, 1983).

In our opinion, leaves with the curved position and angles of about 30° between the stem and the ascending part of the leaf, and between the ascending and the descending part of the leaf may be more efficient than erect leaves (Fig. 15.5). This

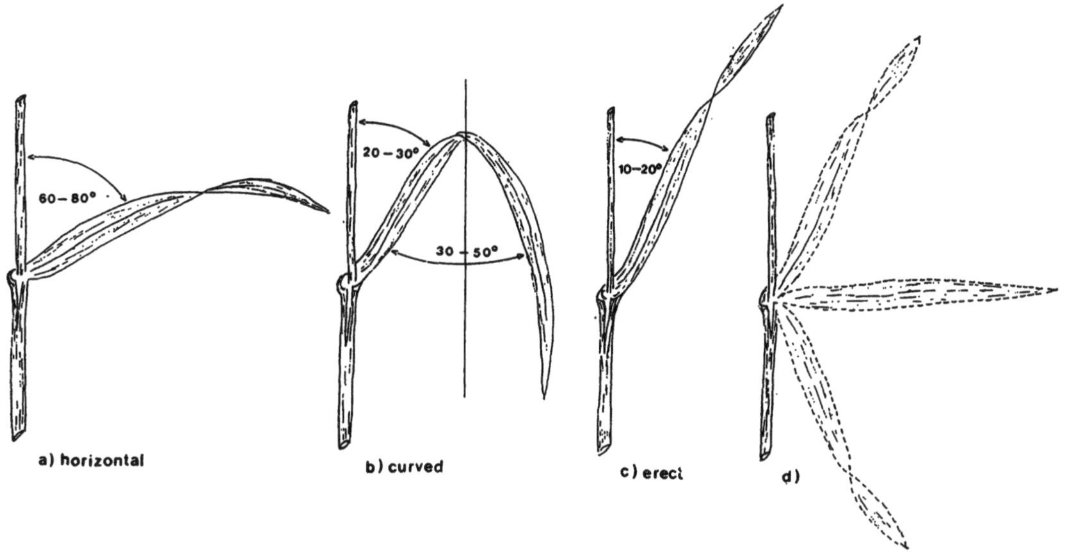

Fig. 15.5 Various positions of wheat flag leaf lamina. d) erect leaf lamina at heading which becomes horizontal at milk stage and pendulous at dough stage of grain.
Source: Borojević and Kraljević-Balalić, 1984

statement is supported by the fact that the curved leaf position is the most stable throughout the stages of development as well as in a dense canopy, while the majority of genotypes with erect leaves lose that position soon after the heading stage, even to become pendulous in the milk stage (Fig. 15.5). Moreover, in genotypes with erect leaves, tip firing is pronounced starting from the heading stage, which is seldom the case with curved leaf genotypes. We have screened a wheat collection of 1,400 entries from 36 countries and found that in the heading stage about 70% of the entries showed the curved position of the flag leaf and the second leaf, about 15% were horizontal, and about 15% were of the erect leaf position (Borojević and Denčić, 1986). It indicated that the very processes of selection favoured the curved type of leaves, because conscious human selection for leaf architecture has begun only recently.

The data on leaf architecture obtained so far seem to indicate that leaf architecture cannot be used alone as a selection criterion for yield but must be combined with several other traits.

Awns

The awns of small grained cereal plants are another morphological criterion that might be related to increased photosynthetic activity. Studies on isogenic lines have shown that awned wheat and barley lines (Grundbacher, 1963; Apel, 1966) have a higher photosynthetic activity than awnless ones as well as the fact that awns may contribute to increasing yield by up to 10%.

Studies by Evans et al. (1972) showed that awned wheat plants significantly outyielded awnless ones under dry conditions but not under wet ones. In connection with this it is interesting to note that in countries closer to the origin of hexaploid wheat, where conditions are rather dry (Afghanistan, Turkey, Russia, Central Europe, etc.) the local and old cultivars were awned, while in countries with a more humid climate (West Europe) the cultivars mostly awnless. Breeding for high-yielding cultivars has changed the situation radically so that now the leading wheat cultivars in Central Europe and the Soviet Union are awnless. It shows again that a single trait is highly unreliable to serve as a selection criterion for yield.

Qualset et al. (1965) and Schaller et al. (1972) showed that full-awned and half-awned lines of the barley cultivar Atlas produced more grain than quarter-awned and awnless lines. Rasmusson and Crookston (1977) studied the effect of multiple awned (three awns per spikelet) and normal awned types (one awn per spikelet) in barley lines and found that all the multiple awned lines obtained by five back-crosses to Dickson yielded less than the control variety Dickson (which is normally awned). The main reason was that the multiple awned lines developed a lower number of kernels per spike and a lower kernel weight. A possible explanation of this phenomenon is that large glumes with three awns could be a competing sink for assimilates during the critical development stage and that could result in weakened florets which would ultimately fail to develop (Rasmusson, 1984).

Mitochondrial complementation

Among the biochemical parameters that could be used as selection criteria, much has been expected from mitochondrial complementation. McDaniel and Sarkisan (1966) as well as other authors later on, found that mitochondria of heterotic combinations of maize, wheat and barley had a higher activity (measured by the ratio ADP:O) than did their parental lines. The same phenomenon was observed in a mixture of mitochondria extracted from the parental lines.

For plant species in which the F_1 is used to produce heterotic hybrids, e.g., maize, sorghum etc., mitochondrial complementation could be used to select heterotic combinations although the correlation between mitochondrial efficiency and heterosis is not high. Conversely, the method is not reliable for plant species in which the F_1 cannot be used; instead, as maintained by Sage and Hobson (1973), it may be used only as one of many yield components. The method of mitochondrial complementation has not found practical application because the necessary biochemical analyses are difficult and long and it is thus simpler and faster to measure the yield itself.

Nitrate-reductase activity

Another biochemical parameter worth mentioning is the employment of nitrate-reductase activity as a selection criterion. Many researchers (Eilrich and Hageman, 1973; Deckard et al., 1977; Dalling and Loyn, 1977) found that increased nitrate-reductase activity was related to increased nitrogen content in grain, as well as increased grain yield. As the method is applicable on 21-day old seedlings, it could be a suitable criterion for selecting superior genotypes. Unfortunately, fluctuations in nitrate-reductase activity in the course of plant ontogenesis and during the growing season make the correlation with yield quite variable. It is thus not surprising that this parameter has not been used extensively as a selection criterion.

Attempts to design efficient laboratory methods that could be used as selection criteria for yield in early segregating generations are certainly commendable and they should be continued. However, these methods should be considered as only one possibility that could help in breeding for yield. They cannot replace any one of the direct methods of selection for yield.

General considerations regarding the breeding for yield

Yield is such a complex character, especially when considered in interaction with environmental factors, that it would be unrealistic to expect one or two selection criteria to suffice for the selection of highest yielding genotypes in early segregating generations.

It has turned out that the measurement of yield of individual F_2 plants and their F_3 progenies is not a reliable criterion for the selection of the highest yielding genotypes. On one hand, the plants are grown in conditions different from those in which the obtained lines are tested (thin and dense stands, respectively). On the other hand, those components of genetic variance are selected which can make the progenies quite different from the original selections. Furthermore, the selection of an enormous number of plants in order to find the most productive ones precludes the possibility of working with a large number of cross combinations. This reduces the chances of success since it is better to work with more cross combinations and a relatively small number of plants in each cross combination.

Therefore, general experience of many breeders is that the measurement of yield of individual F_2 plants and their F_3 progenies (rows) is not a reliable criterion. Superior lines may be expected to be selected with much higher probability if yield is measured in F_4 progenies (rows) obtained from selected F_3 plants, or even better, on the basis of plots (replicates) from selected progenies (rows) in the F_4 and subsequent generations.

Selection for yield via yield components, similar to using selection indices, has been more a subject for theoretical considerations and statistical analyses than a practically applicable method. It requires time-consuming measurement and analysis of plants in early segregating generations. These criteria are not reliable because insufficient attention is paid to correlated responses of individual characters.

Grafius (1978) points out that correlations between characters can either be of a genetic or physiological nature or they can be due to the proportionality in size among plant organs, i.e. the allometry of development. These relationships impose constraints on the form of a plant and are frequently barriers to improved yield. According to Grafius, this approach is based on Sinnot's law which states: "The size of an organ is proportional to the size of the meristem from which it develops" (Sinnot, 1921). The allometry may be associated with (1) organs arising from the same meristem and (2) organs arising from different meristems. Hamid and Grafius (1978) studied the developmental allometry of five traits and its implication to grain yield in barley (Fig. 15.6). On the basis of these and other results, Grafius (1978) proposed two corollaries to Sinnot's law – 1) Plasticity is inversely proportional to ontogenetic proximity, and 2) Number and size of organs tend to have an inverse relationship.

"Traits arising simultaneously from the same meristem will be harder to manipulate independently than those segregated in space and time, even though the latter are subject to competitive inhibition" (Grafius 1978, p. 934). If there are strong correlations between traits as a result of common origin, there is little cause and effect expressed and manipulating them independently would be difficult (see relationship between *D*, *S*, and *Y* in Fig. 15.6). In regard to relationships between *X*, *Y*, and *Z*, they result more from competitive inhibitions than from the effect of common origin and may be less restrictive.

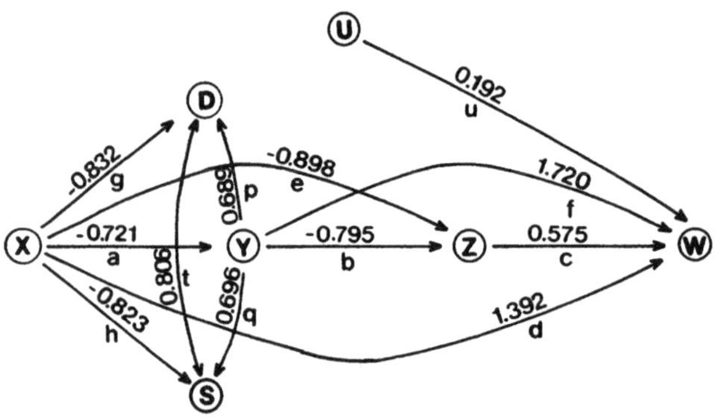

Fig. 15.6 Path coefficient diagram showing the interrelationship of fertile tillers per unit (*X*), seeds, per head (*Y*), kernel weight (*Z*), leaf area (*S*), culm diameter (*D*), and grain yield (*W*). The double-headed arrows refer to simple *r* values, while the single-headed arrows, represented by small letters, are path coefficients and show cause and effect relationships. The unexplained part of the variance of *W* is represented by *U*.

Source: Hamid and Grafius, 1978

It may be concluded that in selecting for yield as well as in constructing a crop ideotype, these corollaries should be taken in consideration.

In our opinion, selection for yield via a plant model (based on visual selection tor several traits integrated in the model in early segregating generations and from fhe F_4 onwards combined with the testing of selected lines for yield *per se*) may be used as an effective method of obtaining superior genotypes.

Rasmusson (1984) states that he has not succeeded in obtaining new barley cultivars with increased genetic yield potential by selection for ideotype characters. He listed correlated responses associated with selection for four ideotype characters during 5 to 10 breeding cycles. Selection for a large spike number was associated with few and small seeds, small stems and leaves and lodging susceptibility. Selection for erect leaves resulted in lodging resistance, but gave late maturity. Selection for semi-dwarf height (sdw gene) was associated with small seeds, late maturity and poor malting quality. And, as mentioned earlier, selection for multiple awns gave few and small seeds.

In explaining these results, Rasmusson (1984) paid attention to component compensation, allometry and pleiotropy. Due to component compensation, increases in spike number had very little influence on the yield potential because reductions in both seed number and seed weight occurred and off-set the increase in spike number. Because of allometric relationships, "the plant breeder is precluded from breeding the ideotype of his choice, i.e., some combinations of characters are not morphogenetically compatible" (p. 102). Negative pleiotropic effects associated with erect leaves and multiple awns were rather unexpected.

Obviously, the correlated responses of multiple traits and the corollaries mentioned by Sinnot and Grafius must be paid heed to when breeding. However, those phenomena should not be considered as static and unchangeable. Only 30 years ago, the dwarf stature of cereals was considered abnormal and such plants were discarded from selection processes. In conditions of primitive production (a thin stand, weed infestation, poor nutrition, etc.), dwarf plants were indeed inferior in yield to plants with a tall stem. As the technology of production intensified, dwarfness came into the focus of interest because it is associated with lodging resistance, better use of fertilizers and machinery, etc. Understandably, due to allometry and pleiotropy, extremes should not be looked for – in the case of cereals, semi-dwarfness has been found to be more favourable than dwarfness.

The situation is similar with erect leaves. The gene pool for erectness, as in the case of dwarfness, is too small to show whether these genes bear negative pleiotropic effects on other traits too. A misconception on a high negative correlation between yield level and quality in wheat has been adhered to for a long time, i.e., high-yielding varieties were not expected to provide a good bread quality. This was refuted by Bezostaia-1 which outyielded a number of other Soviet varieties maintaining at the same time a very good quality. Later on, our variety Partizanka excelled Bezostaia-1 in both yield and quality (Mišić and Mikić, 1976). If genetic correlations between two traits are not complete, it may reasonably be expected that advances may be bred in both traits.

New traits resulting from gene mutations or interactions are only the initial materials which remain to be selected and incorporated into different backgrounds in order to show their values.

Plant breeding is invariably based on genetic laws, physiological processes and interactions between organisms and the environment. The more we know about these processes and relationships, the more we shall pay attention to them in the various phases of the breeding process. However, breeding is a discipline in which it is not possible to make a complete theoretical forecast. Therefore, in addition to the necessary knowledge of genetics, the breeder who has the feeling of the genotype via the phenotype and who lives permanently with plants, will be more successful in work, especially in breeding for yield.

Possibilities of increasing genetic yield potential

Breeding in general strives to find the most suitable ways and methods of increasing genetic yield potential or gain in plants and animals. The necessity to act quickly is dictated by the needs of the enormously growing human population and the constant reductions in agricultural land.

What are the prospects of further increasing the genetic yield potential in plants? Taking into account that more than 90% of the organic matter in a plant is formed in the process of photosynthesis and that only 8% of dry matter in the plant is inorganic; and assuming that the supply of water and nutrients are in optimum and that diseases and pests are absent, it is possible to estimate the maximum productivity of a crop per unit area and unit time. Based on the data of Moon (1940), Loomis and Williams (1963), as calculated by Mac Key (1980), the potential net production of dry matter of a crop may be 67 gr/m2/day or 670 kg/ha/day, providing that solar irradiation, the carbon dioxide content and temperatures are favourable. Calculating the yield of dry matter per 100 days of vegetative period, net production has a potential of 67,000 kg/ha. The yield of grain (at 12% moisture with the ratio grain : straw : root of 5 : 3 : 2) could potentially be 37,000 kg/ha (Table 15.4).

The estimate of dry matter yield and especially of grain yield in Table 15.5 stands indeed for ideal conditions of growing. Of course, it would be exceedingly difficult to provide such conditions. It should be pointed out, however, that the record yields

Table 15.4 Calculation of potential productivity by a crop surface (Basic data from Moon 1940, Loomis-Williams 1963)

Total solar radiation during vegetation period,		
range	$10–80 \times 10^9$ cal/ha/day	
mean	50×10^9 cal/ha/day	
mean of visible radiation (400–700 m)	22×10^9 cal/ha/day	
Total quanta available for photosynthesis	353×10^3 Einsteins/ha/day	
Equation of photosynthesis:		
$CO_2 + H_2O + nhv = (CH_2O) + O_2$		
where $n = 10$, $h =$ Planck's, $v =$ wave frequency,		
Amount of carbohydrate (CH_2O) produced	353×10^2 mol/ha/day	
Gross production of weight (30 g/mole CH_2O)	1,060	kg/ha/day
Respiration loss (33%)	−350	kg/ha/day
Net production (CH_2O) of weight	270	kg/ha/day
Inorganic matter to be added (8%)	+57	kg/ha/day
Reduction 10–15% due to higher mean valoric content of total plant dry matter greater than 3.740 cal/g carbohydrate (CH_2O)	−97	kg/ha/day
Net dry matter production	670	kg/ha/day
or	67	g/m²/day
Dry matter yield/100 days vegetation period	67.000	kg/ha
Grain yield (12% H_2O) at a grain:straw:root ratio of 5:3:2	37.500	kg/ha

Source: Mac Key, 1980

of dry matter achieved under controlled conditions were quite near to the theoretically calculated maximum yields. Plants of the C4 type of phytosynthesis, such as maize and Sudan grass, produce the maximum of 52 and 51 g/m2/day of the above-ground mass, respectively (Table 15.5). If 20% for the dry matter production of the root is added, the sum is 60 g/m2/day; very close to the theoretically calculated 67 g/m2/day (Table 15.4). C3 type plants have a lower production, but rice produced 55 g/m2/day (Table 15.5) in an experiment conducted in controlled conditions of moisture, light, CO_2 and temperature, and the plants were completely protected against pathogens. The experiments mentioned in previous paragraphs show that some of the existing cultivars have very high genetic potentials for yield.

What is the practically used portion of yield potential? The record yields are the best answer. Here are some record yields as reported in the last decade: 23.9 t/ha of maize, 21.5 t/ha of grain sorghum, 14.4 t/ha of rice, 14.1 t/ha of wheat, 7.4 t/ha of soybean, 94.1 t/ha of potato, etc. (see Table 1.1). Those are very high yields which exceed the average ones for the same year and country by 2 to 7 times. Considering the theoretically possible yield of dry matter, record yields hardly reach a half of it or less than a half. In commercial production, in spite of all attempts to create ideal conditions for full phenotypic expression of the genotype in question, the level of genetic potentials could not be achieved because the soil is still a formidable unknown as well as because it is not possible to gain full control of water, temperature, light,

Table 15.5 Record short term top growth rates for some crops

Species	Short term top growth $g/m^2/day$	Author(s)
Plants with C_4-dicarboxyllic acid pathway of photosynthesis:		
Bullrush millet (*Pennisetum typhoides*)	54*	Begg, 1965
Maize (Zea mays)	52	Williams et al., 1965
	52	Murata-Togari, 1975
Napier grass (*Pennisetum purpureum*)	51	Vincente-Chandler et al., 1965
Sudan grass (*Sorghum vulgare*)	51	Loomis-Williams, 1963
	46	Fischer-Wilson, 1975
Sugar cane (*Saccharum officinarum*)	38	Borden, 1942
Plants with C_3-Calvin pathway of photosynthesis:		
Rice (*Oryza sativa*)'	55	Tanaka et al., 1966
	36	Murata-Togari, 1975
Tall fescue (*Festuca arundinacea*)	44	Evans, 1973
Potato (*Solanum tuberosum*)	37*	Lorenz, 1944
Sugar beet (*Beta vulgaris*)	31*	Watson, 1958
	28	Murata-Togari, 1975
Wheat (*Triticum aestivum*)	30	King-Evans, 1967
Soyabean (*Glycine max*)	27	Murata-Togari, 1975
	17	Buttery, 1970

* top and root inclusive
Source: Mac Key, 1980

CO_2 concentration and pathogens. According to Boyer (1982), the largest yield losses, i.e., differences between record and average yields, are due to drought, water excess (unameliorated soils, floods), cold, hail and wind. Diseases and pests cause only 7–10% of the total reduction. This is an indication that the genetic yield potentials of the existing cultivars are not fully exploited but only reach 60 to 80% at best.

It is instructive to review data on the conditions in which some record yields were achieved. According to the "Proceedings of Plant Growth Regulator Working Groups" (August 9–11, 1977, USA) the record yield of maize and wheat have been obtained under the following conditions.

Maize – 21.1 t-ha

This record yield of maize was achieved by Mr. Herman Warsaw in McLean County, Illinois, U.S.A., in 1975. The preparations of the soil for the record yield, regarding quality tillage and application of organic and mineral fertilizers, had already started in 1963. The annual dose of nitrogen was 228 kg/ha. In 1975, pH in the plot was 6.2 and there were added 56 kg/ha of nitrogen, 260 kg/ha of phosphorus, 326 kg/ha of potassium and 50 t/ha of pig manure. The stand was 82,500 plants/ha.

Unfortunately, the information about the record yield does not mention the hybrid used nor the area of the plot.

Wheat – 14.1 t-ha

The yield of 14.1 t/ha (216 bushels/acre) on a plot of 2 hectares was obtained by Mr. R. H. Hageman with the cultivar Gaines in Kittitas County, Washington, USA, in 1965 (see Table 1.1). The plot was reclaimed from pasture, sweet-maize was grown for two years, and 121 kg/ha of nitrogen were added before sowing wheat. The soil had 2–4% humus. Planting was done in October, the seeding rate being 95 kg/ha. The plot was fertilized with 135 kg/ha of nitrogen, 25 kg/ha of phosphorus and 55 kg/ha of potassium. In the period from April to July the plot was irrigated seven times, which is more than usual. The climate of the experimental site is characterized by intensive solar radiation, clear sky without clouds, and cool nights, which allowed the grain filling period to last over 55 days.

The cultivar Gaines belongs to the group of first semi-dwarf wheats with a highly productive spike (Vogel et al., 1963). It had a great impact on the entire CIMMYT programme of Dr. N. Bourlag.

The record yields of wheat achieved in Czechoslovakia and Yugoslavia, although lower, have a very good and meaningful value because they were obtained on large areas in commercial production. Here is a description of conditions in which the cultivar Sava yielded 10.9 t/ha.

Wheat cultivar "Sava" – 10.9 t-ha

That record yield was achieved in the cooperative Šala, Galanta County, Czechoslovakia, in 1976, in a plot of 30 hectares (Table 1.1). The soil was alluvial with 2.7% humus, and the ground water table was at 120 cm. The preceding crop was sugar

Table 15.6 Record yields of winter wheat cultivars and area on which they were obtained in Yugoslavia

Cultivar	t/ha	ha	Farm and place	Year
Sava	9.60	6	Agrokombinat, Subotica	1976
Sava	8.42	107	IPK Osijek, Klisa	1976
Drina	8.65	11	PIRO Šid, Šid	1976
Zlatna Dolina	9.34	6	PIRO Šid, Šid	1976
Biserka	8.50	50	PPK Banat, Kikinda	1977
Novosadska rana-2	8.63	16	PRO Mitrosrem, Sr. Mitrovica	1977
Novosadska rana-1	9.23	12	PIK Tamiš, Seleuš	1984
Jugoslavija	9.87	25	Agrokombinat, Subotica	1984
Jugoslavija	8.00	160	PIK Bečej, Bečej	1984
Zvezda	9.08	58	PD Novi Sad, Sirig	1984
Žitnica	8.75	16	IPK S. Mihalj, Zrenjanin	1984
Baranjka	9.10	25	PD Njegoš, Lovćenac	1984
Skopljanka	8.89	27	PIK Tamiš, Ivanovo	1984

beet which was fertilized with 60 t/ha of manure. The crops prior to this were onions and poppies fertilized with 80 kg/ha of nitrogen, 90 kg/ha of phosphorus and 120kg/ha of potassium. Seedbed preparation was excellent and the planting was made on October 10 and 11, 1975. The seeding rate was 650 grains/m2 or 270 kg/ha. Mineral fertilization included 90 kg/ha of urea, applied before planting with 100 kg/ha of phosphorus and 125 kg/ha of potassium, with 45 kg/ha of urea applied in February, totalling 135 kg/ha of nitrogen. The plot was irrigated before planting from October 1 to 8, 1975, and on April 20, 1976, the irrigation rate being 30 mm in each case. The crop developed vigorously bearing on average 797 spikes/m2. Heading started on May 25 and the crop was harvested on July 17, 1976. The period of grain filling, i.e., from flower to full maturity, was 50 days.

In Yugoslavia, the record yields achieved with several wheat cultivars are somewhat lower, but they have been obtained on large acreages and in several years (Table 15.6).

Genetic changes to increase yield potential

These and other examples given in literature show that yields of several crops increase constantly in many countries (see Table 1.3) and that a plateau or limit has not yet been reached. The limit certainly does exist because the very existence of a species represents a certain limit. Yields above 20 t/ha of grain probably will not be reached with *Triticum aestivum* subsp. *vulgare*. Some new man-made synthetics, such as the already existing *Triticale* or a hybrid between *Triticum* and *Elymus* may move the limit still higher.

There are several possibilities for further increases of genetic potentials for yield and quality. They deserve to be used more intensively in plant breeding.

Changes in harvest index

In the natural associations of a number of plant species, vegetative growth is favoured at the expense of grain because of the competition for light and nutrients. Consequently, the total biomass of these species exceeds the yield of grain two or three times. If the vegetative parts (stem, straw, leaves) are not the main product, then more energy is spent on by-products then on the main one, i.e., grain. That was one of the major conclusions of the modern science which has directed plant breeding to reduce vegetative parts to their optimum size. Consequently, the harvest index was changed so that 50% of the total biomass belongs to grain yield. There have been several notable achievements in that respect. For example, the presently used maize hybrids have the harvest index of more than 40% whereas the index of old maize cultivars was around 25%. Similar improvements have been made with other crops too (Table 15.7).

The change in harvest index resulted principally from the reduction in stem length which in turn increased the resistance to lodging enabling the crops to grow in denser stands and take up more nutrients, i.e., to produce higher yields per unit area. This is associated with the effect of the major genes (Rht_1, Rht_2 and Rht_8) on the reduction of stem height and the increased harvest index in wheat (Fig. 15.7).

Fig. 15.7 Changes in plant height of wheat mainly due to the effect of major reducing height genes (Rht). Varieties from left to right: Bankut 1205 without height reducing genes (RhtO), Akakomughi (Rht8 on 2D chromosome), Saitama 27 (Rht 1S, 4A), Aobakomughi (Rht 2, 4D), Norin 10 (Rht 1, 4A and Rht 2, 4D), Tom Thumb (Rht 3, 4A) and Aibian (Rht 10, 4D)

Table 15.7 Harvest indices of some field crops (% of grain yield in relation to the total biomass of the above-ground parts)

Crop	Average %	Variation in different cultiv.
C_4 type cereals		
Maize, cultivars 1920–1926	24	
Maize, hybrids	42	38–47
Sorghum	41	40–41
C_3 type cereals		
Wheat (old cultivars)	31	23–37
Wheat (new cultivars)	39	28–46
Barley	48	35–52
Rye	27	27–29
Rice	51	43–57
C_3 type legumes		
Bean	59	53–67
Soyabean	32	29–36

Source: A revised table of Zelitch, 1975

A question arises: how much will further increases in the harvest index contribute to the increases in the genetic yield potential? First of all the harvest index (HI) remains to be increased, via stem length reduction, in those plant species in which it is still low (rye, soyabean, etc.) or in which it still can be increased (maize, sunflower, wheat, etc.) (Table 15.7). In spite of progress made in that direction, there is still a limited number of cultivars within species possessing a harvest index of 50%.

Austin et al. (1980) claim that it is possible to increase the HI of wheat from the present level of 50% to about 60%, without a change in the above-ground biomass, achieving a genetic gain in yield of some 25%. Here it should be pointed out that the yield of total biomass should be kept in mind when breeding for increased HI. In wheat, for example, some dwarf lines have a higher HI than do semidwarf cultivars, but their grain yields are lower which make them commercially unimportant (see NS-732 in Table 15.8).

Another possibility to increase the genetic yield potential is to keep the HI at 50% but increase the total biomass which would bring about a corresponding increase in grain yield.

Mac Key (1979), Austin et al. (1980), Kubshrestha and Jain (1982) believe that increases in the total biomass are difficult to achieve. This is because the progress made so far has resulted from dry matter partitioning which did increase the HI but retained the total biomass of the new cultivars on the level of the old ones. This view is difficult to accept because it imposes serious limitations to eventual further progress.

The data in Table 15.8 show that the new semidwarf cultivars, e.g. Siete Cerros, Sava, etc., outyield the old tall cultivar Bankuty 1205 not only by about 50% in grain yield but also by about 5 to 10% in total biomass. In experiments with a large number of varieties developed in different periods, Austin et al. (1985) found that

Table 15.8 Advancement in increasing biomass and grain yield in wheat by breeding (8-year average data, 1970–1977)

Cultivar	Height of stalk cm	Yield in t/ha of		Total biomass		Grain yield %	Harvest index
		straw	grain	t/ha	%		
Bankuty-1205 (Hun.)	112	10.08	4.42	14.50	100	100	0.30
San Pastore (Ital.)	91	9.00	5.91	14.91	103	134	0.40
Bezostaia-1 (USSR)	90	9.31	5.82	15.13	104	132	0.38
Sava (Yug.)	81	8.93	6.90	15.83	109	156	0.44
Siete Cerros (Mex.)	84	9.10	6.62	15.72	108	150	0.42
NS-732 (Yug.)	53	5.88	5.49	11.37	78	124	0.48
Next step	70–80	9.0	9.0	18.00	124	204	0.50

Source: Borojević, 1978

Table 15.9 Characteristics of old and modern varieties of winter wheat grown at Trumpington (UK) in 1983/84

Variety groups (number in groups)	Height to base of ear (cm)	Number of ears per m^2	Date of anthesis (June)	Grain filling period in days	Yield in t/ha					Harvest index
					total	%	straw	grain	%	
Very old (3)	158	386	19	47	19.8	100	13.4	6.4	100	32.3
Old (2)	141	392	17	46	20.3	102	13.0	7.3	121	35.9
Intermediate (2)	104	432	16	49	20.8	105	11.6	9.2	144	44.2
Modern (6)	85	457	14	50	22.8	115	11.5	11.3	176	49.5
Average SE	0.8	14					0.31	0.09		

Source: rearranged after Austin et al., 1985

the semi-dwarf cultivars increase the yield of biomass by 10–15% and the yield of grain by about 50% in respect to old cultivars (Table 15.9).

Therefore, further increases in harvest index and total biomass are possible in wheat and other crops. These increases are bound to bring about further increases in genetic yield potentials of these crops.

Changes in sink capacity

Stem height reduction, expecially in grasses, was accomplished by shortening the length of the internodes while the number of internodes remained rather constant. Consequently, the number of leaves stayed more or less the same and so did the photosynthetic area. It is therefore logical that the old tall cultivars and the new short cultivars do not differ much in the size of the photosynthetic area per plant and their leaf area indexes (LAI), i.e., in their source capacity.

As a result of changed partitioning within shoots, the assimilates saved by stem reduction are translocated to ear development, resulting most frequently in increased grain setting (Evans, 1984). This is possible since in many plant species spikes, ears or panicles are indeterminant, i.e., they develop more flowers than are actually fertilized to produce grain. In consequence, several short straw varieties develop heavier ears at anthesis than do those of comparable tall varieties (Brooking and Kirby, 1981; Thorne, 1982) and a larger number of grains per spike despite their having about the same number of spikelets per spike (Table 15.10).

Table 15.10 Differences between wheat cultivars in source capacity (LAI, LAD) and sink capacity (number of kernels/spike, kernel weight) (10-year data from experiment in Novi Sad, 1968–1977)

Cultivar	LAI m^2/m^2	LAD in days from flow	No. of spikes/ m^2	No. of spike-lets/ spike	No. of grains per spike	1000 grains weight	Grain yield in t/ha
Bankut-1205	6.77	155	675	18.1	19.5	36.7	4.43
Bezostaia-1	7.00	164	634	17.4	23.4	40.4	5.55
Sava	7.47	191	688	18.2	28.0	33.3	6.30
LSD 5%	0.86	13	87	0.76	2.11	1.58	0.455
1%	1.14	17	116	1.89	2.68	2.10	0.606

Source: Borojević, 1980

It should be emphasized that those are indirect changes in sink capacity resulting from changes in stem length and they are not from deliberately breeding for a longer spike, a larger number of spikelets or a larger number of grains.

However, genetic changes should be induced in sink capacity itself, in accordance with the changes made or being made in source capacity. According to Evans (1981), "One key to increased yield potential lies in the greater ability of the differentiating ear to compete with other organs for assimilates. Such competition is influenced by the relative size of the competing organs, by their relative distance from the source of photosynthate, and by the relative directness and capacity of their vascular connections to it" (p. 214). Cook and Evans (1978) performed an experiment by varying the number of grains in two wheat ears competing for assimilates from one leaf at various positions. The results showed the advantage of proximity to the source leaf and even more showed the advantage of greater sink rize.

It should be pointed out that a sizable group of breeding institutes work on the development of genetic materials with increased sink capacity. They remain to be balanced with optimum plant height and leaf architecture (which regularly takes more than one cycle of selection) to produce new commercial cultivars.

Therefore, sink capacity increases, i.e., increase in grain number and mass per spike, panicle, ear or pod, offers prospects of further plant improvements, especially if it is taken into account that green parts of these organs also play an important role in photosynthetic activity.

Increases in photosynthetic efficiency

The fact that only about 1% of solar radiation is used for photosynthesis is an open challenge to scientists who try to increase the efficiency of photosynthesis, and hence the genetic potential for yield (Nichiporovich, 1956, 1970). The success in breeding achieved so far is attributable to the increase in photosynthetic area rather than to any increase in the efficiency of photosynthesis (De Witt et al., 1979). Some wild species from the grass family display a higher efficiency of photosynthesis than do the domesticated species. That phenomenon was interpreted by Tsunoda and Khan (1968) by a thicker mesophyll layer which contains chloroplasts in the wild species.

Although it seems logical to accept the statements of Evans (1981 and 1984) that the selection for greater photosynthetic rates in crop plants could be a direct route to greater yield potential, there is no evidence to support it. Varieties with higher yield potential have not been found to have higher light-saturation rates of CO_2 exchange per unit of leaf area (CER) (Evans and Dudstone, 1970; Khan and Tsunoda, 1970).

It remains to focus attention on indirect ways to increased photosynthetic productivity. Some of these are: duration of photosynthetic activity, disease resistance, prolongation of grain filling period and changes in canopy structure.

Duration of photosynthetic activity

As mentioned earlier in this chapter, of all photosynthetic area parameters (size, index, duration), leaf area duration (LAD) seems to have a positive effect on grain yield (Watson et al., 1963; Welbank et al., 1966; Bingham, 1971; Evans et al., 1975; Austin et al., 1982; Borojević and Williams, 1982; Duvick, 1983). There exist genetic differences between cultivars in leaf area duration, although this parameter is also affected by environmental factors at the time of its development as well as by disease attack.

A long LAD is highly positively correlated with leaf area index (LAI). Since, however, an excessive LAI may have a negative direct effect on yield (Borojević and Williams, 1982), LAD should be harmonized with the length of the vegetative period and the environment in order to preclude their negative effects on grain development.

Duration of photosynthetic activity is certainly a parameter to be taken into consideration, together with changed harvest index and increased sink capacity, when breeding for higher genetic yield potential.

Genetic resistance to pathogens

One of the ways to increase the productivity of photosynthesis is the development of genetic resistance to pathogens. Because of the great importance of genetic resistance to disease and pests, Chapter 14 of this book is dedicated solely to that problem. It is re-emphasized here that breeding for yield inevitably assumes breeding for disease resistance.

Prolongation of grain filling period

A prolongation in the vegetative period may bring an increased production of total biomass and a prolongation of the grain filling period or fruit development may increase the production of grains or fruits. Longer vegetative period is not always positively correlated with grain yield whereas a longer period of grain filling most frequently results in yield increase.

There is at our disposal a large collection of germ-plasm that, for example, contains different genes for vernalization requirement, resistance to low temperatures, response to photoperiod, duration of grain-filling period and other associated characteristics. A suitable choice of parents may give new recombinations of genes that will result in high-yielding varieties with still different phases of development.

Genotypes with earlier spring growth would facilitate earlier spike initiation and allow a longer period of spike development (Bingham, 1971; Kumakov, 1967; Obrascov, 1973). The prolongation of the grain-filling period is of particular importance for the increase of grain dry matter production. For many environments this will be achieved if flowering starts earlier and the time taken to reach maturity remains the same, or vice versa, or as a result of transgressive segregation in both directions. All these changes must be in harmony with the agroecological conditions for which a cultivar is created, particularly with the occurrence of diseases and temperature stresses.

Optimum canopy structure

To achieve maximum yield it is necessary to secure an optimum canopy structure. As many crops are grown in pure stands and dense populations, stems and leaves should not be regarded on an individual plant basis but rather as elements of the canopy structure. For such crops, Donald (1981) proposed the term "communal plants" which may give low yield per plant but, when grown in a pure stand and at a proper density, they are able to give high yields.

The number of stems per unit area serves as an axis on which the parameters of source and sink capacity develop and function. Therefore, the number of stems/m^2 represents the first yield component developed by the plants and it is essential to secure an optimum density of stems/m^2 for each genotype and environment (Borojević et al., 1980). With regard to cereals which have the capacity of tillering, the optimum density may be reached by rather low seeding rates under favourable conditions (Fisher and Kertesz, 1976), by medium seeding rates (Darwinkel, 1978), or by high seeding rates under less favourable environments (Borojević, 1973).

In a dense population, the size and shape of leaves must be different from those in a thin population. It has already been emphasized that the erect leaf position shows an advantage in the interception of solar radiation by the canopy (Loomis and Williams, 1969; Duncan, 1969).

Rice cultivars with erect leaves developed at IRRI (in the Philippines), maize hybrids with erect leaves developed in the U.S.A., Yugoslavia, and elsewhere on the basis of line B 73 with erect leaves (Dr. Russell, personal communication) rank among the highest yielding in their maturity groups, indicating that leaf position does contribute to increased yield potential.

However, since there is quite a variation in the architecture of leaves within a species (Borojević and Denčić, 1986), leaf positions other than the erect one should be considered when trying to enhance photosynthetic activity. Among them, the curved leaf position (Fig. 15.5) seems to be very suitable for dense canopies.

Increased efficiency in the use of mineral nutrients and water

Mineral nutrition accounts for about 8% of the dry matter formed by plants (Loomis and Williams, 1963). Although it may seem small when compared with the 90% of dry matter made in the process of photosynthesis, that process cannot proceed without mineral nutrients and water. In spite of notable successes in breeding for improved above-plant parts, breeding for improved root systems is at its very beginnings. According to some physiological studies, nitrogen uptake depends solely on the density of the root system whereas the utilization of nitrogen taken up by plants may be increased only at the expense of protein content (De Witt et al., 1979).

Although the root system is highly variable, depending on soil structure, moisture and available nutrients, differences have been found among cultivars of the same species. According to Mac Key (1979), cultivars with larger above-ground parts (tall stems and many tillers) have a better developed root system than do semi-dwarfs and dwarf cultivars. There are of course some exceptions, e.g., the Mexican semi-dwarf cultivar Pitic-72 does not fall behind tall cultivars regarding the number of tillers and the number of roots and their distribution along the soil profile (Table 15.11). It indicates that when shortening the stem, to increase harvest index, strict attention should be paid to two things; on the one hand, to avoid a reduction in the root system and on the other, to breed for increased and more efficient root system, for they also contribute to increased genetic potential for yield.

As mineral nutrients, especially nitrogen, take the largest portion of costs of plant production, attention has started to be paid in plant breeding to the so-called nitrogen harvest index (Austin et al., 1980) or the nutrient harvest index (Fischer, 1981).

It has been found that new wheat cultivars with short stems have a higher nitrogen harvest index than do the old tall varieties (Table 15.12). However, it was interesting to note that the grain legumes have a low nitrogen harvest index, on account of a low harvest index of grain. Improvements are nevertheless possible as illustrated by a chickpea variety M-119 as compared against old varieties (Table 15.12).

According to Fischer (1981), the efficiency of nutrient use (E) is a product of nitrogen uptake per unit of supply (uptake efficiency – E_u) and grain yield per unit of uptake (utilization efficiency – E_z):

$$E = E_u \times E_z.$$

Table 15.11 Shoot: root development of spring wheats with tall and short stem grown by hydroponics with polystyrene foam

Parameters (4 replicates)	Brown Schlan-stedt	Cultivars Kolibri	Frontana	Mayo 64	Pitic 62
No. tillers	10.3	5.3	6.3	5.8	9.0
No. crown roots	60.5	32.8	39.3	25.3	44.2
Plant height, cm	109.0	68.2	83.3	62.3	70.9
Root depth, cm	103.3	88.3	109.1	55.3	87.0
Shoot dry weight, g	17.0	6.0	7.9	3.5	5.2
Root dry weight, g	1.04	0.48	0.72	0.28	0.72
% roots in zone 0–20 cm	46	61	43	82	63
% roots in zone 20–40 cm	20	21	18	15	21
% roots in zone 40–60 cm	11	9	13	3	10
% roots in zone 60–	23	9	26	0	6

Source: Mac Key, 1979

Table 15.12 Yield harvest index and nitrogen harvest index in wheat and in chickpea

Cultivars	Yield harvest index	Nitrogen harvest index
Wheat:		
Little Joss (very old)	36	71
Holdfast (old)	36	69
Cappelle-Desprez (inter. height)	42	68
Hobbit (semidwarf)	48	72
Benoist (semidwarf)	48	73
Chickpea:		
JG-62	23	46
M-109	30	51
M-119	35	69

Source: Austin et al., 1980 for wheat and Sinha and Swaminathan 1984 for chickpea.

The efficiency of the utilization of nutrients (E_z) can be calculated as follows:

$$E_z = HI_n/G_n,$$

where HI_n is the harvest index of the particular nutrient (i.e., the percentage of total uptake in the grain at maturity) and G_n is the percentage of concentration of the nutrient in the grain. It has been shown that both the nitrogen harvest index (HI_n) and E_n proved to be higher in high-yielding semidwarf wheat cultivars than in tall ones (Table 15.13).

The increase in nitrogen use efficiency at higher levels of nitrogen input with the shorter high-yielding cultivars, according to Fischer (1981), has resulted less from increased nutrient uptake (E_u) but more from increased efficiency of utilization

(E_z), due to both increasing the nutrient harvest index (HI_n) and decreasing the grain nitrogen content (Table 15.13).

The problem is certainly very interesting and many physiologists have concentrated efforts on studying the genetic aspects of plant mineral nutrition. Two international symposia were organised (Belgrade, 1982; Madison, 1985) dealing with the concentration and content of mineral nutrients in different genotypes, physiological and biochemical aspects of the genetic specificity of plant mineral nutrition and relations between genotypes and nitrogen fixing microorganisms (Sarić, 1987).

Table 15.13 Grain yield, total nitrogen uptake, nitrogen harvest index, grain nitrogen concentration and efficiency of nitrogen utilization (kg grain/kg *N* uptake) for spring wheat cultivars grown under irrigation and high fertility in northwest Mexico, 1870–71

Cultivar group (*n*)	Grain yield (t/ha)	Total *N* uptake (kg/ha)	*N* harvest index (%)	Grain *N* (%)	Efficiency of *N* utilization (kg/ha)
Tall (5)	5.76	200	67.6	2.72	28.3
Low yielding semi dwarf (9)	6.33	206	70.7	2.62	30.8
High yielding semidwarf (10)	7.10	211	72.4	2.45	33.8
S.E.	0.13	4	1.0	0.05	0.6

Source: Fischer, 1981

Studies of N, P, K, Ca, and Mg concentrations (at the stages of heading, milk-wax maturity, and full maturity) in four leaves of 15 wheat cultivars, conducted in laboratory conditions (Sarić et al., 1987) and in the field (Sarić et al., 1986) showed that the cultivars differed significantly in nutrient concentration; the observed variability depending also on leaf age and stage of development. Some cultivars differed significantly in the total biomass production and agricultural yield but these differences could not be correlated with nutrient concentration, not even nitrogen.

Therefore, the breeding for nutrient use efficiency is going to be difficult and again indirect approaches should be applied. As plant physiology advances, it is hoped that some new methods or markers will be developed which could help in the selection for higher efficiency of nutrient use.

It was also observed that genotypes differ in water uptake. Breeding should be employed to find more economic cultivars which would be suitable for growing in semi-arid and arid regions.

Only some possibilities for further increase of genetic yield potential in plants have been mentioned here. Of course, there are other avenues and methods, depending on particular crop and specific environment.

It should also be mentioned that in several crops the grain ought not be the only product looked for. For example, Sudan Sorghum has many uses such as for grain, thatching, fodder, fuel, etc.

In recent years, science has developed several new methods which may not only bring yield increases but also cause radical changes of the plants themselves. These methods of the future will be discussed in Chapters 18 and 19.

References

Adams, M. W., 1967. Basis of yield component compensation in crop plants with special reference to the field bean (*Phaseolus vulgaris*). Crop Sci. 7: 505–511.

Allan, R. E., 1983. Harvest indexes of backcross-derived wheat lines differing in culm height. Crop Sci. 23: 1029–1032.

Anonymus, 1966. New record wheat yields – 216 bushels per acre. Farm J. 90: 42.

Apel, P., 1966. Die Bedeutung der Grannen für die Kornentwicklung. III. Photokynetische Intensität der Ähren verschiedener Gersten- und Weizensorten. Kulturpflanze 14: 163–169.

Austin, R. B., Bingham, Y., Blackwell, R. D., Evans, L. F., Ford, M. A., Morgan, C. L., and Taylor, M., 1980. Genetic improvements in winter wheat yields since 1900 and associated physiological changes. J. Agric. Sci., 94: 675–689.

Austin, R. B., Morgan, C. L., and Ford, M. A., 1985. Grain and straw yields of old and new wheat varieties. Annual Report 1984. Plant Breeding Institute, Cambridge (UK), p. 106–107.

Austin, R. B., Morgan, C. L., Ford, M. A., and Bhagwat, S. G., 1982. Flag leaf photosynthesis of *Triticum aestivum* and related diploid and tetraploid species. Ann. Bot. 49: 177.

Baker, R. J., 1984. Quantitative genetic principles in plant breeding. In: J. P. Gustafson (Editor), Gene Manipulation in Plant Improvement. Plenum Press, New York, p. 147–176.

Bhatt, G. M., 1972. Inheritance of heading date, plant height and kernel weight in two spring wheat crosses. Crop Sci. 12: 95–98.

Bhatt, G. M., 1976. Variation of harvest index in several wheat crosses. Euphytica, 25: 41–50.

Bhatt, G. M., 1977. Response to two-way selection for harvest index in two wheat (*Triticum aestivum L.*) crosses. Australian J. Agric. Res., 28: 29–36.

Bhatt, G. M., and Derera, N. F., 1978. Selection for harvest index among near-homozygous lines of wheat. J. Australian Inst. Agr. Sci., 29: 111–112.

Bhatt, G. M., 1980. Early generation selection criteria for yield in wheat. J. Australian Inst. Agr. Sci., 46: 14–22.

Bingham, J., 1971. Physiological objectives in breeding for grain yield in wheat. Proc. VI. Congress EUCARPIA, Cambridge, pp. 15–30.

Borlaug, N. E., 1968. Wheat breeding and its impact on world food supply. Proc. 3rd. Inter. Wheat Genet. Symp. Canberra, p. 1–36.

Borojević, S., 1971. Building a model of high-yielding wheat varieties. Savremena poljoprivreda (Novi Sad), XIX, 6: 33–47.

Borojević, S., 1973. Canopy structure of different wheat genotypes in relation to the yield of grains. Proc. 4th. Inter. Wheat Genetics Symp. Columbia, Mo., p. 773–780.

Borojević, S., 1978. Genetics, environment and modeling in breeding of organisms. Savremena poljoprivreda (Novi Sad), 11–12: 5–27.

Borojević, S., Čupina, T., and Krsmanović, M., 1980. Green area parameters in relation to grain yield of different wheat genotypes. Zeit. Pflanzenzüchtung, 84: 265–283.

Borojević, S., and Williams, A. W., 1982. Genotype – environment interactions for leaf area parameters and yield components and their effect on wheat yields. Crop Sci., 22: 1020–1025.

Borojević, S., 1983. Genetic changes to increase yield potential in wheat. Proc. 6th Inter. Wheat Genet. Symp., Kyoto, Japan, p. 953–957.

Borojević, S., and Kraljević-Balalić, M., 1983. Determination of optimum density and row spacing for different wheat genotypes. Arhiv za poljoprivredne nauke (Beograd), 43: 309–326.

Borojević, S., and Kraljević-Balalić, M., 1984. Inheritance of leaf architecture at different stages of wheat development. Zeit. Pflanzenzüchtung 93: 89–100.

Borojević, S., 1986. Genetic changes in morpho-physiologic characters in relation to breeding for increased wheat yield. In: V. A. Johnson (Editor), Genetic Improvement in Yield of Wheat. CCSA Madison (USA), Spec. Pub. No. 13, p. 71–85.

Borojević, S., and Denčić, S., 1986. Screening a wheat collection for leaf position at different stages of growth. Plant Breeding, (Zeit. Pflanzenzüchtung), 97: 97–106.

Boyer, J. S., 1982. Plant productivity and environment. Science, 218: 443–448.

Briggs, K. G., and Shebeski, L. H., 1968. Implications concerning the frequency of control plots in wheat breeding nurseries. Canadian J. Plant Sci., 48: 149–153.

Brim, C. A., Johnson, H. W., and Cockerham, C. C., 1959. Multiple selection criteria in soybeans. Agron. J., 51: 42–47.

Brooking, I. R., and Kirby, E. J. M., 1981. Interrelationships between stem and ear development in winter wheat: the effect of a Norin 10 dwarfing gene, gai/Rht_2. J. Agric. Sci., Cambridge, 97: 373–380.

Comstock, R. E., and Robinsson, H. F., 1952. Estimation of average dominance. In: J. W. Gowen (Editor), Heterosis. Iowa State College Press, Ames, pp. 494–516.

Cook, M. G., and Evans, L. T., 1978. Effect of relative size and distance of competing sinks on the distribution of phytosynthetic assimilates in wheat. Australian J. Plant Physiol. 5: 495–503.

Dalling, M. J., and Loyn, R. H., 1977. Level of activity of nitrate reductase at the seedling stage as a predictor of grain nitrogen yield in wheat (*Triticum aestivum* L.). Australian J. Agric. Res., 28: 1–4.

Darwinkel, A., 1978. Patterns of tillering and grain production of winter wheat at wide range of plant densities. Netherlands J. Agric. Sci., 26: 283–298.

Deckard, E. L., Lambert, R. J., and Hageman, R. H., 1973. Nitrate reductase in corn leaves as related to yields of grain and grain protein. Crop Sci. 13: 343–350.

Deckard, E. I., and Busch, R. H., 1978. Nitrate reductase assays as a prediction test for crosses and lines in spring wheat. Crop Sci., 18: 289–293.

De Wit, C. T., van Laar, H. H., and van Keulen, H., 1979. Physiological potential of crop production. In: J. Sneep and A. J. T. Hendriksen (Editors), Plant Breeding Perspectives. Pudoc, Wageningen, p. 47–82.

Donald, C. M., 1981. Competitive plants, communal plants, and yield in wheat crops. In: L. T. Evans and W. J. Peacock (Editors), Wheat Science – Today and Tomorrow. Cambridge Univ. Press, p. 327–339.

Duncan, N. G., 1969. Cultural manipulation for higher yields. In: J. D. Eastin et al. (Editors), Physiological Aspects of Crop Yields. American Soc. Agron., Madison, p. 327–339.

Duvick, D. N., 1983. Genetic contributions to yield grains of U.S. hybrid maize, 1930–1980. In: N. R. Fehr (Editor), Genetic Contributions to Yield Grains of Five Major Crop Plants. CSSA, Spec. Publ. No. 7.

Duvick, D. N., 1984. Progress in conventional plant breeding. In: J. P. Gustafson (Editor), Gene Manipulation in Plant Improvement. Plenum Press, New York, p. 17–31.

Eagle, H. A., and Frey, K. J., 1974. Expected and actual gains in economic value of oat lines from five selection methods. Crop Sci., 14: 861–864.

Edwards, L. H., Kelata, H., and Smith, E. L., 1976. Gene action of heading date, plant height and other characters in two winter wheat crosses. Crop Sci., 16: 275–277.

Eilrich, G. L., and Hageman, R. H., 1973. Nitrate reductase activity and its relationship to accumulation of vegetative and grain nitrogen in wheat (*Triticum aestivum l.*). Crop Sci., 13: 59–66.

Elgin, J. H. Jr., Hill, R. R., Jr. and Zeiders, K. E., 1970. Comparison of four methods of multiple trait selection for five traits in alfalfa. Crop Sci., 10: 190–193.

Evans, L. T., and Dodstone, R. L., 1970. Some physiological aspects of evolution in wheat. Australian J. Biol. Sci., 23: 725.

Evans, L. T., Bingham, J., Jackson, P., and Sutherland, J., 1972. Effect of awns and drought on the supply of photosynthate and its distribution within wheat ears. Ann. Appl. Biol. 70: 67–78.

Evans, L. T., Wardlaw, I. F., and Fischer, R. A., 1975. Wheat. In: L. T. Evans (Editor), Crop Physiology. Cambridge Iniv. Press, p. 101–149.

Evans, L. T., 1981. Yield improvement in wheat: empirical or analytical. In: L. T. Evans and W. J. Peacock (Editors), Wheat Science – Today and Tomorrow. Cambridge Univ. Press, p. 203.

Evans, L. T., 1983. Raising the yield potential: By selection or design? In: T. Kruge, C. P. Meredith and A. Hollaender (Editors), Genetic Engineering of Plants. An Agricultural Perspective, p. 371–389.

Evans, L. T., 1984. Physiological aspects of varietal improvement. In: J. P. Gustafson (Editor), Gene Manipulation in Plant Improvement. Plenum Press, New York, p. 121–146.

Fasoulas, A., 1973. New Approach to breeding superior yielding varieties. Dept. of Genetics & Pl. Breeding, Aristotelian Univ. of Thessaloniki, Greece, Publ. No. 3, 42 pp.
Fasoulas, A., 1976. Principles and methods of plant breeding. Dept. of Genetics & Pl. Breeding. Aristotelian Univ. of Thessaloniki, Greece, Pub. No. 6, 55 pp.
Fasoulas, A., 1981. Principles and methods of plant breeding. Dept. of Genetics & Pl. Breeding, Aristotelian Univ. of Thessaloniki, Greece, Pub. No. 11, 147 pp.
Fasoulas, A. C., 1988. The Honeycomb Methodology of Plant Breeding. Dept. of Genetics and Plant Breeding, Aristotelian University of Thessaloniki, Greece, 167 pp.
Fischer, R. A., and Kohn, G. D., 1966. The relation of grain yield to vegetative growth and post-flowering leaf area in wheat crop under condition of limited soil moisture. Australian J. Agric. Res., 17: 281–295.
Fischer, R. A., and Kertesz, Z., 1976. Harvest index in spaced populations and grain weight in micro-plot as indicators of yield ability in spring wheat. Crop Sci., 16: 55–59.
Fischer, R. A., 1981. Developments in wheat agronomy. In: L. T. Evans and W. J. Peacock (Editors), Wheat Science – Today and Tomorrow. Cambridge Univ. Press, p. 249–269.
Fisher, R. A., 1936. The use of multiple measurements in taxonomic problems. Ann. Eugenics (London), 7: 179–189.
Frankel, O. H., 1947. The theory of plant breeding for yield. Heredity, 1: 109–120.
Frey, K. J., 1971. Improving crop yields through plant breeding. In: Moving off the yield plateau. American Soc. Agron. Spec. Publ. 20: 15–58.
Grafius, I. E., 1964. A geometry for plant breeding. Crop Sci., 4: 241–246.
Grafius, I. E., 1978. Multiple characters and correlated response. Crop Sci., 18: 931–934.
Grundbacher, T. J., 1963. The physiological function of cereal awns. Bot. Rev., 29: 366–381.
Hamid, Z. A., and Grafius, I. E., 1978. Developmental allometry and its implication to grain yield in barley. Crop Sci., 18: 83–86.
Hazel, L. N., 1943. The genetic basis for constructing selection indexes. Genetics, 28: 476–490.
Henderson, C. R., 1963. Selection index and expected genetic advances. NAS, Nat. Res. Council Publ. Washington, No. 982.
Hooker, A. L., 1984. The pathological and entomological framework of plant breeding. In: J. P. Gustafson (Editor), Gene Manipulation in Plant Improvement. Plenum Press, New York, p. 177–207.
Kauffmann, K. D., and Dudley, J. W., 1979. Selection indices for corn grain yield, percent protein and kernel weight. Crop Sci., 19: 583–589.
Kempthorne, O., and Nordskog, A. W., 1959. Restricted selection indices. Biometrics 15, 10–19.
Khan, M. A., and Tsunoda, S., 1970. Evolutionary trends in leaf photosynthesis and related leaf characters among cultivated species and its relatives. Japanese J. Breed., 20: 133–140.
Knott, D. R., and Kumar, J., 1975. Comparison of early generation yield testing and a single seed descent procedure in wheat breeding. Crop Sci., 15: 295–299.
Knott, D. R., 1979. Selection for yield in wheat breeding. Euphytica, 28: 37–40.
Kraljević-Balalić, M., and Borojević, S., 1988. Inheritance of harvest index and related traits in wheat. Proc. 7th Int. Wheat Genet., Symp., Cambridge, U.K. p. 547–551.
Kubschrestha, V. P., and Jain, H. K., 1982. Eighty years of wheat breeding in India: past selection pressures and future prospects. Zeit. Pflanzenzüchtung 89: 19–30.
Kumakov, V. A., 1967. Pokazatelji fotosinteze kod selekcioni priznak u pšenici. Seljhoz. Biologia II, 4: 551–558.
Lin, C. J., 1978. Index selection for genetic improvement of quantitative character. Theoretical and Applied Genetics (TAG), 52 (2): 49–56.
Loomis, R. S., and Williams, W. A., 1963. Maximum crop productivity: an estimate. Crop Sci., 3: 67–72.
Loomis, R. S., and Williams, W. A., 1969. Productivity and morphology of crop stands: patterns with leaves. In: J. D. Eastin et al. (Editors), Physiological Aspects of Crop Yield. American Society of Agronomy, Madison, USA, p. 27–47.
Lush, J. L., 1935. Progeny test and individual performance as indicators of an animal's breeding value. J. Dairy Sci. 18: 1–19.

Lush, J. L., 1947. Family merit and individual merit as basis for selection. American Nat. 81: 241–261, 362–379.
Lupton, F. G. H., Ali, A. M. M., and Subramanion, S., 1967. Varietal differences in growth parameters of wheat and their importance in determining yield. J. Agric. Sci., Cambridge, 69: 111–123.
Mac Key, J., 1979. Some aspects of cereal breeding for reliable and high yields. Inter. Rice Res. Conf., IRRI, Los Banos.
Mac Key, J., 1979. Wheat domestication as a shoot: root interrelation process, Proc. 5th Inter. Wheat Gen. Symp. New Delhi 2, p. 875–890.
Mac Key, J., 1980. Genetic potentials for improved yield. Proc. Workshop Agric. Potentiality Directed by Nutritional Needs. Akadémia Kiado, Budapest, p. 121–143.
Matzinger, D. F., Cockerham, C. C., and Wernsman, E. A., 1977. Single character and index mass selection with random mating in a naturally self-fertilizing species. Proc. Inter. Conf. Quant. Genetics, Iowa State Univ. Press, Ames, p. 503–518.
McNeal, F. M., Qualset, C. D., Baldridge, D. E., and Stewart, V. R., 1978. Selection for yield and yield components in wheat. Crop Sci., 98: 795–799.
McDaniel, R. G., and Sarkissian, I. V., 1966. Heterosis: Complementation by mitochondria. Science, 152: 1640–1642.
McVetty, P. B. E., and Evans, L. E., 1980. Breeding methodology in wheat. II. Productivity, harvest index and height measured on F_2 spaced plants for yield selection in spring wheat. Crop Sci. 20: 587–589.
Mišević, D., 1981. Genetićka varijabilnost i selekcioni indexi za prinos zrna i sadrzaj ulja i proteina u zrnu sintetičkih populacija kukuruza. Doktorska disertacija (Ph.D. thesis) Univ. Novi Sad, (Yugoslavia).
Mišić, T., and Mikić, D., 1976. Breeding for high yield and quality in winter wheat. Proc. 2nd Inter. Winter Wheat Conference (Zagreb), p. 393–401.
Moll, R. H., Stuber, C. W., and Hanson, W. D., 1975. Correlated responses and responses to index selection involving yield and ear height of maize. Crop Sci., 15: 243–248.
Moon, P., 1940. Proposed standard solar radiation curves for engineering use. J. Franklin Inst. 230, 583–617.
Nichiporovich, A. A., 1956. Fotosintez i teoria polučenia visokih urozaev. XV. Timiriazevskoe čtenie. Izd. A.I. SSSR, Moskva.
Nichiporovich, A. A., 1970. Biological basis for plant productivity. Kulturpflanze, 6: 73–109.
O'Brien, L., Baker, R. J., and Evans, L. E., 1978. Response to selection for yield in F_3 of four wheat crosses. Crop Sci., 18: 1029–1033.
Obrascov, A. S., 1973. Physiological aspects of plant breeding for earliness and productivity. Plant Physiol. (Moscow) 20: 175–182.
Pesek, J., and Baker, J. R., 1969. Comparison of tandem and index selection in the modified pedigree method of breeding self-pollinated species. Canadian J. Plant Sci., 49: 773–783.
Pesek, J., and Baker, R. J., 1969. Desired improvement in relation to selection indices. Canadian J. Plant Sci., 49 (6): 803–804.
Qualset, C. D., Schaller, C. W., and Williams, J. C., 1965. Performance of isogenic lines of barley as influenced by awn length, linkage blocks and environments. Crop Sci., 5: 489–494.
Quisenberry, J. E., 1982. Breeding for drought resistance and plant water use efficiency. In: M. N. Christiansen and C. F. Lewis (Editors), Breeding Plants for Less Favorable Environments. Wiley, New York, p. 193–212.
Rasmusson, D. C., 1984. Ideotype research and plant breeding. In: J. P. Gustafson (Editor), Gene Manipulation in Plant Improvement. Plenum Press, New York, p. 95–119.
Rasmusson, D. C., and Crookston, R. K., 1977. Role of multiple awns in determining barley yields. Crop Sci. 17: 135–140.
Rosielle, A. A., and Frey, K. J., 1975. Application of restricted selection indices for grain yield improvement in oats. Crop Sci. 15: 544–547.
Rosielle, A. A., Eagles, H. A., and Frey, K. J., 1977. Application of restricted selection indexes for improvement of economic value in oats. Crop Sci., 17: 359–362.
Sage, G. C. M., and Hobson, G. E., 1973. The possible use of mitochondrial complementation as an indicator of yield heterosis in breeding hybrid wheat. Euphytica 22: 61–69.

Sarić, M., Mišić, T., Vulić, B., and Momčilović, V., 1986. Genetic aspects of mineral nutrition of wheat. II. Concentration of N, P, K, Ca and Mg in leaves of plants grown in field. Savremena poljoprivreda (Novi Sad), 34: 289–384.
Sarić, M., 1987. Progress since the first international symposium: "Genetic aspects of plant mineral nutrition", Beograd 1982 and perspectives of future research. Plant and Soil, 99: 197–209.
Sarić, M., Krstić, B., and Stanković, Z., 1987. Genetic aspects of mineral nutrition of wheat. I. Concentration of N, P, K, Ca and Mg in leaves. J. Plant Nutrition, 10: 1539–1545.
Schaller, C. W., Qualset, C. O., and Rutger, Y. N., 1972. Isogenic analysis of the effect of the awn on productivity of barley. Crop Sci. 12: 531–535.
Shebeski, L. H., and Evans, L. T., 1973. Early generation selection for widerange adaptability in the breeding program. Proc. Fourth Inter. Wheat Genet. Symp. Columbia, Mo. USA, p. 587–594.
Simmonds, N. W., 1983. Plant Breeding: The state of the art. In: T. Kosuge, C. P. Meredith and A. Hollaender (Editors), Genetic Engineering of Plants. An Agricultural Perspective, p. 5–25.
Sinnott, E. W., 1921. The relation between body size and organ size in plants. American Naturalist, 55: 385–403.
Smith, H. F., 1936. A discriminant function for plant selection. Ann. Eugenics, (London) 7: 240–250.
Sneep, J., 1977. Selection for yield in early generations of self-fertilizing crops. Euphytica, 26: 27–30.
Tallis, G. M., 1962. A selection index for optimum genotype. Biometrics 18: 120–122.
Thorne, G. N., 1971. Physiological factors limiting the yield of arable crops. In: P. F. Wareing and J. P. Cooper (Editors), Potential Crop Production. Heinemann, London, p. 143–158.
Thorne, G. N., 1982. Distribution between parts of the main shoot and the tillers of photosynthate produced before and after anthesis in the top three leaves of main shoots of Hobbit and Maris Huntsman winter wheat. Ann. Appl. Biol. 10: 553.
Tsunoda, S., and Khan, A. H., 1968. Differences among strains of rice in the photosynthetic tissues. I. A comparative leaf anatomy of indica and japonica. Tohoku J. Agr. Res. 19: 1–7.
Vogel, O. A., Allen, R. E., and Peterson, C. J., 1963. Plant and performance characteristics of semi-dwarf winter wheats producing most efficiently in Eastern Washington. Agron. J., 55: 397–398.
Watson, D. J., Thorne, G. N., and French, J. A., 1963. Analysis of growth and yield of winter and spring wheats. Ann. Bot. New Series, 27: 1–22.
Welbank, P. J., French, S. A., and Witts, K. J., 1966. Dependence of yields of wheat varieties on their leaf area durations. Ann. Bot. 30: 291–299.
Williams, J. S., 1962. The evaluation of a selection index. Biometrics 18: 375–393.
Yap, T. C., and Harvey, B. L., 1972. Inheritance of yield components and morpho-physiological traits in barley, *Hordeum vulgare* L. Crop Sci. 12: 283–286.
Young, S. S. J., and Tallis, G. M., 1961. Performance index for lifetime production. J. Animal. Sci. 20: 506–509.
Zelitch, I., 1975. Improving the efficiency of phytosynthesis. Scientific American 188: 626–633.

16. USE OF HETEROSIS

In a large number of crosses between genetically diverse lines or varieties, hybrid vigour or heterosis occurs in the F_1 generation. Namely, F_1 plants are more vigorous and higher yielding than their parents. This phenomenon is exploited to obtain higher yields in commercial production, especially with cross pollinated plants in which it is difficult to maintain homogeneous and stable cultivars and in which, due to random mating, each reproduction differs from the preceding one.

The use of heterosis may be classified as a method of developing new cultivars by hybridization. But, taking into account that breeding is used only as far as the F_1 generation plus the fact that there is a number of specific methods of producing and using heterotic F_1 combinations, it may also be considered to be a specific path to the development of new cultivars. The main features of this method will be discussed in this chapter.

Heterosis is considered to be the state of maximum heterozygosity which is most easily achieved by crossing genetically diverse inbred lines. Such lines are not easy to obtain, especially in some cross-pollinated plants. Therefore, the development and utilization of heterotic hybrids is a fairly complex process which comprises the following procedures:

(1) the choice of materials for the development of inbred lines,
(2) the development of inbred lines,
(3) testing inbred lines for combining ability, and
(4) the production of heterotic hybrids.

Choice of materials for the development of inbred lines

Local populations and varieties, and newly developed cultivars or their mixtures serve as source materials for the development of inbred lines. It is essential for the source material to have a large genetic variability which warrants that a large number of genetically diverse inbred lines will be obtained.

In the case of maize, which is counted among the most important crops whose hybrid varieties are used in commercial production, the source materials for the development of inbred lines were all local populations, mostly yellow dents, that used to be grown in the U.S.A. Yugoslav maize breeders also used local populations from all maize-growing regions of the country as well as several maize varieties, e.g., Maksimir Yellow Dent, Osijek Yellow Dent, Novi Sad Yellow Dent, etc. As a great portion of the local populations of maize has been used up, efforts are now focused on discovering new source materials for new inbred lines. Synthetic varieties and various hybrid populations possessing new gene recombinations are selfed to make inbred lines (see Chapter 12.).

Development of inbred lines

The development of inbred lines of self pollinated plant species offers no problem because varieties themselves are either already pure lines, or such lines can be made in a short time by the method of individual selection.

In the case of cross-pollinated plants which tolerate self pollination, inbred lines are produced by selfing through six or more generations. Individual plants are selected from the source population and seeds are sown in separate rows or small plots. The progenies are observed in the course of the vegetative season and the best plants are selfed. If species with bisexual flowers are dealt with, e.g., sunflower, rye, etc., the head or the spike is isolated before the beginning of flowering and the seeds obtained are the result of selfing. In monoecious plants (maize, watermelon) and dioecious plants (hops, asparagaus), the male and the female sexual organs must be isolated separately before the beginning of flowering and when pollen reaches maturity it is applied to the stigmata of the female parts of the same plant or other plants.

Seeds of the plants from the first generation of selfing (S_0) are sown in separate rows or plots using the pedigree method. S_1 plants are quite diverse due to segregation of traits after selfing. The segregation takes place because the parents are heterozygous for most traits. Superior plants within each progeny are selected for the next cycle of selfing and inferior progenies are discarded.

Seeds obtained by further selfing are sown by the pedigree method. S_2 progenies are already more uniform, i.e., the formation of lines commences as does differentiation amongst the progenies. The effect of inbreeding becomes visible because most progenies have smaller and less developed plants. Degenerative processes take place in individual plants, e.g., stunting, leaf yellowing, albinism, partial sterility of ovaries and anthers, susceptibility to diseases, etc. (i.e., inbreeding depression).

The best plants are selected from the best progenies, i.e., lines, and they are further selfed. The seeds obtained are sown by the pedigree method. In the S_3 generation, plants within a line are fairly uniform while the lines further diversify. Of course, a large number of lines still segregates in a number of traits. The inbreeding effect intensifies, reducing plant height and yield and causing degeneration in some plants. Some inferior lines must therefore be definitely discarded.

In further generations, the best plants are continually selected, selfed and grown by the pedigree method. After 6 to 8 generations of selection and selfing, the negative effect of inbreeding ceases and lines are obtained which are very uniform in most traits (over 96% of homozygosity). Compared with the source material, the lines have a lower stem, fewer leaves and a lower yield, i.e., they are quite different from the original material.

As the process of inbreeding provokes a radical differentiation among lines, it is possible to obtain hundreds of different inbred lines from a single source population. The higher the genetic diversity of the source material, then greater is the number of genetically different inbred lines that will be derived in the process of inbreeding.

The development of inbred lines by selfing is suitable for plant species which produce a fair number of seeds when selfed, e.g., maize and sunflower. However, there are species in which it is difficult to produce inbred lines, either due to incompatibility as in rye or for other reasons such as the very long time between generations in forest trees. Attempts have recently been made to produce diploid homozygous lines of the latter species via haploids. Such doubled haploids may be used for the development of hybrids (see Chapter 18).

Testing inbred lines for combining ability

Once developed, inbred lines should be mutually crossed in order to find those combinations which will ensure the best heterosis in the F_1. Heterosis is fully expressed when one allele is fixed in one line and a different allele in the other line, as indicated by the formula of the quantitative explanation of heterosis offered by Falconer (1967):

$$HF_1 = \Sigma\, dy^2 .$$

This formula shows that heterosis is the product of dominant effects (d) of a large number of genes but, even more so, the product of square differences (y^2) in gene frequency between the crossed lines. The differences are largest when the lines are completely homozygous, but it takes more than 6 generations of selfing to achieve it (see Table 4.3). Such lines form only one kind of gamete and the crossing of different lines produces the most favourable combination of such gametes or genes. An advantage of using F_1 is that such exact combinations of genes may continuously be produced, which is not the case with varieties or populations where superior combinations of genes are few and they cannot affect significantly the phenotypic value. This is why inbred lines started to be produced because genetic variability created between lines in the process of selfing is much higher than the variability between varieties of the same species.

Heterosis being the state of maximum heterozygosity, much higher heterozygosity is achieved by crossing inbred lines rather than varieties. If capital and lower-case letters are used to designate different allelic genes in varieties and inbred lines, then, when varieties are crossed, the state of heterozygosity in the F_1 generation will be:

Variety A × Variety B

A b C d e F G i J etc. *A B C d E F G I J* etc.

F_1 *AA Bb CC dd Ee FF GG Ii JJ* etc.

The F_1 generation is heterozygous in few alleles and homozygous in most alleles. Consequently, heterosis will be less expressed than if two inbreds derived from the above varieties are crossed:

Line 1 from variety A × Line 15 from variety B

$A\,B\,c\,D\,e\,F\,g\,i\,j$ etc. $a\,b\,C\,d\,E\,f\,g\,I\,j$ etc.

F_1 $Aa\ Bb\ Cc\ Dd\ Ee\ Ff\ gg\ Ii\ jj$ etc.

When genetically different inbred lines are crossed, the state of heterozygosity is attained in the F_1 generation for most alleles, i.e., traits, resulting in luxuriant growth of the entire organism. Therefore, wherever it is technically feasible and economically justifiable, exploitation of line heterosis is given priority over the exploitation of varietal heterosis.

As heterosis is not an invariable result of crossing two inbred lines, because lines may be too genetically similar, the value of a line is estimated on the basis of the degree of the heterosis it achieves in combination with other lines. To check combining ability of inbred lines, they should be diallely crossed, each line should be crossed with all other lines. As it is important for the expression of heterosis which line will be used as female and which as male, the whole crossing scheme should be reciprocated. For example, if 40 inbred lines are to be tested for combining ability, it is necessary to make $n/n - 1$ cross combinations, i.e., 40 (40 − 1) = 1,560. If a group to be tested consists of 100 inbred lines, which is not much, 9,900 cross combinations must be made. As this task would be too large to handle, a simpler and yet reliable method of testing inbred lines for combining ability has been developed.

Davis (1927) proposed a common tester to be used instead of crossing each line with all others. A tester may be a variety or a hybrid which must have a large number of positive traits on a wide genetic basis in order to exploit the genetic variability of the inbred lines as much as possible and thus be able to show how certain traits of the inbred lines are combined. Testers for maize, for example, are either a standard single- or double-cross hybrid or a very good inbred line grown in the region.

The crossing of inbred lines to a tester provides information on the general combining ability of the inbred lines but it remains unknown how they combine mutually. It has been observed that lines which combine well with a tester regarding yield level, are in most cases good combiners with other lines producing a heterotic yield in the F_1 generation. Conversely, the lines which bring a low yield when crossed to a tester may not produce a heterotic yield when crossed to other lines (see Chapter 9).

Inbred lines which bring a high yield with a tester are selected for further work and lines inferior to the tester are discarded. A large number of lines is eliminated that way and the remaining ones are diallely crossed to assess their specific combining ability (Sprague and Tatum, 1942).

Lines which exhibit the highest heterotic effect regarding yield in the F_1 generation and which also show other favourable traits may be used as components for producing commercial hybrids. If F_1 seeds of single-cross hybrids ($A \times B$) are used directly for production, as in the tomato and onion, then the test of specific combining ability completes the research programme and the best components are handed over to seed producers to organize commercial production.

In some plants, double-cross hybrids are preferred to single-cross hybrids because of cheaper and safer seed production. However, it means that single-cross hybrids must be additionally tested for combining ability since hybrid ($A \times B$) may combine

Table 16.1 Yield in bu/acre of 10 single-crosses obtained from 5 inbred lines of maize

Hybrid	Yield	Hybrid	Yield
(23×24)	41.7	(24×27)	72.1
(23×26)	62.6	(24×28)	69.3
(23×27)	70.8	(26×27)	64.2
(23×28)	64.4	(26×28)	60.4
(24×26)	65.6	(27×28)	59.6

Source: Anderson, 1938

Table 16.2 Realized and expected yield of 15 double-cross maize hybrids obtained from 10 single-crosses in Table 16.1

Hybrid combinations	Realized yield	Expected yield
crosses of lines: 23, 24, 26, 27		
(23×24)×(26×27)	68.8	67.8
(23×26)×(24×27)	62.4	60.6
(23×27)×(24×26)	62.0	60.2
crosses of lines: 23, 24, 26, 28		
(23×24)×(26×28)	65.0	65.5
(23×26)×(24×28)	59.8	58.0
(23×28)×(24×26)	56.0	58.5
crosses of lines: 23, 24, 27, 28		
(23×24)×(27×28)	71.1	69.2
(23×27)×(24×28)	58.1	59.4
(23×28)×(24×27)	58.0	60.4
crosses of lines: 23, 26, 27, 28		
(23×26)×(27×28)	68.2	65.0
(23×27)×(26×28)	65.0	62.7
(23×28)×(26×27)	65.7	63.4
crosses of lines: 24, 26, 27, 28		
(24×26)×(27×28)	70.2	66.5
(24×27)×(26×28)	62.0	64.7
(24×28)×(26×27)	62.7	64.4
LSD	5.26	3.41

For example, expected yield of double-cross (23×24)×(26×27) is calculated from the yield of single-crosses:

(23×26) = 62.6
(23×27) = 70.8
(24×26) = 65.6
(24×27) = 72.1

271.1 : 4 = 67.8

Source: Anderson, 1938

better with hybrid $(C \times D)$ than with hybrid $(E \times F)$. To avoid another cycle of crossings, investigations have been conducted which demonstrated that yields of double-cross hybrids may be reliably predicted on the basis of yields of single-cross hybrids. For example, on the basis of the yields of the single-cross hybrids (Table 16.1) it is possible to estimate the yield of the double-cross hybrid (Table 16.2).

Fifteen double-cross hybrids can be made from 10 single-cross hybrids. Anderson (1938) conducted the crossings and found that the actual yields were in very good agreement with the calculated yields (Table 16.2). This procedure saves much time and expenses and it still provides reliable data on combining ability of double-cross hybrids.

It should be mentioned at this point that double-cross hybrids of maize are not intensively used any more because most inbred lines presently in use have a good seed yield (1.5–2.0 t/ha) making the price of F_1 seeds of single-cross hybrids competitive on the market.

Production of heterotic hybrids

Although heterosis is manifested in a large number of plant species, there is still a relatively small number of crops in which hybrid vigour is commercially exploited. The main reason for this are the difficulties in hybrid seed production associated with the constitution of sexual organs of these plant species.

In maize, which is a monoecious species, detasseling is a simple and effective procedure of removing male organs from the female plant which facilitates the production of sufficient quantities of seeds. In other plants, such as the sunflower, rice, wheat, etc., which have bisexual flowers, emasculation would be a complicated and costly operation. Of all methods of hybrid seed production examined in these plants, the method of male sterility has been the most successful.

Male sterility

Male sterility is a phenomenon when plants do not produce viable pollen grains, which is due either to the stunting of male sexual organs (stamens) or to meiotic irregularities in apparently normal stamens leading to the development of abnormal microspores and non-functional or sterile pollen grains.

Male sterility is genetically controlled and there are three types:

(1) genetic or nuclear male sterility, controlled by one or more gene loci in chromosomes,
(2) cytoplasmic male sterility, controlled by extra-chromosomal factors or plasmagenes, and
(3) cytoplasmic-nuclear male sterility, controlled by both kinds of factors.

Cytoplasmic males sterility has been found to be most useful in plant breeding.

Cytoplasmic male sterility (cms)

This type of sterility occurs spontaneously in maize, sugar beet, carrot, onion, pepper and other species (von Wettstein, 1924; Rhoades, 1933). Sources of cytoplasmic male sterility that are presently in use originate mostly from interspecific crosses (wheat, rice, tobacco, beans, peppers, *Brassica* species, etc.) or even intergeneric crosses (*Aegilops* species × *Triticum* species) (Jain, 1959; Laser and Lersten, 1972).

When a cytoplasmic male sterile (cms) plant is crossed with a male fertile plant, male sterile progeny is obtained. It is an indication that plasmagenes contained in cytoplasm are responsible for this trait. Plasmagenes are transmitted to progenies only via egg cells, since the male transmits only the nucleus via pollen and thus it does not affect progenies with respect to male sterility. This again indicates that cytoplasmic male sterility itself cannot be used in breeding heterotic hybrids because crosses between cms lines and fertile lines would result in sterile F_1 hybrids. It is not a problem for onions and other plant species whose F_1 generation is grown for vegetative parts, but it is a serious problem for the sunflower, maize, and other plant species which are grown for seed. Therefore, the male line or pollinator must contain genes for male fertility restoration, so-called *Rf* genes, to ensure the development of seeds in F_1 plants.

Thus, production of hybrids by cms lines proceeds as follows:

(1) introduction of the cms trait into the line to be used as the female,
(2) maintenance of the cms line ("*A*" lines) and its fertility maintainer or analogue ("*B*" line),
(3) introduction of *Rf* genes in the pollinator ("*R*" line), and
(4) hybrid seed production by growing "*A*" line and "*R*" line in alternate rows.

The cms trait is incorporated into suitable lines by the method of back crossing. After at least 6 back crosses of the cms line to the male line, an "*A*" line is produced which has all parental traits plus the cms trait. The "*A*" line is maintained by being crossed to the source line ("*B*" line), i.e., its isogenic line or fertile analogue.

Since all normal lines, i.e., those that are not cytoplasmic male sterile, do not possess *Rf* genes for fertility restoration, it is necessary to find lines which do possess *Rf* genes. They are commonly found in the same stock from which the cms material has been derived but in some instances they must be transferred from other species, as in the case of the transfer of *Rf* genes from *Triticum timopheevi* to *T. aestivum*. *Rf* genes are also incorporated into the pollinator (*R* line) also by the method of back crossing. This is important for crops grown for seed, whilst for those species grown for their vegetative parts (sugar beet, onion, grasses) the pollinator is not required to possess *Rf* genes.

If the pollinator contains an *Rf* gene or genes in the chromosomes, it means that the cms line contains, in addition to sterile cytoplasm, nuclear genes (*rf*) which are not capable of restoring fertility.

A diagram of hybrid seed production based on cytoplasmic male sterility is presented in Fig. 16.1.

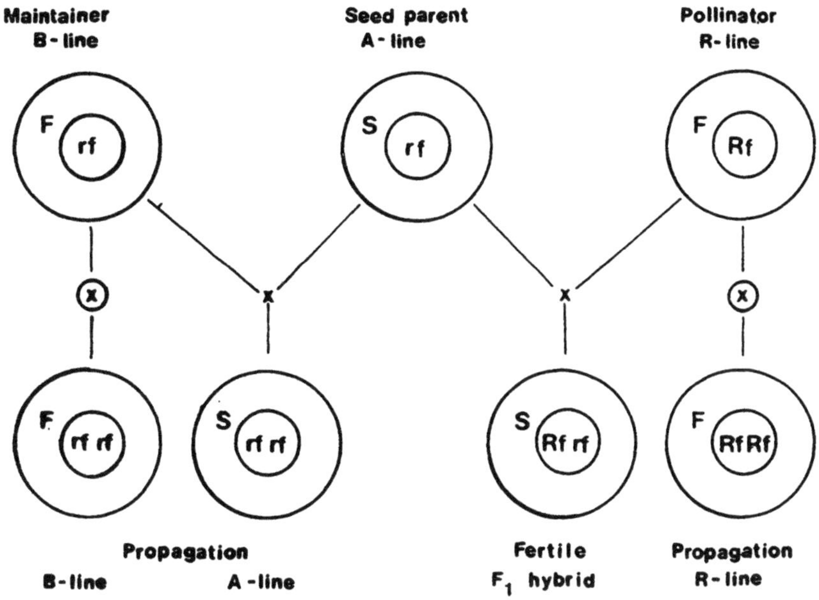

Fig. 16.1 Production of hybrid seeds on the basis of cytoplasmic male sterility (cms). Letter *S* stands for sterile cytoplasm (carrier of plasmogenes for male sterility), *F* stands for fertile cytoplasm Rf genes for male fertility restoration, *rf* genes which do not restore male fertility, *x* stands for crossing, ⊗ for selfing

Production of hybrid seed by the cms method involves numerous difficulties. Male sterility itself is not completely sterile and a certain number of fertile plants occur in "*A*" lines. Also, *Rf* genes may be affected by environmental factors resulting in incomplete fertility and decreased heterotic effect in F_1 plants. Because of changes introduced in "*A*" and "*R*" lines, they are frequently less resistant to pathogens than are their normal analogues, etc.

Nuclear male sterility

Nuclear or genetic male sterility is mostly controlled by a single recessive gene (sorghum, beans), but it may also be controlled by several genes (the barley, tomato).

Since chromosomal genes are concerned, segregation takes place in the progenies of female and male lines. Therefore, the procedure of hybrid seed production differs from that of the cms type (Fig. 16.2).

A difficulty encountered in the use of this type of male sterility is that fertile plants must be removed from seed plot before flowering, to prevent them from pollinating male sterile plants which would decrease the heterotic effect (Fig. 16.2). Furthermore, as the gene for male fertility is dominant (*Ms*), the number of plants must be twice the number than is actually necessary because all *Msms* plants have to be removed.

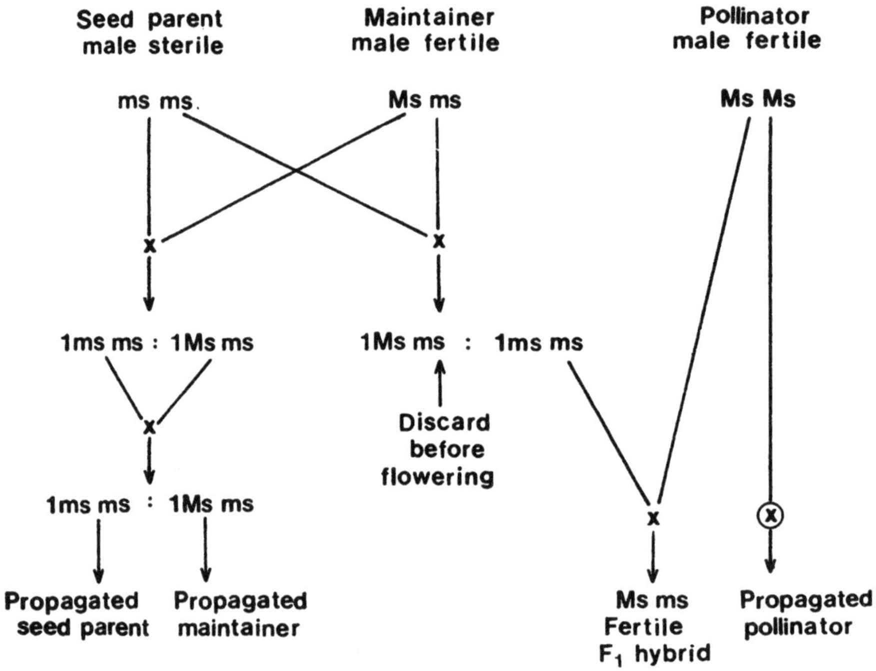

Fig. 16.2 Production of hybrid seeds on the basis of nuclear gene sterility. *Ms* – dominant gene for male fertility, *ms* – recessive gene for male sterility

To overcome these obstacles in the use of nuclear male sterility, Ramage (1975) applied in barley a so-called *balance system of tertiary trisomics* or BTT system which involves the production of trisomic plants for one extra chromosome which consists of segments of two non-homologous chromosomes. This extra chromosome is the carrier of *Ms* genes and the two normal chromosome carry *ms* genes. The extra chromosome is not transmitted to progenies through the male. Progenies of trisomic plants segregate to disomics (diploid) which are male sterile (*msms*) and make about 70% of the total number of plants, and trisomics which are fertile (*Msmsms*) and make about 30%. Ramage succeeded in incorporating into the extra chromosome a marker gene for a violet colour of the stem which allows for the differentiation of sterile and fertile plants. The method has not been accepted because of its complexity as well as because BTT plants are weak and develop fewer grains than normal plants.

Functional male sterility

This type of male sterility covers the occurrence of plants which produce normal pollen but the mechanism of anther opening is abnormal and the pollen cannot be released. This phenomenon may be used to advantage in hybrid seed production.

Plants with functional male sterility may be used as females, without a need to remove anthers, and a line to be used as a male need not contain genes for fertility restoration.

Functional male sterility has been found in potatoes, eggplants, blackberries, etc. Several tomato hybrids have been developed in Bulgaria on the basis of this type of male sterility (Daskalov and Georgiev, 1976).

Chemical induction of male sterility

In plants with bisexual flowers, in which it would be difficult to obtain cytoplasmic or nuclear male sterility and mechanical removal of anthers would be uneconomical, efforts have been made to incapacitate pollen by chemicals without damaging stigmata and ovaries.

It has been found that chemicals such as Ethrel (2-chlorethanphosphoric acid), and Mendok (2,3-dichloroisobutyrate) kill pollen in wheat, rice, sugar beet, etc. Moreover, gibberellic acid, which is a growth stimulator, has been found to incapacitate pollen in the lettuce, onion, sunflower, etc. These chemicals are called *gametocides*.

The development of the technique of male sterility induction by gametocides made room for an inexpensive and quick commercial production of hybrid seed, without a long procedure of introducing cms genes and *Rf* genes. However, gametocides that have been used earlier exhibited a number of negative side effects. First of all, they could not incapacitate all pollen grains, treatment dosage and scheduling had to be set precisely, and the treatment had to be repeated several times in the course of the vegetative period. Furthermore, gametocides tended to cause damage in female sexual organs, deformities in plants and a prolonged growing period. Thus gametocides have not found a wide application yet but efforts are invested in developing better gametocides and improving application methods.

Use of heterosis in commercial practice

Maize

The phenomenon of heterosis has found its widest application in this crop and it has actually revolutionized its entire maize production. Heterosis in maize increases yield by up to 30% in relation to standard varieties or varieties from which inbred lines have been derived.

The production of hybrid maize based on crossing inbred lines commenced in the U.S.A. in the course of the 1930s thanks to the pioneer efforts of East and Hayes (1912), Shull (1909) and Jones (1917). Before World War II, maize hybrids were grown on 55% of the world's maize acreage. Right after the war, the entire acreage was practically sown with hybrid seeds. Double-cross hybrids have predominated at the beginning to give way to single-cross hybrids which are presently grown on 80–90% of the acreage (Duvick, 1977). In Europe, especially in France, Italy, Yugoslavia, Rumania, the Soviet Union, etc., maize varieties have rapidly been replaced by hybrids in the post-war period which brought significant yield increases (Tavčar, 1959; Trifunović, 1978).

Hybrid seed production requires the maintenance of inbred lines which have been found experimentally to produce heterotic combinations. Each inbred line is maintained by means of spatial isolation and hand pollination.

In practice, F_1 maize hybrid seeds are produced by alternately sowing four rows of the female line to two rows of the male line. Female plants are detasseled (emasculated) as soon as tassels start to appear (this activity must be completed before flowering, i.e., pollen shedding). As the maize plant is pollinated by means of wind, the female's silk can be pollinated only by pollen of the male line (Fig. 16.3). Seeds which develop on the ears of the female are F_1 hybrid seeds and those on the male are seeds of that inbred line.

Detasseling is conducted mechanically because the operation is neither difficult nor expensive. Attempts have been made to introduce cytoplasmic male sterility in female lines. However, it considerably prolongs the development of testing inbred lines and brings additional risks. For example, a large number of maize hybrids grown in the U.S.A. at the beginning of the Seventies possessed the so-called Texas (T) cytoplasm for male sterility. A race of *Helminthosporium maydis* (*Cochiobolus heterostrophus*) which developed in 1970 was so virulent on all hybrids possessing

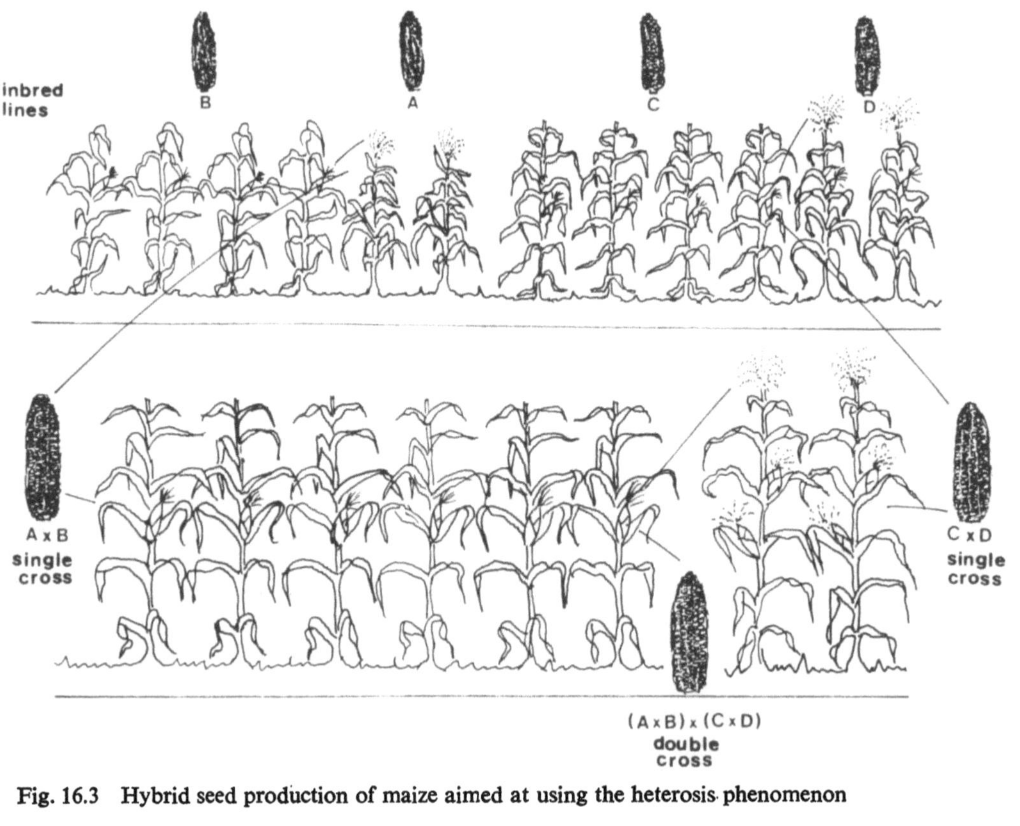

Fig. 16.3 Hybrid seed production of maize aimed at using the heterosis phenomenon

T cytoplasm that a 15% yield reduction was experienced over vast acreages. The use of T cytoplasm was abandoned overnight. The other types of sterile cytoplasm, such as S and C, are not convenient for introduction and are not widely used.

Hybrid seeds from ears of the female plants produce in the next season heterotic F_1 single-cross plants which outyield the inbred lines several times over (inbreds yield 1.5 to 2.0 t/ha) and the variety from which the inbred was derived by about 30%.

To produce seeds of double-cross hybrid – $(A \times B) \times (C \times D)$, F_1 seeds of one single-cross hybrid, e.g., $(A \times B)$ are sown in six female rows and F_1 seeds of another hybrid $(C \times D)$ in two male rows. The female rows are detasseled before pollen shedding and are pollinated by pollen coming from the male rows to produce seeds of the double-cross hybrid. These seeds are used for the production of double-cross plants in the subsequent season (Fig. 16.3).

Fig. 16.4 Single-cross hybrid NSSC-70 produced by crossing inbred line NS-796 and American line C 103
Source: Courtesy of late Prof. R. Savić, Faculty of Agriculture Novi Sad, Yugoslavia

Besides single and double-cross hybrids, other kinds of maize hybrids are also used, such as three-way crosses – $(A \times B) \times C$, and top crosses – $A \times$ variety or $(A \times B) \times$ variety.

Highest yields are obtained from single-cross hybrids (Fig. 16.4), then from three-way crosses, etc.

Hybrid maize is adapted to most intensive cultural practices producing average yields between 7.0 and 9 t/ha (Fig. 16.5) and record yields of over 20 t/ha (see Table 1.1).

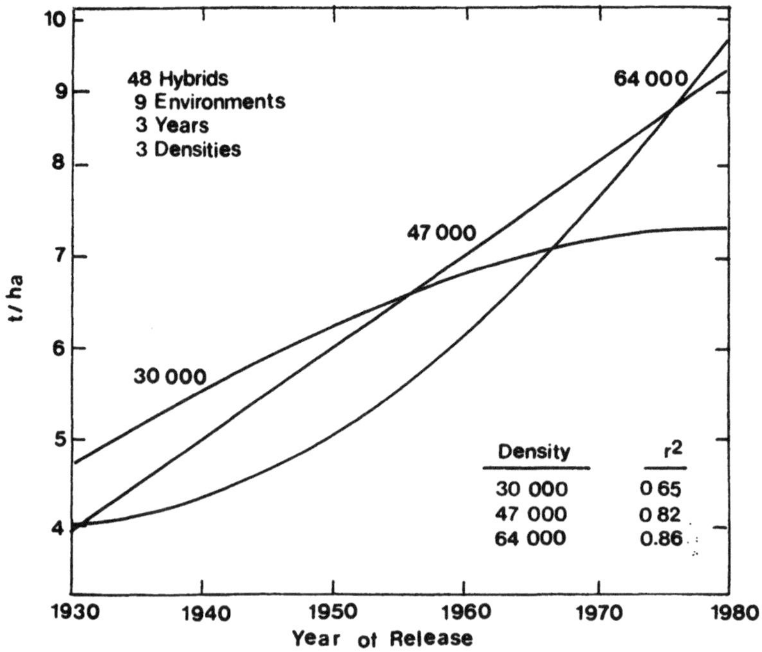

Fig. 16.5 Regressions of hybrid grain yield on year of hybrid release, for 47 maize hybrids and one open pollinated variety tested on three plant populations. The hybrids were released at intervals from 1934 to 1978
Source: Duvick, 1984

Sorghum

The phenomenon of heterosis is largely used in grain sorghum. Sorghum hybrids are superior to commercial cultivars in yield, lodging resistance and, in some instances, in resistance to drought and diseases.

As the sorghum flowers are bisexual and very small, the exploitation of heterosis is based exclusively on the use of cytoplasmic male sterility.

Sunflower

The sunflower is the second crop in importance after maize for the use of hybrids in commercial production. Although inbred lines of sunflower started to be developed in the Soviet Union after World War I (Yagodkhin, 1937), sunflower hybrids could not be introduced into commercial production because male sterile lines did not exist and there was no other way to produce hybrids.

The work on sunflower hybrids was intensified after Leclercq's (1969) discovery of cytoplasmic male sterility in the interspecific cross *Helianthus petiolaris* × *H. annuus*. After that, it only remained to introduce into inbred lines the genes for fertility restoration (*Rf* genes) which were discovered in some wild sunflower species (Vranceanu, 1974).

In only one decade, sunflower hybrids based on *cms* and *Rf* genes have been developed in Rumania (Vranceanu, 1974), France (Leclerq, 1969), Yugoslavia (Skorić, 1980) (Fig. 16.6), and the U.S.A. The introduction of hybrids brought considerable increases in sunflower acreage. France, the U.S.A. and Canada, which used to grow sunflowers on a limited acreage, have become important producers of sunflower oil.

Fig. 16.6 Novi Sad hybrid sunflower. Left – cms seed parent inbred line, right – pollinator parent with *Rf* genes, in the middle F_1 hybrid.
Source: Škorić, 1988

It should be emphasized that this major breakthrough in sunflower production was preceeded by the extraordinary achievements of V. S. Pustovoit (1966) who succeeded in increasing the oil content in the seeds of sunflower varieties from 33% to 50% (Pustovoit, 1968).

Sunflower hybrids are not superior to varieties in oil percentage but, thanks to higher grain yields, they produce more oil per unit area. The number of sunflower hybrids is still relatively low and they are not as adapted to various environmental conditions as are some varieties (e.g., Peredovik). As diseases are the major limiting factor in sunflower production (e.g., *Phomopsis* sp., *Sclerotinia sclerotiorum*, etc.), much effort has been invested into finding and introducing genes of resistance to pathogens into the cultivated sunflower as well as to increasing the number of diverse hybrids.

Rye

In rye, heterosis is expressed in vigorous growth and increased yield. Still, there are no high correlations between various traits. It happens in some combinations that heterosis is expressed for plant height and spike length but not for grain yield (Borojević, 1957).

Varietal heterosis rather than line heterosis is exploited in rye. The development of inbred lines has been attempted in Sweden and the U.S.A. but they were difficult to obtain and more difficult to maintain since rye is pronouncedly a cross pollinated species.

Rye has bisexual flowers which require either the manual emasculation of individual flowers or the application of a gametocide which has not been successful so far. Instead, heterosis is exploited by mixing seeds of varieties previously found to be heterotic. If two varieties A and B are mixed at the ratio 50:50, it can theoretically be expected to obtain about 50% F_1 seeds i.e., $A \times B$ and the rest are $A \times A$ or $B \times B$ seeds. About 66% hybrid seeds may be expected from a mixture of 3 varieties, 75% from four varieties, etc. (Bredeman and Heuser, 1931; Hmeljev, 1952; Borojević, 1957).

Since it is a case of varietal heterosis, yield increase in relation to the better parent can be up to 25% but, considering that 100% F_1 hybrid seed cannot be produced, the actual increases range between 10 and 15%. Nevertheless, this yield increase due to heterosis is practiced in most rye-growing countries because it incurs minimum expense.

In recent years, the rye acreage has been considerably reduced on account of increases in wheat yields and the introduction of *Triticale* (see Chapter 18). The extent of studies on heterosis in rye has been reduced accordingly.

Wheat

Although heterosis has been observed in crosses of different wheat varieties, it has not been put to use because of difficulties involved in producing sufficient amounts of F_1 seed, associated with the bisexual nature of wheat flowers. Therefore, it is not possible to practically exploit heterosis in wheat without the introduction of male sterility. Cytoplasmic male sterile lines ("*A*" lines) have been available for a long time but there remains to find male pollinators possessing the capacity of male fertility restoration (*Rf* genes) in order to be able to produce fertile F_1 plants.

First results in the development of hybrid wheat were given by Wilson (1967) and Smith and Johnson (1963) in the U.S.A. Male lines possessing genes for male fertility restoration in the F_1 generation have been found in the species *Triticum timopheevi*.

However, numerous difficulties have been encountered in trying to practically exploit heterosis in wheat. The principal difficulty is the finding of *Rf* genes which would ensure a complete fertility restoration in F_1 plants. If the number of *Rf* genes is greater than two, it is not easy to make a good "*R*" line which would restore fertility at 100%. "*R*" lines should produce much pollen for as long as possible in order to secure maximum production of F_1 seeds. It is difficult to achieve in self pollinated species such as wheat and this is why the amount of hybrid seeds produced is regularly low. The "*A*" line should be made to flower as long as possible, with flowers wide open, and the stigma capable of receiving pollen for a long time.

The effect of environmental factors on hybrid seed production poses a separate problem. High temperatures and low relative air humidity affect both pollen and the stigma negatively causing poor seed set in some years and regions.

As much as the studies on hybrid wheat are an ingenious scientific endeavour, these efforts are unjustified from the practical point of view. The process of developing *cms* lines is long. It takes 4 to 5 years, even when a greenhouse is available, to back cross *cms* plasmagenes into a standard variety. However, this period is sufficiently long to develop new cultivars by conventional breeding methods, which are higher yielding than the variety converted into an "*A*" line. The same stands for the introduction of *Rf* genes and the development of "*R*" lines. After that it is necessary to test combinations of "*A*" and "*R*" lines for heterotic effects in grain yield because only such lines can be offered to seed producers for hybrid seed production.

Table 16.3 Grain yield of F_1 wheat hybrids and their parents in the experiment at Cambridge 1985

	Yield (% of male parent)	
Male parent Aquila	With fungicide 8.6 t/ha	Without fungicide 5.7 t/ha
Female parent		
Gawain	117 (105)*	120 (108)
Norman	112 (97)	113 (85)
Rendezvous	111 (96)	123 (111)
3489/24	112 (99)	102 (98)
3547/46	113 (103)	102 (97)
3615/9	110 (103)	109 (93)
LSD for 0.05	6.8	12.0

* Yields of the female parents are shown in parentheses.

Source: Bingham and Lupton, 1987

Another obstacle in the production of wheat hybrid seed is the volume of production. While in maize the seed rate is about 10 kg/ha, hybrid wheat requires 150 to 200 kg/ha which for the moment presents an insurmountable problem to large scale seed production.

In the period after World War II, wheat breeding by conventional methods has achieved notable successes. In many countries, high-yielding varieties have been obtained which, in combination with intensified cultural practices, brought about significant increases in average yields and total production. As wheat is an allohexaploid, due to its homeologous chromosomes, a considerable number of genes is contained in several doses which may give rise to so-called homeologous heterosis, i.e., heterosis fixation in the homozygous state (Mac Key, 1970). This also explains why heterosis is better expressed by hexaploid wheat than by diploid barley as well as why wheat breeding has been so successful.

Nevertheless, studies on hybrid wheat have not been abandoned although emphasis has been placed on the use of chemical gametocides. Large chemical companies offer gametocides with which the achieved level of hybridity is generally over 95% and the yields of F_1 seed are 85% of that of the male parent. Using these gametocides at the Plant Breeding Institute (PBI) in Cambridge, several hundred F_1 hybrid combinations have been tested to find out those with the best heterotic effect. According to Bingham and Lupton (1987), the yields of F_1 hybrids were more than 10% higher than those of the higher yielding parent, particularly in the fungicide-treated plots (Table 16.3). However, the advantage of the best pure line cultivar was smaller. At the same time the authors state that hybrid wheat exploitation will be mainly dependent on the ecnonomics of seed production and equivalent yield increases required to cover these costs.

We may hope that further studies and new techniques will allow heterosis in wheat to be exploited as extensively as in the rice.

Rice

As rice and wheat are the staple cereals for human food, possibilities of increasing rice yields are also searched for. The People's Republic of China, where rice production receives much attention, was the first to succeed in growing hybrid rice on a large scale. These rice hybrids were made on the basis of cytoplasmic male sterility (cms) and over 90% of them belong to the *indica* type (Virmani et al., 1981). Hybrid varieties are distinguished by a more powerful root system, more intensive tillering, a larger panicle and larger grains, outyielding standard varieties by up to 25% in irrigation. According to Mao Chang-Xiang (1987), studies on hybrid rice were commenced in P. R. China in 1964 and by 1973 cms maintainer and restorer lines needed to start hybrid seed production were developed.

The commercial production of hybrid rice started in China in 1976 at about 150,000 ha, and this increased to 8.9 million hectares by 1986, i.e., to 26% of the total rice acreage of 32 million hectares. The average yields of hybrid rice were 6.47 t/ha against 5.25 t/ha achieved with standard varieties, i.e., 23% more (Yuan Long-Ping, 1986). In Jiangsu Province the average yield of hybrid rice was 8.0 t/ha on 767,000 ha (Mao Chang-Xiang, 1987). This large expansion of hybrid rice was made possible by permanent increases in hybrid seed yield which increased from 0.9 t/ha in 1973 to 1.9 t/ha in 1986 (Mao Chang-Xiang, 1987) on account of superior cms and restorer lines and the introduction of supplementary pollination using pole and rope-pulling methods (Virmani and Edwards, 1983).

The success in hybrid rice growing in P.R. China moved the neighbouring and other rice-growing countries to pay more attention to the development of hybrid varieties of rice. In this context, a research cooperation involving the Hybrid Rice Research Center in Changsha and other Chinese institutes and the International Rice Research Institute (IRRI) in Manila (Philippines) deserves to be mentioned. IRRI focuses on the development of promising cms lines which are then used in China and other countries for the development of hybrid varieties (Virmani, 1987). To promote further work on hybrid rice in other countries and to coordinate research efforts, the International Symposium on Hybrid Rice was convened in 1986 in Changsha, Hunan Province, P.R. China. Among recommendations to promote further investigations on hybrid rice, the priority areas for further research are of particular interest to mention.

The three-line method ("*A*", "*R*", and "*B*" lines) currently in use is expected to remain effective for the next decade with necessary improvements of cms systems and restorer lines.

The two-line method, however, shows great promise with at least two approaches. First, to produce photoperiod-sensitive genetic male steriles so that any normal line can serve as a restorer and second, to use chemical emasculation which allows any two lines to be used as parents to produce enough hybrid seeds.

A one-line method using apomixis or asexual seed production is the ultimate genetic tool for developing true-breeding hybrids with permanently fixed heterosis. This is stated as a long-term goal in the Recommendations from the Symposium in Changsha.

So rice has become the first self pollinated cereal in which the phenomenon of hybrid vigour is successfully expoited over large areas.

Vegetables

Heterosis in tomatoes has been extensively used for a long time in Bulgaria and the Netherlands, etc. It is manifested in increased number of fruits per plant, bringing yield increases from 20% to 30%. Although the tomato is a selfpollinated species and each flower must be emasculated and pollinated with a corresponding component, the whole operation is profitable because each fruit contains a large number of seeds so that two workers in one shift can produce enough hybrid seeds for sowing one hectare. Several tomato hybrids based on functional male sterility have been made in Bulgaria (Daskalov and Georgijev, 1976).

Heterosis is also extensively used in the onion, carrot, kale and cabbage. As these crops are grown for vegetative parts, the female line may possess any type of male sterility and the *maintainer* need not possess *Rf* genes. Consequently, F_1 seed production is economical with these crops which do not require high seed rates in the first place.

Heterosis is also used in the cucumber, melons and other species from the Cucurbitaceae family. Parental components are sown in alternate rows and female plants are emasculated. The operation is simple because anthers are large, but, due to a long flowering period, newly developed flowers must be emasculated at two-day intervals. Production of hybrids based on male sterility is another possibility with these plant species.

Fruits

In a number of fruits, especially in the apple and walnut, heterosis is expressed as a more vigorous growth of trees, earlier maturation and larger fruits. Heterotic effect is observed already on seedlings which may be used for scions or vegetative propagation, to ensure the exploitation of heterosis over a period of several years.

Other plant species

Heterosis is practically used in the poppy and tobacco. As these species have a large number of seeds in a single fruit and low seed rates on account of very small seeds, manual production of F_1 seeds is profitable.

The use of heterosis is especially advantageous in species which reproduce both by seed and vegetatively. Considerable success has been achieved in growing Canadian poplars and some ornamental and decorative species.

It is reasonable to expect that new methods will be developed to increase the number of crops in which heterosis will be exploited in commercial production. The achievements made so far place the use of heterosis among the major successes of plant genetics and breeding.

In addition to the theory explaining that hybrid vigour is based on heterozygosity which is superior to homozygosity, new hypotheses have emerged in recent years showing that homozygosity may have advantages over heterozygosity and so one can speak of inbred vigour.

On the basis of long-term experiments on *Nicotiana rustica* carried out at the University of Birmingham (Jinks and Pooni, 1976, 1981; Pooni and Jinks, 1980, 1982) using biometrical genetic analyses, Jinks (1983) came to the conclusion that the dispersion of dominant increasing alleles, which may be linked in repulsion, is the most likely cause of heterosis, rather than overdominance. If this explanation is correct, states Jinks (1983), it is safe to predict that it will be possible to obtain pure breeding lines superior to heterotic F_1s from such crosses.

Pooni et al. (1985, 1986), on the basis of random samples of inbred lines extracted from single and double crosses of *Nicotiana rustica*, demonstrated that it is possible to select inbred lines superior to heterotic F_1s or better parents if a sufficiently large population is grown in late generations, allowing for transgressive segregants to occur. The chances of recovering these extreme inbred lines are remote unless the alleles of the like effect are completely dispersed amongst the parental varieties.

According to Fasoulas (1988), whether hybrid vigour is superior to inbred vigour or not is determined by the load of deleterious genes. Heterozygosity does increase the immediate survival and fitness of organisms but it maintains a heavy load of deleterious genes which affect adversely the response to selection. Therefore, the main task of artificial selection is to remove deleterious and other unfavourable genes accumulated by natural selection and replace them by favourable alleles. Fasoulas (1988) states: "In fact, hybrids are superior to inbred lines only in the presence of deleterious genes. Once efficient selection removes deleterious genes, pure line cultivars become more important than hybrids" (p. 113).

These are some new ideas in explaining advantages of heterozygosity and homozygosity which may be useful in developing new approaches and methods of plant breeding.

References

Allard, R. W., 1960. Principles of Plant Breeding. John Wiley. New York, 485 pp.

Anderson, D. C., 1938. The relation between single and double cross yields in corn. J. American Soc. Agron. 30: 209–211.

Bingham, J., and Lupton, F. G. H., 1987. Production of new varieties: an integrated research approach to plant breeding. In: F. G. H. Lupton (Editor), Wheat Breeding, Its Scientific Basis. Chapman and Hall, London, p. 487–538.

Borojević, S., 1957. (Heterosis of rye produced by crossing varieties). Poljoprivredna znanstvena smotra (Zagreb), 16: 29–65.

Bredemann, G., and Heuser, W., 1931. Beiträge zur Heterosis bei Roggen. Zeit. f. Züchtung, R.A. Bd. XVI. Hf 1.

Daskalov, C., and Georgiev, C., 1976. New trends and results of the use of heterosis in tomato. Proc. 17th. Congress of EUCARPIA. Budapest, p. 303–305.

Davis, R. L., 1927. Report of the Plant Breeder. Rept. Puerto Rico Agr. Expt. Station, p. 14–15.

Duvick, D. N., 1977. Genetic rates of gain in hybrid maize yields during the past 40 years. Maydica 22: 187–196.

Duvick, D. N., 1984. Progress in conventional plant breeding. In: J. P. Gustafson (Editor), Gene Manipulation in Crop Improvement. Plenum Press, New York. p. 17–31.

East, E. M., 1936. Heterosis. Genetics 21, No. 4.

East, E. M., and Hayes, H. K., 1912. Heterozygous in evolution and plant breeding. U.S. Dept. Agric. No. 243.

Falconer, D. S., 1967. Introduction to Quantitative Genetics. R. MacLehose, Glasgow, 1960. The Ronald Press, New York, 3rd ed. 1964, reprinted 1967, 365 pp.

Fasoulas, A. C., 1988. The Honeycomb Methodology of Plant Breeding. Dept. of Genetics and Plant Breeding, Aristotelian University of Thessaloniki, Greece, 167 pp.

Hmeljev, I. B., 1952. Ispoljzovanie smjesi semjan pri mežsortovom opyljenii ozimoi rži (in Russian). Selek. i semen. No. 1.

Hughes, W. G., Bennett, M. D., Bodden, J. J., and Galanapoulou, S., 1974. Effects of time of application of Ethrel on male sterility and ear emergence in wheat *Triticum aestivum*. Ann. Appl. Biology 76: 243–252.

Jain, S. K., 1959. Male sterility in flowering plants. Bibliographia genetica 18: 101–166.

Jinks, J. L., and Pooni, H. S., 1976. Predicting the properties of recombinant inbred lines derived by single seed descent. Heredity 36: 253–266.

Jinks, J. L., and Pooni, H. S., 1981. Properties of pure breeding lines produced by dihaploidy, single seed descent and pedigree breeding. Heredity 6: 391–395.

Jinks, J. L., 1983. Biometrical genetics of heterosis. In: R. Frankel (Editor), Heterosis, Springer-Verlag, Berlin, p. 1–46.

Jones, D. F., 1917. Dominance of linked factors as a means of accounting for heterosis. Genetics 2: 466–479.

Laser, K. D., and Lersten, N. R., 1972. Anatomy and cytology of microsporogenesis in cytoplasmic male sterile Angiosperms. Botanical Review 38: 425–454.

Leclerq, P., 1969. Une stérilité male cytoplasmique chez le tournesol. Ann. Amélioration Plantes 19: 99–106.

Lukyanenko, P. P. (Editor), 1968. Geterozis v rastenievodstve. Izd. Kolos, Leningrad. 390 pp.

Mac Key, J., 1970. Significance of mating system for chromosomes and gametes in polyploids. Hereditas 66: 165–176.

Mao Chang-xiang, 1987. Recent advances in hybrid rice in China. Hybrid Rice News Letter, No. 1.

Pooni, H. S., and Jinks, J. L., 1980. Non-linear genotype *x* environment interactions. Heredity 45: 389–400.
Pooni, H. S., and Jinks, J. L., 1982. Comparative analysis of dispersion and association crosses to detect linkage and epistatic components of variation. Heredity 49: 210–220.
Pooni, H. S., Jinks, J. L., and Johannes, D., 1985. Predicting the properties of random inbreds and second cycle hybrids using progeny families and hierarchical selfs of a four parent cross. Heredity 55: 111–119.
Pooni, H., Jinks, J. L., and Johannes, D., 1986. Comparison of inbred lines extracted from two single crosses and their double cross by single seed descent. Heredity 56: 263–276.
Pustavojt, V. S., 1966. Izabranie trudi. Izd. Kolos, Moskva, 367 pp.
Pustavojt, G. V., 1968. Current work on sunflower breeding at VNIIMK. Proc. 3rd Inter. Sunflower Conf., Crookston, Minnesota. p. 46–52.
Ramage, R. T., 1975. Techniques for producing hybrid barley. Barley News Letter 18: 62–65.
Rhoades, M. M., 1933. The cytoplasmic inheritance of male sterility in *Zea mays*. J. Genetics 27: 71–93.
Schmidt, J. W., and Johnson, V. A., 1963. Studies on hybrid wheat. 10th Spring Wheat Workers Confer. Jan. 30–31, St. Paul, Min. p. 1–10.
Shull, G. H., 1909. A pure line method of breeding. Am. Breeders Assoc. Rept. 5: 51–59.
Shull, G. H., 1948. What is heterosis. Genetics 33: 439–446.
Sprague, G. F., and Tatum, L. A., 1942. General vs. specific combining ability in single crosses of corn. J. American Soc. Agron. 34: 923–932.
Škorić, D., 1980. Desired model of sunflower hybrid and newly developed NS-sunflower hybrids. Helia (Bucharest), No. 3: 19–25.
Tavčar, A., 1959. Oplemenjivanje bilja. Nakladni zavod Hrvatske, Zagreb.
Trifunović, V., 1978. Maize production and maize breeding in Europe. In: Maize Breeding and Genetics. John Wiley, New York, p. 41–58.
Virmani, S. S., Chaudhary, R. C., and Khush, G. S., 1981. Current outlook on hybrid rice. Oryza, 18: 67–84.
Virmani, S. S., and Edwards, I. B., 1983. Current status and future prospects for breeding hybrid rice and wheat. Advances in Agron. 36: 145–214.
Virmani, S. S., 1987. IRRI reports promising cms lines. Hybrid Rice New Letter, No. 2.
Vranceanu, A. V., 1974. Sunflower. Editure Acad. Rep. Soc. Romania, Bucharest. 322 pp.
Wettstein, F. von, 1924. Über Fragen der Geschlechtsbestimmung bei Pflanzen. Naturwissenschaften 12: 761–768.
Wilson, J., 1967. Problems in hybrid wheat breeding. Euphytica, Suppl.
Wittwer, S. H., 1979. Future technological advances in agriculture and their impact on the regulatory environment. BioScience, 29: 603–610.
Yagodkin, I. G., 1937. Application of the method of inbreeding and diallel crossing in sunflower growing. Selekcija i semenovodstvo 1: 21–27.
Yuan Long-ping, 1986. Hybrid rice in China. Hybrid Rice News Letter, No. 1.

17. MUTATIONS IN PLANT BREEDING

The importance of gene mutations in the development of genetic variability has been discussed in Chapter 3. Despite a low frequency of spontaneous or natural mutations (about 10^{-6}), when we consider a long period of time and the large number of genes in each organism, spontaneous mutations gain prominence in plant breeding as well as in the evolution of species.

Use of spontaneous mutations

Sengbush (1942) in Germany found a mutant of the yellow lupin (*Lupinus luteus*) which did not contain in its seeds the alkaloid which used to make the yellow lupin unsuitable for animal feed. To date, numerous cultivars of sweet lupin have been developed which are so used.

In fruit trees, large numbers of spontaneous mutations have been recorded and used either directly as new cultivars or in breeding programmes. As many fruit species are reproduced vegetatively, somatic mutations may successfully be used. Stanković (1960) itemized a large number of spontaneous mutations that are used in pomiculture as new cultivars. He mentioned changes in fruit form with apples and pears, date of maturity in peaches and cherries, etc.

A particularly large number of spontaneous mutations is found in ornamental plants. An old tulip cultivar, Murillo, has produced over 60 mutants of different colour that are used in breeding (Doorenbos, 1954). Preil (1986) reported that a fair number of ornamental species in *in vitro* culture tended to produce somatic mutants which can be used in breeding.

The best known spontaneous mutation in field crops is the *opaque* gene in maize which governs the lysine content and is used in breeding high-lysine hybrids. A dwarf mutant registered in the sorghum cultivar Standard Milo has served as a basic stock in breeding short, mechanically combinable types of grain sorghum in many countries (Poehlman, 1983).

Use of induced mutations

Recent decades have witnessed an intensive work on the induction of mutations by irradiation, chemicals and other mutagenic agents. The frequency of induced mutations (10^{-3}) almost doubles those occurring naturally and they have been looked upon as a powerful tool for the development of new cultivars. However, the available mutagens cause not only changes in genes but also cause chromosomal aberrations, many of which are deleterious in their effect on the trait and the entire organism. Consequently, there has been a limited number of induced mutations directly usable as new cultivars.

Gustafsson (1947) obtained three mutants with increased grain yield from the barley cultivar Gull Corn exposed to *X*-rays, and also a mutant with short straw and a mutant with a shorter vegetative period from the cultivar Bonus. The last two mutants, named Pallas and Mari, used to be grown commercially in Sweden and England.

Gregory (1956), following *X*-ray treatment of peanuts (*Arachis hypogaea*), obtained more productive lines as compared with the source cultivar.

Scarascia-Mugnozza (1967) and Bozzini et al. (1973) have obtained several durum wheat mutants with a short straw by means of induced mutation. This trait could not be introduced by any other method because the species *Triticum durum* does not include short-strawed genotypes. These mutants were more resistant to lodging which not only allowed the application of increased doses of mineral fertilizers but virtually revolutionized the production of durum wheats.

As such work has extended in scope and more attention has been paid to the choice of materials to be treated, certain numbers of mutants have been developed which are either directly usable or can be included in breeding programmes as sources of new genes (Sigurbjörnsson, 1983; Donini et al., 1984) (Table 17.1).

Table 17.1 Number of cultivars developed by using induced mutations (According to Mutation Breeding Newsletter Nos 1—20)

	Direct use of mutant	Mutant used in cross-breeding	Total
(a) Seed propagated species:			
Barley	23	34	57
Bread wheat	21	8	29
Durum wheat	5	7	12
Rice	34	13	47
Oat, rye, pearl millet, maize, grasses	10	8	18
Common bean	6	4	10
Peanut	1	9	10
Pea	6	2	8
Chickpea, soyabean, lupin	10	2	12
Others	30	12	42
(b) Vegetatively propagated species:			
Fruit trees	20	1	21
Sugar cane, peppermint, potato	10	–	10
Ornamentals	216	7	223
Total	392	107	499

Source: Donini et al. 1984

Induced, as well as spontaneous mutations are of particular importance in breeding vegetatively propagated plants. Van Harten and Broertjes (1986) cite examples of ornamental and fruit plants where commercial mutants have increased in number especially since 1970 (Table 17.2).

Table 17.2 Commercial Mutants of various Vegetatively Propagated Crops (April, 1985)

Plant Group	Number of commercial mutants					Total number of mutants	Crops of which (induced) mutants were commercialized (number)
	Before 1950	1950–1960	1960–1970	1970–1980	1980–		
ORNAMENTALS							
Tuber and bulb crops	1	1	13	36	5	56	*Dahlia (45)*, *Gladiolus (1)*, *Lilium* (2), *Polyanthes* (2), Tulip (6)
Pot plants	–	–	7	54	12	73	*Achimes* (8), *Azalea (10)*, *Begonia (21)*, *Calathea (1)*, *Ficus (1)*, *Guzmania (1)*, *Hoya* (4), *Kalanchoë* (3), *Streptocarpus (24)*
Cut flowers	–	–	18	121	40*	179*	*Alstroemeria (24)*, Carnation (12), Chrysant* (130), *Euphorbia fulgens (1)* Rose (12)
Other ornamentals	–	–	–	20	4	24	*Abelia (1)*, *Azalea (1)*, *Bougainvillea (5)*, *Forsythia* (2), *Malus (1)*, *Populus (1)*, *Portulaca (11)*, *Weigelia (2)*
FRUIT CROPS	–	–	4	14	4	22	Apple (4), Apricot (1), Black Currant (1), Cherry (7), Olive (1), Peach (4), Fig (1), Grapefruit (1), Pomegranate (2)
OTHER CROPS	–	–	–	14	3	17	Peppermint (3), Sugar cane (11), Grasses (1), Matgrass (2)
Totals	1	1	42	260	68*	372	

* In *Chrysanthemum* induction of mutations is every-day practice and the exact number of commercial mutants cannot be determined anymore. It must exceed several hundreds or even many by now.
Source: Harten and Broertjes, 1986

As regards the traits for which induced mutants have been obtained, they are mostly qualitative traits controlled by major genes (macromutations) and to a lesser degree the quantitative traits controlled by minor genes (micromutations). Donini et al. (1984) listed such mutants in cereals (Table 17.3).

Table 17.3 Characters of cereal mutant cultivars developed without cross-breeding

Characters	No. of occurrences in released varieties of:				
	Barley	Bread wheat	Durum wheat	Rice	Total
A. Plant morphology					
Reduced plant height	12	6	5	7	30
Stiff straw	6	5	0	0	11
Erectoid type	1	0	0	0	1
Lodging resistance	8	3	4	5	20
Increased tillering	2	1	0	1	4
Awned	0	2	0	0	2
B. Maturity					
Earlier, up to 4 days	3	1	0	0	4
Earlier, 5–10 days	3	0	0	2	5
Earlier, over 10 days	0	0	0	4	4
Earlier, not specified	2	7	0	5	14
Later	0	1	0	0	1
C. Seed characteristics					
Seed size	1	1	0	5	7
Seed colour	0	6	0	0	6
Seed quality (baking, cooking, malting, protein)	4	4	0	4	12
D. Resistance to disease					
Mildew	2	2	0	0	4
Rusts	0	9	1	0	10
Blast (*P. orykae*)	0	0	0	2	2
Others	1	0	0	1	2
E. Other characters					
Adaptability	1	1	1	0	3
Photoperiod-insensitivity	1	0	0	3	4
Improved winter hardiness	2	2	0	2	6
Improved threshability	1	0	0	2	3
Less shattering	0	0	0	1	1
Improved drought tolerance	0	0	0	2	2
Quick grain filling	0	1	0	0	1
Resistance to field sprouting	0	1	0	0	1
F. Increased yield					
Between 3–10%	3	2	1	0	6
Higher than 10%	0	1	0	1	2
Higher but not specified	7	9	4	15	35

Source: Donini et al., 1984

The role of induced mutation in plant breeding

The induced mutants used as commercial cultivars mentioned above indicate that relative success has been achieved in vegetatively propagated ornamental plants which presently take over 50% of the commercially used mutants (Table 17.1). This is to be expected because somatic mutants too may be used with these species. Secondly, the main objective is not the yield but such things as changes in colour and forms which are easier to achieve. With seed propagated species, the number of induced mutants used as cultivars is modest, and only a few of them have succeeded in remaining in commercial production. Moreover, the majority of positive induced mutants with reduced plant height, shorter vegetative period, greater disease resistance, higher protein content, etc., were not competitive due to inferiority in yield performance. This is not surprising since many genes, especially the major ones, e.g., Rht (reducing height) and Ppd (photoperiod), exhibit a pleitropic effect. However, as pointed out by Mac Key (1984), directly induced mutations are in fact raw products not yet exposed to any kind of fitness evaluation.

Therefore, the number of induced mutants used indirectly in cross-breeding should be expected to be considerably larger then the number of mutants used as cultivars. However, the data presented in Table 17.1 show that the former number is almost four times smaller than the latter. Perhaps these data are incomplete or the number of mutant cultivars has been overestimated.

As new genetic variability is essentially based on mutations, spontaneous or induced, their contribution to plant breeding could be enlarged by introducing more systematic approaches to: the selection of material for treatment, the methods of mutant detection and methods of mutant selection and growing.

To facilitate the work on induced mutations, the International Atomic Energy Agency (IAEA) in Vienna, in cooperation with the Food and Agriculture Organization (FAO) in Rome, published in 1969 and 1977 manuals on mutation breeding with instructions on all aspects of induced mutations and their use in plant breeding (Sigurbjornsson, 1977).

Later on comments will be given on the main principles and methods of work on induced mutations which should be adhered to in order to increase the efficiency of the work.

Breeding objectives

A conventional breeding programme based on gene recombinations with long and extensive experiences cannot be expected to be successful unless clear breeding objectives have been defined. The same is true for mutation-breeding programmes because induced mutations occur randomly and they can neither be foreseen nor governed. As it is not realistic to expect mutagen-treated organisms to produce mutations in any given number of genes, especially not in the positive direction from the agronomic point, it is necessary to define specific breeding objectives. It means that a plan should be made on how to improve one or a few traits in a cultivar or line to make it more useful either directly or in a hybridization programme.

Legend to Fig. 17.1

Select parent variety according to breeding objectives
Select mutagen and doses
Treat sufficient number of seeds
Make laboratory test to check % of survival

M_1 Space plant several thousand seeds in the field. Observe changes and select plants individually for M_2. The plants represent cereals which tiller.

M_2 Look for desired mutants and select individually for M_3. The rest of plants may be bulked for a sample to be sown next generation for selection of quantitative characters.

M_3 The same as in M_2.

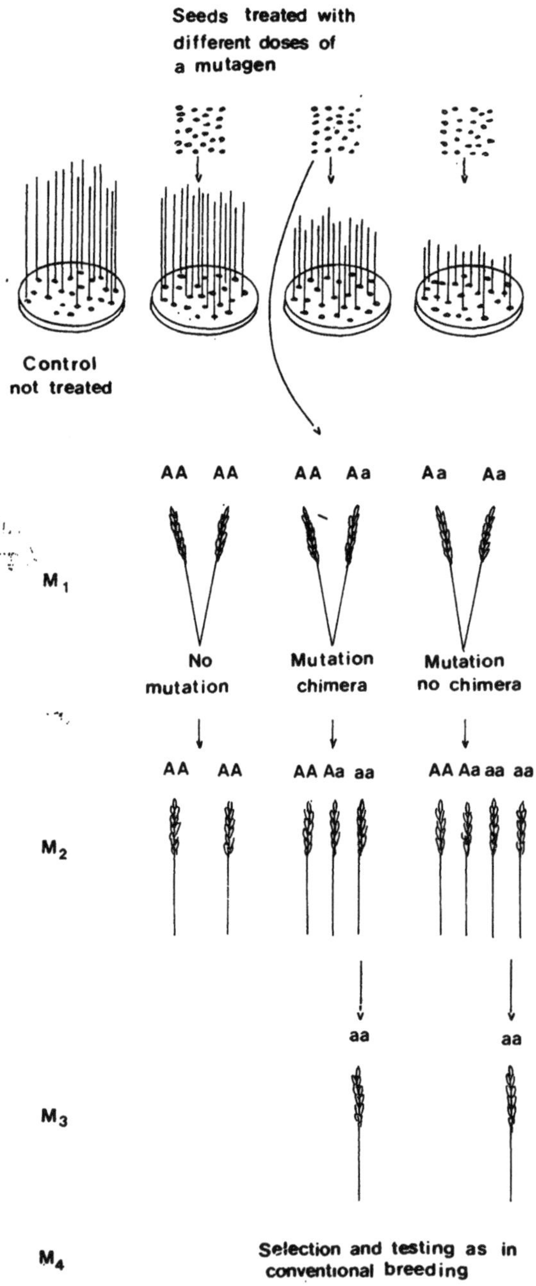

Fig. 17.1 A diagram of induced mutation (*A* to *a*) treating seeds by a mutagen

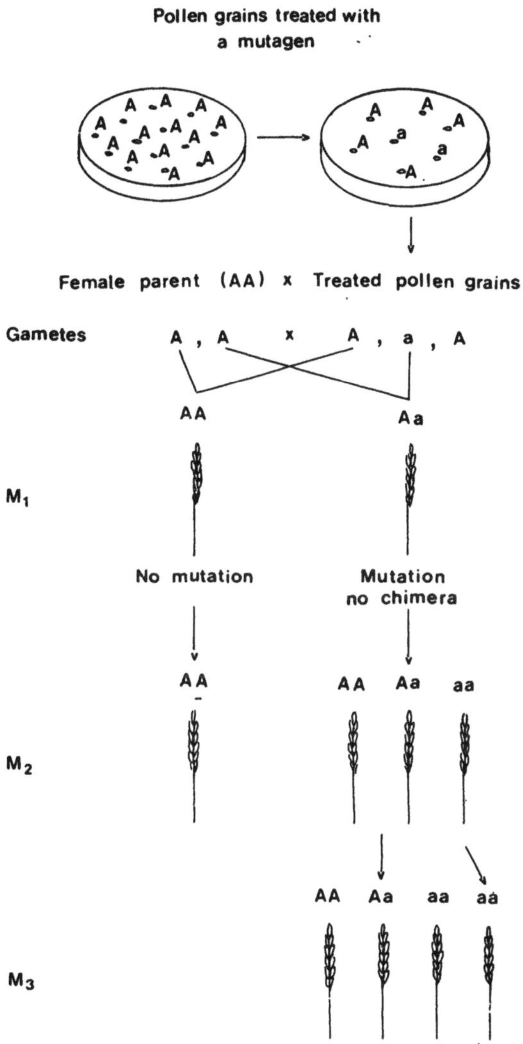

Fig. 17.2 A diagram of induced mutation (*A* to *a*) treating pollen grains with a mutagen

Choice of source material

It ensues from the above that the material to be treated should be good commercial cultivars or advanced lines which need improvement in a single or a few traits, for example, reduced stem height to increase resistance to lodging, resistance to a certain pathogen, removal of a harmful alkaloid, etc.

When choosing materials for treatment, one should bear in mind that differences may exist in the frequency and in the spectrum of mutations among various genotypes within a species and between species. This has been the subject of a number of papers (Gustafsson, 1947; Mac Key, 1954; Konzak, 1959; Scossiroli, 1960; Sparrow, 1961; Borojević K., 1965; Favret, 1965; Brock, 1965, etc.).

To be able to distinguish between the induced variability and the variability already present in the source material, genetically pure seeds, i.e., foundation or inbred line seeds should be treated. Untreated seeds from the same sample should serve as the control in a back-up experiment to be conducted simultaneously with the growing of the M_1 and M_2 generations.

Mutagen doses

It is recommended, especially to beginners, to use three doses (plus controls) of each mutagen, i.e., radiation (*X*-rays, gamma rays) or chemical (ethyl methane-sulphonate, ethylene imine, etc.). The mutagen doses should be about 20% higher or lower from that found by laboratory or greenhouse tests. Three doses should be applied each year to each M_1 to be sure that at least one dose will give a sufficient number of expected mutants.

Since the frequency of induced mutations is about 10^{-3}, to have a high probability of success in recovering mutants, it is necessary to treat several thousand seeds taking into account the percent of survival in the M_1 generation. This will ensure that a sufficiently large population of an M_2 generation is available for the selection of desired mutant(s). A diagram of the main stages in the development of induced mutants is shown in Figures 17.1 and 17.2.

Understandably, there are other technical details which must be observed in order to increase the success in induced mutations.

The general principles and techniques mentioned are also applicable to asexually propagated plants, but considering specific characteristics of individual species, special instructions have been provided in the I.A.E.A. Manual 1977, p. 159–169.

Conclusion

The work on induced mutations has rendered a great contribution to genetic studies and it has become a method of plant breeding. As more had been expected of the method than it could have returned, there were periods of disappointment and slowdown in experiments.

Objectively, mutation breeding can never become an independent plant-breeding method because it focuses on individual genes which then remain to be recombined with other genes to produce acceptable commercial cultivars. Mutation breeding should be considered as a part of the system of recombination breeding which remains as the main method for developing new plant cultivars. In that sense, perhaps it should not be referred to at all as a breeding method because not any institution, will undertake to base a breeding programme on induced mutations alone.

References

Bozzini, A., Bagnara, D., Mosconi, C., Rossi, L., and Scarascia-Mugnozza, G. T., 1973. Trends and results of durum wheat mutation breeding at Casaccia. Proc. Symp. Genet. and Breed. of durum wheat, Bari. p. 339–347.

Borojević, K., 1965. The effect of irradiation and selection after irradiation on the number of kernels per spike in wheat. In: The Use of Induced Mutations in Plant Breeding (Report FAO/IAEA Tech. Meeting, Rome, 1964), Pergamon Press, Oxford, p. 505–513.

Brock, R. D., 1965. Induced mutations affecting quantitative characters. In: The Use of Induced Mutations in Plant Breeding (Report FAO/IAEA Tech. Meeting, Rome 1964), Pergamon Press, Oxford, p. 443–450.

Donini, B., Kawai, T., and Micke, A., 1984. Spectrum of mutant characters utilised in developing improved cultivars. In: Selection in Mutation Breeding. Proc. of a consultants meeting, FAO/IAEA, Vienna, June 21–25, 1982. International Atomic Energy Agency, Vienna, p. 7–31.

Doorenbos, J., 1954. Notes on the history of bulb breeding in the Netherlands. Euphytica 3: 1–11.

Favret, E. A., 1965. Induced mutations in breeding for disease resistance. In: The Use of Induced Mutations in Plant Breeding (Report FAO/IAEA Tech. Meeting, Rome, 1964), Pergamon Press, Oxford, p. 521–536.

Gregory, W. C., 1956. Induction of useful mutation in the peanut. In: Genetics in Plant Breeding, Brookhaven Biol. Symp. 9.

Gustafsson, A., 1947. Mutation in agricultural plants. Hereditas 33.

International Atomic Energy Agency, 1977. Manual on Mutation Breeding. 2nd ed., IAEA, Vienna, 288 pp.

Konzak, C. F., 1959. Radiation induced mutations for stem rust resistance in oats. Agron. J. 51: 518–520.

Mac Key, J., 1954. Neutron and *X*-ray experiment in wheat and a revision of the speltoid problem. Hereditas 40: 65–180.

Mac Key, J., 1984. Selection problems and objectives in mutation breeding. In: Selection in Mutation Breeding. Proc. of a consultants meeting, FAO/IAEA, Vienna, June 21–25, 1982. International Atomic Energy Agency, p. 35–48.

Poehlman, J. M., 1983. Breeding Field Crops. (2nd ed.). AVI Westport, Conn. (USA), 486 pp.

Preil, W., 1986. In vitro propagation and breeding of ornamental plants: Advantages and disadvantages of variability. In: W. Horn, C. J. Jensen, W. Odenbach and O. Schieder (Editors), Genetic Manipulation in Plant Breeding, Walter de Gruyter, Berlin, p. 378–403.

Scarascia-Mugnozza, G. T., 1967. Mutacioni indotte e miglioramento genetico delle piante agrarie. Atti XI. Conv. Soc. Ital. Genetica Agraria 20: 361–375.

Scossiroli, R. E., 1960. Caratteri quantitativi e mutabilitá poligenica. Atti. Ass. Genet. Ital., Padova, 5: 9–38.

Sengbush, von, R., 1942. Süszlupinen und Oll-lupinen. Landwirtschaftliche Jahrbücher, 91: 719–880.

Sigurbjörnsson, B., 1977. Mutations in plant breeding programmes. In: International Atomic Energy Agency, Manual on Mutation Breeding, IAEA, Vienna, p. 1–7.

Sigurbjörnsson, B., 1983. Induced mutations. In: D. R. Wood (Editor), Crop Breeding. American Society of Agronomy, Madison, Wis. p. 153–176.

Sparrow, A. H., 1961. Types of ionizing radiation and their cytogenetic effects. Mutation and Plant Breeding, National Academy of Science, 891: 55–119.

Stanković, D., 1960. Oplemenjivanje voćaka i vinove loze. Zavod ra udźberike Beograd, 211 pp.

Van Harten, A. M., and Broertjes, C., 1986. Mutation breeding: A stepping-stone between Gregor Mendel and genetic manipulation (a treatise for vegetatively propagated crops). In: W. Horn, C. J. Jensen, W. Odenbach and O. Schieder (Editors), Genetic Manipulation in Plant Breeding, Walter de Gruyter, Berlin, p. 3–15.

18. CHROMOSOME ENGINEERING

When it was discovered that the doubling of an entire genome chromosome set (x) resulted in an increase of cell volume and, consequently in an increase of plant parts, especially vegetative ones, methods of induced polyploidy and its practical use were developed.

Polyploids are organisms which have a multiple number of chromosomes in relation to the haploid number. When reffering to polyploids it is important to distinguish between the numbers of chromosomes in somatic cells, i.e., the sporophyte (the $2n$ number) and the basic number of chromosomes that make up the genome (the x number). In bread wheat *Triticum aestivum*, for example, the chromosome number is 42 ($2n = 42$) but the species is hexaploid having six sets of chromosomes with the basic number 7 ($x = 7$) in each A, B and D genome which makes $2n = 6x = 42$. In the diploid species the values of n and x are identical ($n = x$). Both natural and artificial polyploids develop in two ways:

(1) by increasing their own chromosome number from $2x$ to $3x$, $4x$, $8x$, etc.; such organisms are called *autopolyploids*;

(2) by increasing the chromosome number by hybridization between two different species; such organisms are called *allopolyploids*.

Natural polyploids will not be discussed here because they are already used in plant breeding. This chapter will be concerned with methods of induced polyploidy, i.e., chromosome manipulation or chromosome engineering, and with the use of various polyploids in genetic studies and plant breeding.

It has been found that the treatment of the vegetative apex of plants with colchicine ($C_{22}H_{25}O_6$) and some other chemicals leads to the doubling of the chromosome number in somatic cells. Colchicine prevents the formation of the spindle fiber at mitosis and the chromosomes, in consequence to that, do not move to the equatorial plane but disperse within the cell. After the division of the centromeres, the chromosomes divide but each set does not move to the opposite poles and all chromosomes remain in one cell. Such a cell has the doubled number of chromosomes ($2n = 4x$) instead of the diploid number ($2n = 2x$) and in further mitotic division it produces filial cells with $2n = 4x$ chromosomes.

The techniques of colchicine treatment of plant parts is fairly simple. The vegetative apex of monocotyledons is treated with a 0.25% solution of colchicine for half an hour to one hour; in the case of dicotyledons, a young bud is placed on agar containing a 1–2% solution of colchicine and kept there for several hours. After that, the treated parts are planted in sterile pots. Considering the differences between species in their stages of development, the method of induced polyploidy is specific to each species and the reader who needs details may refer to the manuals written by Eigsti and Dustin (1957), Enken (1966), and several others.

The main method of producing allopolyploids is hybridization between different species. This does not render polyploids directly and it is necessary to apply the colchicine treatment to double the chromosome number, or to conduct back crossings with the parent possessing the larger chromosome number, or to introduce a third variety to serve as a bridge.

Success in polyploidy induction varies from one species to another but better results are achieved:

- with species possessing a small chromosome number than with species possessing a large chromosome number,
- with allogamous plants than with autogamous plants,
- with species grown for forage than with species grown for seed production.

Autotetraploids

Natural autotetraploids are formed by the union of gametes with non-reduced chromosome numbers ($n = 2x$). Artificial autotetraploids are made by the colchicine treatment. Compared with diploids, autotetraploids have larger plant parts, which is due to the larger number of chromosomes and thus the larger cell volume.

Tetraploid rye ($2n = 4x = 28$). The first variety of tetraploid rye was made by Dorsey (1936) who used the method of high temperature shock. Muntzing (1941) produced several tetraploid rye varieties by the colchicine treatment. These varieties had thicker stems, wider leaf blades, longer spikes, and their seeds were almost two times heavier than the seeds of diploid rye (Fig. 18.1). The change in seed size brought about a change in the ratio between carbohydrates and proteins so that tetraploid rye has about 2% more proteins and better baking properties than does diploid rye. A serious shortcoming of tetraploid rye is a high percentage of sterility per spike, 20% higher than that in diploid rye when grown in field conditions. Consequently, the number of seeds is smaller and the yield of tetraploid rye does not exceed that of diploid rye. The sterility in tetraploid rye results from the formation of trivalents and univalents, as well as the lack of the necessary balance between the chromosomes and the cytoplasm (Muntzing, 1951).

Because of fairly high sterility, tetraploid rye has not been widely accepted by growers. It is necessary to perpetually keep selecting the best spikes and keep developing new varieties of tetraploid rye to reduce the sterility rate. Another disadvantage of tetraploid rye is that it ought not be grown in the vicinity of a plot with diploid rye because completely sterile triploid plants result from mutual pollination. Tetraploid rye is very good for green forage because it develops large vegetative parts.

Tetraploid red clover ($2n = 4x = 28$). The development of polyploids is more profitable with crops grown for forage. Tetraploid red clover (Table 18.1), vetch, and some other forage crops have been commercially grown for years.

The disadvantage of tetraploid red clover, like some other polyploid forage crops, is a lower seed set than in corresponding diploids (Dewey, 1980). Besides some irre-

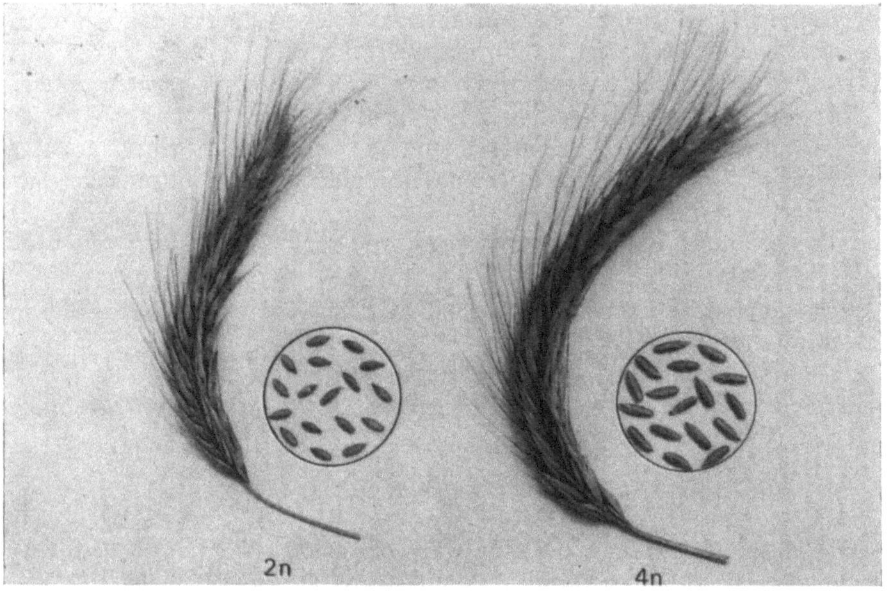

Fig. 18.1 Spike and grains of diploid $2n = 2x = 14$ rye (left) and tetraploid rye $2n = 4x = 28$ of the cultivar Petkus grown in Novi Sad

Table 18.1 Green bulk yield of diploid and tetraploid varieties of red clover from 30 trials in Sweden

Green bulk t/ha	Varieties			Ulva in % when Ultuna 100%
	2x Ultuna	2x Merkur	4x Ulva	
First year	26.0	31.2	30.4	117
Second year	20.2	23.9	33.3	165
Second year as % of the first year	78	77	110	

Source: Hagberg, 1962

gularities in meiosis, one of the reasons is that tetraploid plants have larger but fewer flowers than diploid plants, making it more difficult for honeybees and bumblebees to reach pollen and so pollination tends to be less efficient. Therefore, selection could go in two directions – to select tetraploids with somewhat smaller flowers or to select bees with a longer proboscis adapted to reach more pollen.

Tetraploid ryegrass. Several varieties of autotetraploid perennial ryegrass (*Lolium perenne*) and Italian ryegrass (*L. multiflorum*) have been developed in the Netherlands. The tetraploid varieties grown in commercial production have a 72–84% larger thousand grain weight than do the diploid varieties, 8–18% larger yield of green bulk, but approximately the same yield of dry matter. Furthermore, tetraploid varieties do not form as dense a canopy and they are not as resistant to trampling as are the diploid varieties, although animals prefer to graze tetraploid varieties (Bogaert, 1975).

Because of considerable genetic changes and the disbalanced ratio of the nucleus and the cytoplasm, the cycle of mitosis in tetraploids is retarded and they mature later than diploids. Also, the number of mitotic divisions may be reduced so that tetraploids frequently do not have an outstanding growth rate in spite of their enlarged plant parts.

The common shortcoming of artificial tetraploids is partial sterility which occurs in spite of an even number of chromosomes. In autotetraploids there are four homologous chromosomes (diploids have two of them) which form a quadrivalent in the pachytene stage of meiosis. If in anaphase I, one pair of chromosomes moves to one pole and the other pair to the other pole, the sexual cells contain the same number of chromosomes and they are capable of fertilization. However, it may so happen that some chromosomes do not form the quadrivalents but trivalents and univalents. In most induced autotetraploids, the percentage of chromosomes associated in quadrivalents ranges only between 30 and 60% (Cauderon, 1986). The result is the formation of gametes with an uneven number of chromosomes and those which do not have all chromosomes are not capable of normal fertilization. If a gene is present in four doses instead of two doses, it may interact differently with other genes including genes located in chloroplasts and mitochondria (Morris, 1983). This is the reason for the occurrence of partial sterility which lowers the practical value of some tetraploids, excepting those which may be vegetatively propagated or those which are not grown for seed.

The induction of autotetraploids has been practiced for more than half of the century but it can hardly be said that it has produced significant results. Only a few such crops have managed to establish a firm position in commercial production although attempts have been made with a large number of species. Obviously, the change from the diploid to the tetraploid state requires a much longer period for biochemical and genetical balance to be achieved within the cells and plant parts as well as for adaptation to the environment. Moreover, it is necessary to enlarge the genetic base for inducing autotetraploids upon which selection pressure may act.

Autotriploids

Organisms in which each complete genome i.e., each chromosome is present three times, are called triploids ($3x$). They are formed when an ovule with a non-reduced chromosome number ($n = 2x$) is fertilized by pollen with a normal reduced chromosome number $n = x$. Nowadays, when several tetraploids are available, triploids are made by crossing tetraploids and diploids.

Triploid organisms develop more vigorously than do their diploids equivalents. In some species, they are better than the tetraploids and are commercially used, e.g., sugar beet. Because of the odd number of chromosomes, however, the meiosis of triploids is absolutely irregular and they are completely sterile. With some crops such as grapes, water melons, etc., sterility is an advantage which greatly promotes the commercial use of triploids as they have no seed. In the case of species which are vegetatively reproduced, such as grasses, fruit trees and decorative plants, commercial triploids are quite frequent.

Sugar beet ($2n = 2x = 18$, $3x = 27$, $2n = 4x = 36$). The sugar beet which is presently produced and used is a plant which may freely be called a scientific product.

In the 18th century, *Beta vulgaris* contained only 5–7% sugar. In the first half of the 19th century, Achard and von Koppys increased the percentage of sugar to 9% by the method of mass selection. In the second half of the 19th century, Vilmorin moved the percentage of sugar up to 11% by the method of family selection. Nowadays, the content of sugar exceeds 17% (cit. lit. Hendriksen, 1979).

A further step forward was made in 1934 when Soviet scientists Bordonos and Savitsky produced monogerm sugar beet seed (one seed in a cluster instead of several as in multigerm seed) which rendered the practice of sugar beet thinning unnecessary and allowed for precision sowing.

Fig. 18.2 Roots of diploid $2n = 2x = 18$, (left), triploid $3x = 27$, and tetraploid sugar beet $2n = 4x = 36$ (right). (Courtesy of Prof. P. Dokić, Faculty of Agriculture, Novi Sad)

The next step was the development of tetraploid sugar beet by the colchicine treatment (Schwanitz, 1938) and the proof (Peto and Boyes, 1949) that triploids are more productive than tetraploids (cit. lit. Hendricksen, 1979). Triploid sugar beets develop a larger leaf area which promotes root development and increases the yield of roots by 20 to 30% and the content of sugar by 1–2% in relation to diploid sugar beets

(Fig. 18.2). In that way, the triploid sugar beet varieties, at first multigerm and later monogerm, entered into commercial production on a large scale, especially in Europe while in the U.S.A. diploid sugar beet varieties remained predominant.

As the triploid plants are completely sterile, the production of triploid seed is ensured by alternate sowing of diploid and tetraploid plants or by mixing their seeds. Since the pollen of the diploid plants is more fertile than that of the tetraploid plants, the population consists of more triploid seeds (hybrid seeds) than diploid and tetraploid seeds which are also present.

It was necessary to discover the phenomenon of male sterility, which was first found by Owen in 1942, in order to fully exploit the advantages of triploid sugar beets.

These achievements made room for the present large-scale production of triploid monogerm sugar beet hybrids which is a further advance in the production of sugar.

The following components are needed for the production of triploid hybrids:

(1) a diploid cytoplasmically male sterile variety – *diploid cms*, (*S*) *xxzz*;
(2) a male sterility maintainer – *0 type diploid*, (*N*) *xxzz*;
(3) a tetraploid pollinator – *tetraploid.*

Cytoplasmic male sterility has been found in *Beta vulgaris*, *B. maritima*, and other species so that presently a fair number of cms sources is used in breeding triploid sugar beets. It is assumed that the male sterility is controlled by the cytoplasmic factor (*S*) and the normal fertility by the cytoplasmic factor (*N*) and at least two recessive genes, *x* and *z*.

The O-type for each cms line is found by crossing a cms plant (*S*)*xxzz* with a normal plant. If the progeny of the cms plant is male sterile, it may be considered to be a male line with the formula (*N*)*xxzz*.

The tetraploid parent is obtained by treating germinated diploid seeds with a 0.2% solution of colchicine for 15 hours at the temperature of about 30 °C.

To produce superior commercial triploid cultivars, it is necessary, before hybrid production, to perform selection for reliable male sterility, monogermity (single-seeded fruit) and seed yield at the diploid level. At the tetraploid level, it is necessary to ascertain that all plants are truly tetraploid and prolific in the production of fertile pollen. Understandably, selection for increased yields of roots and sugar, disease resistance, etc. must be continuously pursued. For technical details on triploid sugar beet production, the reader may refer to the publication "Plant Breeding Perspectives" edited by J. Sneep and A. J. T. Hendriksen (1979) which gives a diagram of the production of the tetraploid pollinator, the male sterile parent, and the triploid sugar beet hybrid (p. 211–215).

Seedless watermelons ($3x = 33$). A Japanese geneticist Kihara (1951) made seedless watermelons by crossing diploid ($2n = 2x = 22$) and tetraploid ($2n = 4x = 44$) watermelons. Because of the completely irregular meiosis, triploid plants ($3x = 33$) do not form seeds which, in this particular case, is an advantage (Fig. 18.3).

Although production of triploid watermelons is expensive, because the crossing of tetraploid × diploids must be performed manually each year, it may be profitable if the market is sufficiently large.

Fig. 18.3 Seeded diploid watermelon $2n = 2x = 22$, (top corners) and seedless triploid watermelon $3x = 33$ obtained from Micado Seed Growers, Co., Chiba City, Japan and grown in Novi Sad

Triploid and tetraploid grapes ($3x = 57$, $2n = 4x = 76$). Triploid and tetraploid varieties of the grape have also been made by means of colchicine. Tetraploid grapes have large berries and a reduced number of seeds, which is a commercial advantage. Their shortcomings are sparse bunches and a reduced yield. Triploid grapes are completely seedless which makes them ideal table cultivars.

Allopolyploids

Allopolyploids are organisms arising as a consequence of duplicating different genomes after hybridization of two species or genera. The duplication occurs after the fusion of F_1 gametes possessing the non-reduced chromosome number. A safer method of obtaining allopolyploids is the treatment of the F_1 generation with colchicine. Furthermore, allotetraploids or amphidiploids may be produced by backcrossing the F_1 generation with the parent possessing the higher chromosome number. The production of allopolyploids will be illustrated by the development of a man made synthetic species, *Triticale.*

The cross of wheat and rye produces F_1 plants with 28 chromosomes whose traits are mostly intermediate. But, rye dominates in some traits (hairy neck, spike length), wheat also in some (stem diameter). The F_1 plants are completely sterile because all chromosomes are univalents and the gametogenesis is irregular.

♀ *Triticum aestivum* subsp. *vulgare* × ♂ *Secale cereale*
(bread wheat) (rye)

genomes A B D R
↓ ↓ ↓ ↓
chromosomes 7′ 7′ 7′ 7′
↙
F_1 7′ + 7′ + 7′ + 7′ = 21′ (wheat) + 7′ (rye)

♀ F_1 (21′ + 7′) × ♂ 21′ (wheat′)
↓
BC_1F_1 21″ (wheat) + 7′ (rye)

gametes ♀ ♂
21 + 1 21 + 3
21 + 6 21 + 4
21 + 2 21 + 7
↙
×
↗
21 + 7 21 + 0
etc etx
↓
21″ (wheat) + 7″ (rye) = 56 chromosomes
octoploid Triticale

′ = univalents
″ = bivalents

When the F_1 plants are backcrossed to wheat, the resulting progenies possess 42 chromosomes derived from wheat and 7 chromosomes derived from rye. As the wheat chromosomes form bivalents at meiosis, the univalents from rye do not exhibit a strong negative effect on the meiotic process and the plants of the first backcrossing are partially fertile. Different gametes develop in the course of gametogenesis of the plants of the first backcrossing. The wheat chromosomes, being bivalents, are arranged regularly at the poles, 21 chromosomes at each, but the rye chromosomes being univalent, are arranged irregularly. If an ovule possessing 21 + 7 chromosomes is fertilized by a pollen grain possessing also 21 + 7 chromosomes, the plant subsequently developed will have all wheat and all rye chromosomes, i.e., 56 chromosomes in total, and this is a new synthetic species called *Triticale*.

Octoploid triticale (genomes – *AABBDDRR*, $2n = 8x = 56$). Wishing to combine the positive traits of common wheat with the positive traits of rye, Müntzing (1939) in Sweden, Pisarev (1966) in the Soviet Union, and others hybridized vulgare wheat and rye. Fertile F_1 hybrids have been obtained in several instances, resulting from the mating of two sexual cells possessing the non-reduced chromosome number, or more frequently, by the colchicine treatment which produces amphidiploids at a higher frequency.

This hybrid has been expected to combine the adaptability to an adverse environment coming from rye with the yield and quality coming from wheat. However, the result fell short of expectation. In Sweden, the octoploid *Triticale* yields less than wheat in the wheat-growing regions and less than rye in the rye-growing regions. Attempts at further improvement of the new species are under way (Müntzing, 1979).

Hexaploid Triticale (*AABBRR*, $2n = 6x = 42$). In the last couple of decades, large efforts have been invested in breeding hexaploid *Triticale* by crossing turgidum wheats ($2n = 4x = 28$) with rye ($2n = 2x = 14$) and treating F_1 plants with colchicine. The first encouraging results were obtained by Jenkins and Larter in Canada, Kis in Hungary, Zillinsky in Mexico, Pisarev and Shulyindin in the Soviet Union, Sanchez-Monge in Spain, Wolski in Poland, etc. (An extensive literature on the development of *Triticale* has been supplied by Rigin and Orlova, 1977, Müntzing, 1979, and Wolski, 1988).

The hexaploid *Triticale* has a long and attractive spike with more than 30 spikelets (Fig. 18.4), a high protein content (15–17%), and satisfactory baking properties. However, the endosperm of first varieties was underdeveloped (two doses of wheat chromosomes and one dose of rye chromosomes), and the yield was not quite satisfactory. Thus, the hexaploid *Triticale* varieties have not been very popular among growers.

Fig. 18.4 Spikes of tetraploid wheat – $2n = 4x = 28$ (left), hexaploid triticale – $2n = 6x = 42$ (middle) and diploid rye – $2n = 14$ (right) grown in Novi Sad

As a result of an increased number of breeders working on the hexaploid *Triticale*, the introduction of new turgidum wheats and rye varieties into the crossing programmes and the intensive selection for improved quality, new cultivars have been made which compete more successfully with wheat in yielding capacity. The cultivars Mapache, Beagle, Rahum, etc., developed at the CIMMYT in Mexico, produce high yields (Zillinsky, 1974). The hexaploid *Triticale* cultivar Lasko, developed by T. Wolski in Poland, took the first place by yield compared with wheat cultivars in the 16th International Winter Wheat Performance Nursery for 1984, giving 5.75 t/ha, while the most productive wheats were 9D-27-262 from Rumania and NS 2704 from Yugoslavia averaging 5.69 and 5.66 t/ha, respectively.

The cultivar Lasko was released in 1982, but it was not recommended for cultivation in Poland due to its insufficient winter-hardiness (Wolski and Tyminiecka, 1983). However, according to Varughese et al. (1987), Lasko is the widest grown triticale variety in the world. It is used as a standard variety in the International Eucarpia Triticale Yield Nursery. Recently, Polish breeders have developed several triticale varieties with good winter-hardiness, among them Grado, Dagro, Bolero, and Malno (Wolski, 1988). By improving the grain filling and increasing the yield level of triticale, the protein content became lower but it is still on the level of wheat varieties. However, the lysine content is higher than in wheat and close to that in rye.

As a result of breeding this new man-made crop in various institutes around the world, several new spring and winter cultivars have been developed. The area under triticale has increased considerably to exceed 1,200,000 hectares by 1986 (Varughese et al., 1987). Among the countries growing more than 300,000 hectares per year are Poland, France, the U.S.S.R., Australia, and the U.S.A. According to an estimate of Wolski (1988), the winter triticale will occupy about one million hectares in Poland by 1990.

Hybrids between various species and genera offer an avenue to further advances in genetic potentials for yield. For example, 42-chromosome hybrids have been made between *Elymus*, which possesses over 100 spikelets per spike (Fig. 18.5), and wheat by a Soviet scientist N. V. Cicin. Mujeeb-Kasi et al. (1983) reported on the most recent attempts to incorporate desired genes from *Elymus* species into *Triticum aestivum*. Modern methods of chromosome and genetic engineering may allow new breakthroughs in that field.

Hybrid ryegrass. The hybrid ryegrass produced by crossing *Lolium perenne* and *L. italicum* (Breese et al., 1975) is considered a successful synthetic amphidiploid. It has a rapid initial growth and a high nutritional value, inherited from the Italian ryegrass, as well as high resistance and fertility, inherited from the perennial ryegrass.

Rapeseed (*Brassica napus*, $2n = 4x = 38$) is also a synthetic made by crossing *B. campestris* ($2n = 2x = 20$) and *B. oleracea* ($2n = 2x = 18$). *Brassica napus* has developed naturally in the same way but the crosses between natural and synthetic allotetraploids resulted in an increased genetic variability which is used for breeding new rapeseed cultivars (Olsson, 1963).

Besides the allopolyploids mentioned in the previous paragraphs, several others have also been developed but they are used either as experimental materials or they have only a theoretical significance for genetic studies.

Allotetraploids are usually not as vigorous as autotetraploids. They undergo much larger changes which reduce to an extent the effect of the chromosome number on the development of various plant parts.

It has also turned out that the production of polyploids is more successful in crops that reproduce by cross-pollination than by self-pollination. In a self-pollinated crop a high proportion of genes is in a homozygous state, while a cross-pollinated crop is in a heterozygous state which provides a much larger genetic base from which to select better adapted genotypes.

An increase of an entire set of chromosomes (i.e., a genome) brings positive changes only to a certain limit. An excess of chromosomes may produce a depressive effect, depending on the species in question. While in strawberries octoploids and decaploids are more luxuriant than diploids, plants at high ploidy levels have the same cell volume as diploids in many other species. Natural wild species above octo-

Fig. 18.5 The increased number of spikelets per spike and number of grains per spikelet may offer a chance for further increasing genetic yield potential. From left to right: *Triticum aestivum* subsp. *vulgare*, *Secale cereale*, Triticale and *Elymus*, grown in Novi Sad

ploid level are quite rare although a 20-ploid *Gallium grande* and a 38-ploid *Poa litiroza* have been found (Brown, 1972).

Auto- or allopolyploids may still be induced in a large number of forage, medicinal and horticultural plants. The presently used methods of polyploidy induction must be further improved in order to intensify the use of this phenomenon in plant breeding.

Aneuploids

Aneuploids are organisms in which individual chromosomes are increased or reduced in number as compared with disomics ($2n$). The designation for most commonly used aneuploids is as follows:

monosomic: $2n - 1$, organism which lacks one chromosome,
nullisomic: $2n - 2$, lacks both chromosomes of one homologous pair,
trisomic: $2n + 1$, has one chromosome present three times,
tetrasomic: $2n + 2$, has one chromosome present four times.

There are combinations of monosomic for one chromosome and trisomic for another chromosome, etc. Aneuploids cannot be used directly in production. However, when series of monosomics and trisomics are made for each chromosome, they may be very useful for substitution or addition of individual chromosomes in already developed cultivars.

Series of monosomics and trisomics have been developed for several agricultural crops (Table 18.2). Detailed analyses and literature references on the development, characteristics, and use of aneuploids have been supplied by Sears (1954, 1972), Khush (1973), and Riley and Law (1984).

Monosomics and trisomics reveal the role of certain chromosomes in the expression of traits and the development of an organism as a whole. Monosomics allow breeders to localize the position of genes for certain traits on chromosomes. If one wishes to establish which chromosome in a variety carries a gene or genes for resistance to leaf rust, he will cross it with another variety which has the series of monosomics, i.e., he will cross it with every monosomic in the series. If the resistance is dominant and controlled by a single gene, 75% resistant and 25% susceptible plants may be expected in the F_2 generation of the crosses with each monosomic except one. The cross which gives progeny with all resistant plants indicates that the resistance gene is located on the chromosome for which this plant is the monosomic. In general the crossing pattern may be represented as follows:

Location of chromosome carrying a dominant gene:

a) a disomic parent possessing a dominant gene is crossed with the monosomic series of another parent;

b) in the F_1 generation, both disomic and monosomic plants show a dominant character and the critical chromosome cannot be identified;

c) in the F_2 generation, the disomic plants segregate into dominant and recessive ones at the ratio 3 to 1, respectively. The progeny that does not segregate is monosomic and it possesses the chromosome which carries the dominant gene.

If a recessive gene is intended to be located, the chromosome carrying the gene may be determined already in the F_1 generation:

a) a disomic parent possessing a recessive gene is crossed with the monosomic series of another parent;
b) in the F_1 generation, the plant with the recessive character is monosomic and it possesses the chromosome which carries the recessive gene.

Table 18.2 Complete sets of monosomic and trisomic lines are available in the following crop species

Wheat	(*T. aestivum*)	$2n = 42$, with 21 monosomic lines
Oats	(*Avena sativa*)	$2n = 42$, with 21 monosomic lines
Tobacco	(*Nicotiana tabacum*)	$2n = 48$, with 24 monosomic lines
Tomato	(*Lycopersicon esculentum*)	$2n = 24$, with 21 trisomic lines
Spinach	(*Spinacia oleracea*)	$2n = 12$, with 6 trisomic lines
Chilli	(*Capsicum annuum*)	$2n = 24$, with 12 trisomic lines
Rye	(*Secale cereale*)	$2n = 14$, with 7 trisomic lines
Rice	(*Oryza sativa*)	$2n = 24$, with 12 trisomic lines
Sorghum	(*Sorghum vulgare*)	$2n = 20$, with 10 trisomic lines
Barley	(*Hordeum vulgare*)	$2n = 14$, with 7 trisomic lines

Source: Kush, 1973, Riley and Law, 1984

Numerous genes on individual chromosomes have been located by monosomic series. Abundant literature on the subject has been collected by Morris (1973).

For example, it has been found that the Lr9 gene for resistance to leaf rust of wheat is located on the long arm of chromosome 6B (designated as $6B^L$), the Hg gene for hairy awns on the short arm of chromosome $1A^S$ (Sears, 1954, 1966; Tsunewaki et al., 1966), etc.

Chromosome substitution and addition

The development of monosomic series enabled the substitution of individual chromosomes in one variety to be made by chromosomes from another variety. If two varieties are crossed, chances are small that a progeny will be found possessing all the chromosomes of one parent and only one chromosome from the other parent whose trait is intended to be transferred. This was made possible by the development of monosomic series. These new and exciting possibilites will be illustrated by a few examples.

Sears (1956) decided to try to transfer the resistance to leaf rust of *Aegilops umbellulata* to the bread wheat. As these two species do not cross, he used *Triticum dicoccoides* as a bridge:

♀ *T. dicoccoides AABB* × ♂ *A. umbellulata UU*

↙

F_1 × bread wheat *AA BB DD* (*T. aestivum*)

↙

BC_1F_1 × bread wheat

↙

BC_2F_2 × bread wheat

↓

BC_2F_3

One plant with 21″ (wheat) + 1′ (*Aeg.*)
(The symbol ″ stand for bivalents and ′ for univalents)

This crossing scheme produced one plant resistant to leaf rust that possessed 42 wheat chromosomes and one bivalent or two *Aegilops* chromosomes. This addition, however, showed some drawbacks, e.g., a brittle spike axis, sterile pollen, etc. It was thus necessary to separate that part of the *Aegilops* chromosome which carried the rust resistance from the part that carried the negative traits. The progenies of the original plant were exposed to *X*-rays and irradiated pollen was used for pollination. Out of 6,091 plants produced, 40 of them had the desired part of the *Aegilops* chromosome translocated to chromosome *6B* of the bread wheat. The other part of the *Aegilops* chromosome disappeared as a fragment and the negative traits disappeared together with it. Sears called the new variety *Transfer*, which for many years served as a source of resistance to leaf and stem rust.

A more recent example is the transfer of eye-spot (*Pseudocercosporella herpotrichoides*) resistance from *Aegilops ventricosa* to *Triticum aestivum* by means of *T. turgidum* subsp. *carthlicum* as a bridge (Doussinault et al., 1983, Cauderon 1986):

♀ *Aegilops ventricosa* × ♂ *Triticum turgidum carthlicum* (*Tr. persicum*)
DD M^vM^v — *AA BB*
$2n = 28$ — $2n = 28$

↙

♀ amphidiploid F_1 × ♂ *T. aestivum* cv Marne
AA BB DD M^vM^v — *AA BB DD*
$2n = 56$ — $2n = 42$

↙

♀ *AA BB DD* M^v F_1 × ♂ cv Marne
$2n = 49$

↙

♀ F_1 × ♂ cv Marne

↓

selfing and selection

↙

♀ strain *VPM*1 × ♂ cv Moisson
AA BB DD — *AA BB DD*
$2n = 42$ — $2n = 42$

↓

Tr. aestivum subsp. *vulgare*, cv Roason

The cultivar Roason has gene(s) for resistance from *A. ventricosa* on chromosome *7D* and it also contains the *5B/7B* translocation.

The addition of chromosomes from one species to another has been applied in several instances. The premier cases were the development of addition lines by incorporating a rye chromosome into a wheat chromosome set, carried out by O'Mara (1940, 1951), Evans and Jenkins (1960). Wheat and rye are first crossed, the F_1 is treated with colchicine to double the chromosome number and produce fertile plants ($21''w + 7''r$) which are then back crossed to wheat. In subsequent generations, selection is applied for individual rye chromosomes until the disomic addition ($21''w + 1''r$) is obtained for each rye chromosome pair.

The addition lines of rye, oats, wheat and tobacco that have been made so far have not rendered suitable commercial varieties because genotypes with one or more pairs of chromosomes added from another species are unstable and the chromosome is frequently lost.

Inter-varietal and inter-specific chromosome substitutions are of much greater significance in plant breeding then is the addition of chromosomes.

If a variety possessing a monosomic series may serve as a recipient variety, and it is intended to incorporate into it the resistance to leaf rust (e.g., Lr 9), a donor variety which possesses the gene (e.g., Transfer) should be chosen and crossed with the recipient. In the F_1 generation, plants resistant to leaf rust are selected and crossed again with the recipient variety. After 4 to 6 back-crosses and selfing, disomic plants are obtained which have most chromosome pairs from the recipient variety and one pair of chromosomes from the donor variety carrying the gene for resistance to rust.

So far, the intervarietal chromosome substitutions involved the variety Chinese Spring because Sears (1954) developed a complete monosomic series and supplied seeds to all interested breeders. As Chinese Spring has no commercial value, its substitutions lines were more important for genetic and theoretical studies of the role of individual chromosomes in controlling certain traits.

At the Plant Breeding Institute in Cambridge, Dr. Law and his associates have used as the recipient or recurrent variety a good commercial variety Capelle Desprez and more recently a semi-dwarf variety Hobbit "Sib" (Law et al., 1987). Substitution lines of Capelle Desprez have been used to study the genetics of vernalization requirements, day-length response, plant height, disease resistance, baking quality, and yield components. One of these studies is demonstrated in Figure 18.6 for substitution lines of the variety Mara into Capelle Desprez which clearly shows that regarding plant height, the chromosome *2D* and translocated chromosomes $5B^S$ — $7B^S$ are responsible for the dwarfism of Mara as compared with Capelle Desprez. Many valuable substitution lines have been produced but, interestingly enough, in regard to yield no one proved to be superior to the control variety. The same has been experienced in transferring particular genes by the conventional back crossing procedure where these genes very rarely contributed to yield increase in the recurrent variety (Fig. 18.6).

The above examples illustrate intervarietal substitutions, i.e., substitutions between two varieties of the same species. However, it is possible to carry out chromosome

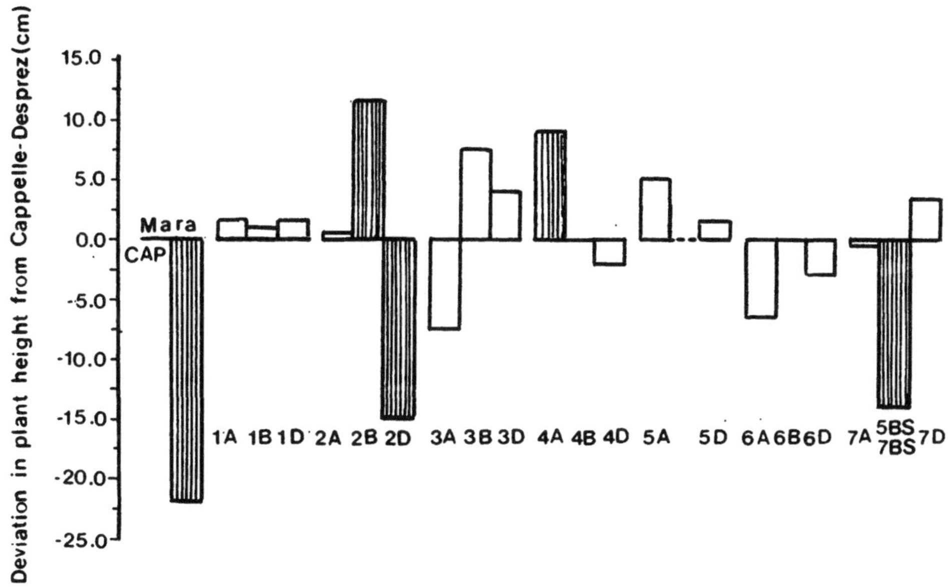

Fig. 18.6 Deviations in final plant height from the height of the variety Cappelle Desprez for each of the substitution lines of Mara into Cappelle Desprez and for the donor variety Mara. Final plant height of Cappelle Desprez is 103 cm.
Source: Law et al., 1987

substitutions between different species and even different genera. In wheat, progress in that respect has been made possible by the findings of Riley and Chapman (1958), Riley (1968) and Okamoto (1957) that the pairing of homologous i.e. homeologous chromosomes is controlled by a dominant gene (Ph) located on the long arm of the chromosome *5B*. Sears (1976) discovered a mutant Ph gene which allows the pairing of non-homologous chromosomes. Therefore, if the action of the chromosome *5B* is prevented, or if the plant is a nullisomic for the long arm of *5B*, the pairing of homeologous chromosomes will take place and a substitution will be possible. Riley (1968) transferred the resistance to *Puccinia striiformis* from *Aegilops comosa* to the vulgare wheat in the following way:

♀ *T. aestivum* subsp. *vulgare* × ♂ *Aegilops comosa*
(Chinese Spring)
AA BB DD *MM*

↙

F_1 × Chinese Spring

↓

several back crossings

↓

a plant with 21″*w*. + 1″ (2*M* from *Aeg.*),
resistant to *Pucc. striiformis*

The resulting plant had 42 vulgare wheat chromosomes and one bivalent or pair – the *2M* chromosome – from *Aegilops*. As the addition line was unstable and the *Aegilops* chromosome would have been lost in further generations, it was crossed with *Aegilops speltoides* which supressed the action of the chromosome *5B*:

♀ the plant with 21″ + 1″(*2M*) × ♂ *A. speltoides*,

several back crossings to Chinese Spring and selection for resistance to the rust gives a vulgare line with 20″ + *2MD* resistant to *P. striiformis*.

The product of this hybridization programme was the variety Compair which has 20 bivalents from vulgare wheat and the chromosome *2M* from *A. comosa* which pairs with the chromosome *2D* from Chinese Spring. Compair has regular meiosis and may serve for the transfer of resistance to *P. striiformis* without encountering difficulties regularly imposed by interspecific hybridization.

Law et al. (1983) achieved a significant increase in protein content by substituting the chromosome *2M* from *Aegilops comosa* for either chromosome *2A*, *2B*, or *2D* in vulgare wheat (Table 18.3).

Table 18.3 Mean yields, percentage proteins and yield components of lines in which chromosome *2M* of *Aegilops comosa* has been substitutet for either chromosome *2A*, *2B*, or *2D* in the wheat variety Chinese Spring

Genotypes	Yield per plant	% protein (= $N \times 5.7$)	250 Grain weight	Grain Number per Ear	Tiller Number
Chinese Spring *2A/2M*	5.0	18.92***	7.2**	21***	9.9***
Chinese Spring *2B/2M*	6.2	18.81***	7.5*	24***	8.8**
Chinese Spring *2D/2M*	6.9	21.09***	8.4***	22***	9.3**
Chinese Spring	7.3	16.82	6.9	36	7.5

Significantly different from Chinese Spring

* *P* 0.05–0.01
** *P* 0.01–0.001
*** $P < 0.001$

Source: Law et al. 1983

The best known chromosome substitutions are those between wheat and rye of which some produced wheat varieties of high commercial value. Zeller (1973), Zeller and Hsam (1983), Rajaram et al. (1983) provided detailed reviews of wheat-rye substitutions and translocations.

The most successful substitutions and translocations are those of the chromosome *1B* of wheat and the chromosome *1R* of rye. A wheat variety "Zorba" was developed at Weihenstephan (FRG) which has a complete rye chromosome *1R* pair substituting for the wheat chromosome *1B* pair. This substitution proved to be resistant against some races of stem rust (Sr 31), leaf rust (Lr 26), stripe rust (Yr 9), and powdery mildew (Pm 8). As a result of crossing to normal wheat varieties, the short arm of rye chromosome *1R* has been translocated to wheat chromosome *1B*, forming the *1BL/1RS* translocation but still carrying the rye genes for disease resistance.

At the same time at Salzmunde (GDR) a wheat cultivar Salzmunder Bartweizen was developed with *1B/1R* translocation (the breeder was G. Riebesel). From this cultivar several cultivars with *1B/1R* substitution have been obtained, such as Orlando, Almus, etc. A sister of Salzmunder Bartweizen, called Neuzucht, was sent to Krasnodar (USSR) where the breeder P. P. Lukyanenko used it in a crossing programme and developed cultivars Caucasus, Aurora, Skorospelka, etc., containing *1B/1R* substitution or translocation. It is interesting to note that Caucasus and Aurora, although having good disease resistance and excellent spike structure and fertility, did not become widely grown cultivars in the Soviet Union, mainly due to their long vegetation period, but they have made a great contribution to developing high-yielding cultivars in many countries.

Neuzucht, Caucasus and Aurora were used as parents in developing the cultivars Amika, Istra and Solaris in Czechoslovakia, Burgas in Bulgaria and Lovrin 10 in Rumania. In Yugoslavia, the cultivars Aurora and Caucasus were the main genetic stock in the crossing programme by which a breakthrough was made above the yield level that had been reached by the cultivar Sava. Several registered cultivars possess the *1B/1R* translocation, among them Jugoslavija, Balkan and Zvezda which are presently in the group of leading cultivars in commercial production (Borojević, 1983; Mišić, 1987).

According to Rajaram et al. (1983) the CIMMYT advanced spring wheat lines such as Veery *S*, Bobwhite *S* and Alondra, carry the *1B/1R* translocation obtained from Caucasus, Aurora, or Weique Red Mace (USA), as one of the parents. These lines proved to have high yield potential, good rust resistance and wide adaptability in various growing environments.

It has to be mentioned that some of *1B/1R* translocation cultivars, e.g., Clement, Beno and some others, are inferior in bread making properties such as having poor stand in fermentation, reduced volume or wet and smeary dough (Zeller and Hsam, 1983).

Among other wheat-rye substitutions, there is the cultivar Amigo which carries the *1AL/1RS* translocation developed by E. E. Sebesta (USA). In Regional Wheat Trials in 1984, four out of five best advanced lines had the cultivar Amigo as one of the parents.

Substitutions or translocations of wheat chromosomes by other rye chromosomes (*2R* to *7R*) have also been attempted and locations of some rye genes have been determined, but these investigations were not as important for plant breeding as *1B/1R* substitutions or translocations (Zeller and Hsan, 1983).

In any case, aneuploidy is extensively used in modern plant breeding. Its particular advantage is that it allows the development of new cultivars differing in only one or two chromosomes from the recipient cultivar. It is especially advantageous for interspecific hybridization because a random combining of chromosomes from two species is usually inefficient on account of the occurrence of negative traits together with positive ones which may render the entire programme of interspecific hybridization worthless.

It should nevertheless be mentioned that aneuploidy is a time-consuming method of obtaining new plant materials, especially new cultivars for commercial production. It is mostly used as a step or a transitory stage in plant improvement by hybridization.

Haploids

Haploid organisms possess in their somatic cells the same chromosome number (n) as a gamete of that species. The earliest detected and studied plant haploids were found in *Datura stramonium* (Blakeslee et al., 1923). Later on, haploids were found in tobacco, tomato, maize, wheat and other species.

A haploid plant develops from an unfertilized ovule (gynogenesis), other gametophytic cells (apogamy), a male gamete (androgenesis), or by polyembryony (Lacadena, 1974). As haploids may contain different genomic formulae, according to the ploidy level of the species they sprang from, diploid and haploid designations should be reviewed for their origin (Table 18.4). The frequency of natural haploids is fairly low, about 1 per 1,000 in the case of parthenogenesis and about 0.1 per 1,000 in the case of androgenesis.

Table 18.4 Types of crops and derived haploids in relation to genome composition

'Crops	Genome formula	Derived haploids	Genome formula
monoploids[1]	$2n = x$	none	
autodiploids[2]	$2n = 2x$	monohaploids	$2n = x$
autotriploids[3]	$2n = 3x$	different, mostly aneuploid	
autotetraploids[4]	$2n = 4x$	autodihaploids	$2n = 2x$
autohexaploids[5]	$2n = 6x$	autotrihaploids	$2n = 3x$
allodiploids[6]	$2n = x_A + x_B$	monohaploids	$2n =$ either x_A or x_B
allotetraploids[7]	$2n = 2(x_A + x_B)$	allodihaploids	$2n = x_A + x_B$
allohexaploids[8]	$2n = 2(x_A + c_B + x_C)$	allotrihaploids	$2n = x_A + x_B + x_C$

1. Natural monoploid cultivars include cv. Kleiner Liebling of *Pelargonium zonale* and a clone of *Thuja plicata* f. *gracilis*, both propagated vegetatively for many years before the discovery of their monoploidy.
2. Many crops.
3. *Musa acuminata* (seedless banana), *Citrullus vulgaris* (seedless water melon), *Beta vulgaris* (sugar beet triploid varieties), maintained either vegetatively or by tetraploid-diploid matings.
4. *Solanum tuberosum* (potato), *Medicago sativa* (lucerne), *Dactylis glomerata* (cocksfoot).
5. *Phleum pratense* (timothy).
6. *Solanum ajanhuiri* (a diploid potato).
7. *Gossypium hirsutum* and *G. barbadense* (cotton), *Nicotiana tabacum* (tobacco), *Arachis hypogaea* (groundnut), *Triticum durum* (macaroni wheat), *Brassica napus* (rapeseed).
8. *Triticum aestivum* (bread wheat), *Avena sativa* (oats).

Source: Riley et al., 1979

As a rule, haploid plants are shorter than diploids, weaker, later and completely sterile because of univalent chromosomes. Thus they are eliminated in all species except those which reproduce vegetatively.

However, it happens with some haploid plants that the meiotic division does not take place in macrosporogenesis and microsporogenesis and this results in a spontaneous development of a fertile diploid plant.

After the development of the colchicine treatment for chromosome doubling, haploids have been treated to obtain diploid organisms, the so-called *dihaploids* or *doubled haploids*.

Dihaploid organisms are completely homozygous for all allelic genes which is not the case with either self-pollinated organisms such as wheat, soyabean, tomato, etc. or lines resulting from long inbreeding, such as maize, sunflower, etc. This finding has been immediately recognized as a possibility of developing completely homozygous diploids via haploids which may be useful in genetic studies as well as in plant breeding.

To overcome the dependence on the spontaneous occurrence of haploids which has a low frequency, *in vitro* and *in vivo* methods of increasing haploid frequency and methods for a reliable production of dihaploids have been developed. The topic has been paid due attention, first at a special symposium, "Haploids in Higher Plants", held in 1974 in Guelph, Canada, and then at other scientific meetings such as "Genetic Manipulation in Plant Improvement" held in 1984 in Columbia, USA, "Genetic Manipulation in Plant Breeding" held in 1986 in Berlin, etc.

Methods of obtaining haploids and doubled haploids

The method of anther culture has become the most successful *in vitro* method of obtaining haploids from pollen. So far, haploids have been produced in over 60 plant species, 90% of them from the *Solanaceae* and *Gramineae* families (Sunderland 1974, 1983; Hu Han and Shao, 1981; Jensen, 1986). However, the successful regeneration of whole plants from anther callus has been achieved only in a limited number of plant species, tobacco, potato, cabbage, tomato, rice, wheat, petunia, etc. (Vasil et al., 1979; Jensen, 1986) (Table 18.5)

According to Ouyang Tsun-wen et al. (1973), Hu Han (1978), and other researchers at the Institute of Genetics in Peking, the wheat anther culture produces a relatively high percentage of diploid plants resulting from spontaneous chromosome doubling, rendering the haploid treatment with colchicine unnecessary in the process of the development of doubled haploids.

Difficulties in the production of haploids by anther and pollen culture arise from insufficient knowledge of the effects of the culture medium, optimum temperature, light and other factors affecting callus development and plant regeneration. At present, a satisfactory frequency of haploid production has been achieved only in tobacco (Table 18.6).

Table 18.5 Selected crop species and recommended procedures for obtaining haploids

Species + = haploids reported ++ = recommended procedure +++ = easy procedure	Mode of origin and production: Spontaneus – selection	Wide species crosses	Interploid crosses	Genetic control mechanism	Chromosome elimination	Anther-pollen culture	Ovary-ovule culture	References
Cotton		+		++				48
Potato		++				++		17, 34, 48, 63
Tobacco		+				+++		14, 17, 24, 53
Tomato						++		48, 66
Brassica	+					++		18, 19, 38
Flax	++							48, 63
Asparagus						++		17, 48, 63
Capsicum						++	+	13, 17
Alfalfa			++			+		44, 48, 63
Sugar beet	+						++	3, 11, 55, 58
Petunia						++		17, 59
Gerbera						+	++	45, 65
Pelargonium	+					+		48
Barley				+	++	++	+	23, 30, 33, 35, 56, 60
Maize				++		+		6, 7, 54
Rice						++	+	1, 65
Wheat						++	+	2, 5, 28, 43, 65
Sorghum						+		44, 48
Thuja	+							34

Source: Jensen, 1986 and also for references*

It has been found that different plant genotypes respond specifically to the culture medium in regard to callus formation or plant regeneration (Schaeffer et al., 1979; Lazar et al., 1983). Therefore, it is spoken about anther culturability as a way of increasing number of plants from anthers. Moraes-Fernandes (1983), Borojević K. et al. (1986) and others found rather large differences in the number of calli and regenerated plants. These differences depended on the genotypes of wheat growing on the same culture medium, the F_1 anthers in the majority of cases giving better results than the anthers from particular varieties. Still the number of regenerated plants was low and not sufficient for breeding work.

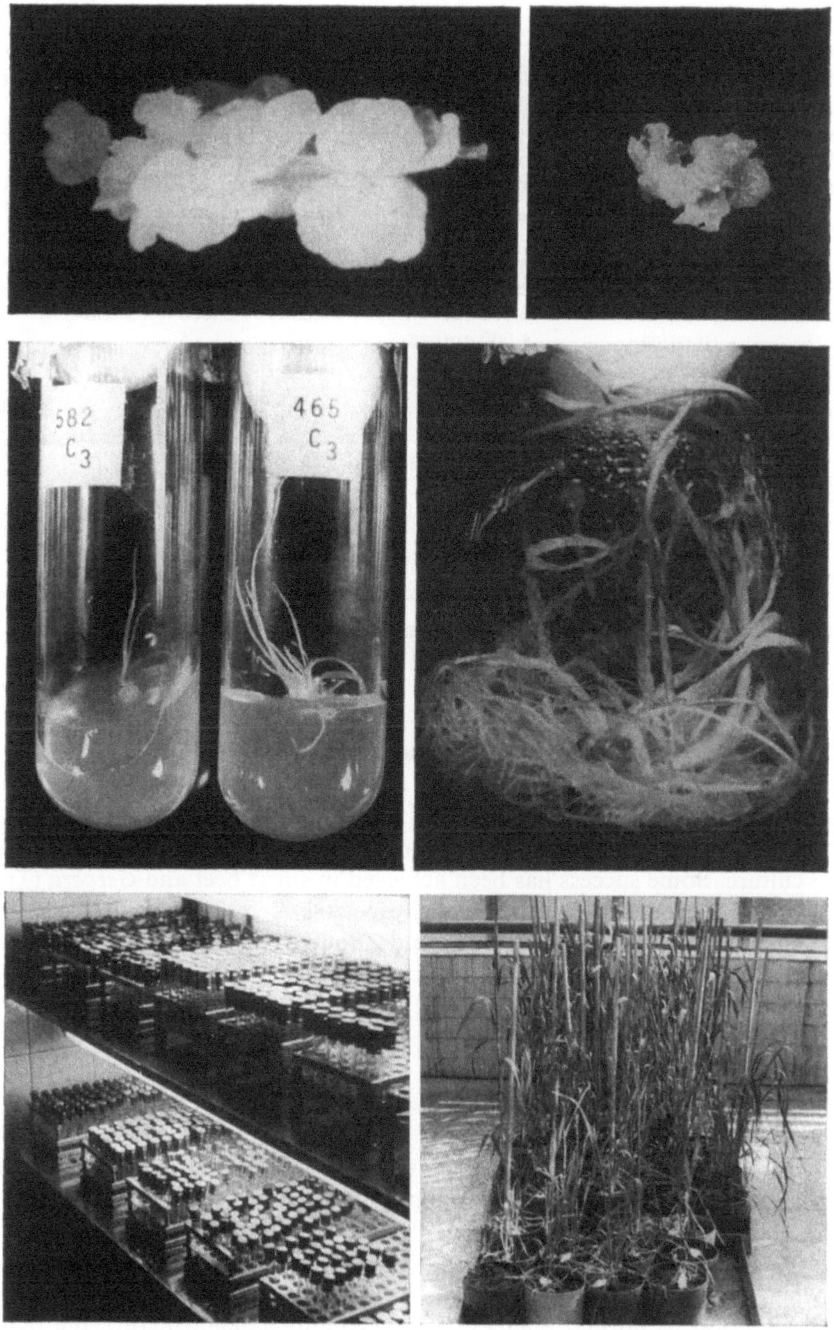

a b c d e f

Fig. 18.7 Production of haploids and doubled haploids of common wheat by anther culture in vitro. a – androgenic calli in anther culture, b – an androgenic callus with embrionic structures, c – androgenic calli on medium for plant regeneration, d – regenerated haploid plant on rooting medium, e – growing chamber with anther cultures, f – haploid plants (in front) and spontaneous diploids obtained by anther culture in vitro. (Courtesy of Eng. S. Šesek, Faculty of Agriculture, Novi Sad)

Table 18.6 Potential and frequencies of haploid production in 3 systems of production: anther-pollen culture, wide species crosses (incl. chromosome elimination) and ovule-ovary culture. (The estimates are taken from the 14 review papers listed in references)

Method of production	Species	Potential cells per culture unit	Potential embryons		Haploid frequencies per 1000* (hypothetical estimates)
			Induced	Plant	
Anther-pollen culture	Tobacco	40,000 Per	20,000	2,000	50
	Brassica	1,000 anther	150	5	5
	Barley	1,500	100	1	0.6
Chromosome elimination	Barley	Per 30 spike	25	5	per 100** 10–17
Wide species crosses	Potato	Per 300 berry	200	50	10–15
Ovule	Barley	30 (Spike)	2.5	0.2	0.1–0.8
Ovary culture	Sugar beet	100 (Ovary)	30–50	10–25	10–25

Source: Jensen, 1986 and also for review papers in references.

Method of ovary and ovule culture

As haploids may develop naturally from an unfertilized ovule or other cells in the ovary, attempts have been made to produce haploids of agricultural plants by ovary and ovule culture. Some success has been achieved in sugar beet and *Gerbera* (Transvaal daisy) while attempts with other crops. especially cereals, have produced poor results (D'Halluin and Keimer, 1986; Yang and Zhou, 1982) (Table 18.5). The method is not a prospective one because it is expensive and it requires much labour.

Method of wide species crossing

Since *in vitro* methods of haploid production have been quite inefficient for most crops, attempts have been made to develop *in vivo* methods.

In a number of species from *Solanum*, *Medicago*, *Trifolium*, *Raphanus*, *Brassica* and other genera, the embryo may develop parthenogenetically from an ovary (gynogenesis) when crossed with another species. When a diploid is crossed with another diploid, in addition to normal plants ($2x$ embryo and $3x$ endosperm), the ovule is sometimes not fertilized by a generative nucleus of the pollen grain. Instead, a secondary nucleus of the embryo sac ($2x$) is fertilized by the secondary nucleus of the pollen grain (x) resulting in a monohaploid possessing the haploid number of chromosomes in the embryo (x) and the triploid endosperm ($3x$) (Fig. 18.8).

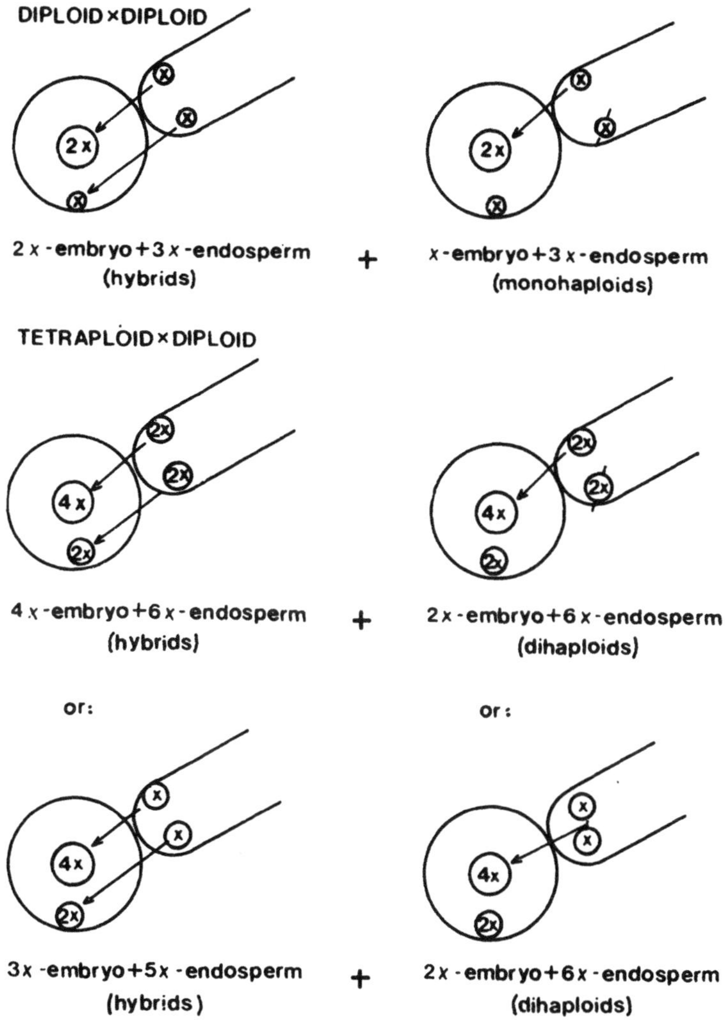

Fig. 18.8 Possible modes of origin of parthenogenetic haploids from diploid×diploid and tetraploid×diploid matings respectively. Arrows indicate fertilization. An oblique line through a nucleus means the loss of that nucleus (above and centre right). A line connecting nuclei (below right) indicates fusion of those nuclei.

Source: Riley, 1979

When a tetraploid and a diploid are crossed, as a result of parthenocarpy, a dihaploid possessing $2x$ embryo and $6x$ endosperm instead of $4x + 6x$ is obtained (Fig. 18.8). For example, autotetraploid *Medicago sativa* is pollinated by diploid *Medicago falcata* to induce dihaploids (Bingham, 1969). In the case of the strawberry, which is an octoploid, tetraploid *Potentilla anserina* is used a s a pollinator (Janick and Hughes, 1974). In the case of *Solanum tuberosum*, diploid *Solanum phureja* is used as a pollinator (Hougas et al., 1964; Jensen, 1986).

There is also a phenomenon of semigamy, when the generative cell of a pollen grain enters the embryo sac but does not fuse with the ovule. The result is a chimeric haploid plant.

The frequency of haploids may be increased by a late pollination (Chase, 1969), chemical treatment (Lacadena, 1974), or under the effect of alien cytoplasm (Tsunewaki et al., 1968).

A special *in vivo* method of haploid production is the *chromosome elimination method* which has been particularly effective with *Hordeum bulbosum.*

When the cultivated barley (*Hordeum vulgare*) is crossed with wild barley (*Hordeum bulbosum*), the resulting seeds have underdeveloped endosperm. It is possible to grow plants from such seeds by embryo culture but chromosomes derived from *H. bulbosum* are gradually eliminated at early stages of development and plants that survive are in fact *H. vulgare* haploids (Simpson and Snape, 1981). Chromosome elimination takes place no matter which species is used as male or female. The elimination does not occur if only diploid *H. vulgare* is crossed with tetraploid *H. bulbosum* (Kasha, 1974).

As in the method of anther culture, the method of crossing with *H. bulbosum* indicated that differences in the frequency of producing haploids (i.e., doubled haploids) occurred depending on the genotype involved (Baenziger et al., 1984). Snape et al. (1986) crossed two recipient lines of wheat and their derived substitution lines with *H. bulbosum* to find that both donor lines failed to set seed, while both substitution lines were crossable and gave 7.6 to 18% seed set. The former was due to the crossability genes Kr1 (*5B*) and Kr2 (*5A*) that supress crossing with *H. bulbosum*, while the latter substitution lines which included Chinese Spring *5B* had the allele kr1 that confer crossability. Petrović et al. (1986) crossed a large number of Yugoslavian winter and spring wheat cultivars with two genotypes of *H. bulbosum*. Crossability was different depending on the wheat genotype, but even more interesting was a high percentage of seed set (40–80%) without the embryo in two spring cultivars, giving only 1 to 2 seeds with the embryo. This is a confirmation of the results obtained by Sitch and Snape (1986) that genes which control seed set are independent of the genes controlling embryo formation.

Use of haploids and doubled haploids in plant breeding

Haploids, and doubled haploids obtained through them, have all allelic genes in the homozygous state. This is otherwise impossible to achieve either in pronouncedly self-pollinated species or in self-compatible cross-pollinated species by long inbreeding. Such a genetic constitution allows us to study the effects of individual genes, determination of recessive mutations, the study of phylogenesis of species, and other phenomena which are difficult to study in diploid organisms.

It would certainly be advantageous to increase the use of haploids in plant breeding. With plant species which allow a routine and mass production of haploids and doubled haploids, it would be the fastest way to make homozygous diploid lines necessary for the development of heterotic hybrids. In the case of cross-pollinated species, for which it is difficult or impossible to make inbred lines due to self-incompatibility or other barriers, the production of haploids seems to be the only possibility to obtain homozygous diploid lines for development of hybrids. With forest trees and other perennial species which would require a very long time to develop inbred lines, the production of haploids by anther culture allows a fast development of homozygous lines for the production of commercial hybrids.

Production of doubled haploids may accelerate the process of selection of self-pollinated plants such as wheat, barley, soyabean, etc. in which six or more generations are necessary after hybridization for new genotypes to stabilize.

Furthermore, doubled haploids may be used for the production of cytoplasmically male sterile (cms) lines of sunflower, rice, sugar beet, onion, etc. to be used for developing heterotic hybrids.

However, doubled haploids that are truly usable in plant breeding and comparable in efficiency with the conventional breeding methods must fulfill certain criteria which Snape et al. (1986a) defines as follows:

"(a) easy, consistent production of large numbers of doubled haploids of all genotypes in the breeding programme,
(b) doubled haploids should be genetically normal and stable,
(c) the doubled haploid population should contain a random sample of the parental gametes."

So far, no method of producing doubled haploids fulfills those criteria. To some extent the exception may be tobacco (Dunwell, 1986) and potato (Sneep et al., 1979) where doubled haploids are rather routinely produced.

The frequency of doubled haploids in barley, wheat and many other plants obtained either by anther culture or by using the *Hordeum bulbosum* method is very low and cannot compensate for the number of homozygous lines which are obtained in segregating generations by the conventional breeding methods.

Another limitation is that the percentage of doubled haploids is different when different genotypes or cultivars are used; in some, practically no haploid plants are obtained. It would not be reasonable to expect actual genetic differences to occur in response to the application of different methods. It is rather that culture media are not adapted to the stages of development of particular genotypes.

Due to the effect of the culture medium, haploids, as well as by colchicine treatment derived doubled haploids, sometimes show somaclonal variation and deviate from original genotypes. For normal breeding procedures that may be considered a disadvantage, particularly since a great part of somaclonal variation has no value for breeding purposes (see Chapter 19).

Crossing programmes are performed to obtain a rich pool of gene recombinations, and to select the best recombinations of genes from the parents used. If the frequency of haploids is only 1 to 2% green plants per spike (Snape et al., 1986a), then the frequency of those with desirable gene recombinations is negligible.

It was expected that the use of doubled haploids would speed up breeding work considerably when F_1 gametes are subjected to anther culture. Instead of waiting for six years after a cross, homozygous lines could be obtained in two years. The time factor is important, but since the doubled haploid population does not contain a random sample of the parental gametes, the method cannot be considered efficient.

Furthermore, it is definitely advantageous to quickly produce homozygous lines but not any line. When the conventional breeding method is used, selection already starts in the F_2 generation. It is not a random selection but is the selection of desirable phenotypes. When the method of doubled haploids is used, it may happen that none of the obtained 1–2% of homozygous plants meets the requirements set by the breeding objective. If only this method is relied upon, a large number of desirable lines is irretrievably lost. This cannot happen with the conventional breeding methodology.

When the doubled haploid anther culture and the conventional breeding method are compared, the time factor is mentioned in favour of the former method (Baenziger et al., 1984). However, nobody has come to the main point of comparing the two methods regarding the efficiency in the production of cultivars. Kasha and Reinbergs (1981) supplied an example of several barley cultivars obtained by the doubled haploid breeding method but these cultivars were not superior to cultivars obtained by the conventional method in yield potential. Baenziger et al. (1983) reported that the doubled haploids derived from the wheat cultivar Kitt yielded significantly less than the cultivar or lines obtained from it by the single seed descent method (SSD). Other information from the literature refer more to the variability and instability of doubled haploid derived lines than to their practical use (Baenziger et al., 1984).

With self-pollinated plant species such as wheat, barley, soyabean, tomato, etc., which do not offer problems in the production of F_1 seeds and sowing in the field, and in which the process of self-pollination itself leads to the development of homozygous lines, doubled haploids may hardly replace the method of conventional breeding or may hardly be a cheaper solution.

With cross-pollinated and perennial plant species for which it is either very difficult or quite impossible to produce homozygous lines by inbreeding, the production of doubled haploids, if it becomes routine and economical, will gain large importance in breeding programmes. According to Griffing (1975), doubled haploids may be used in recurrent selection for population improvement, primarily this is due to the absence of dominance and epistasis. For the same reason doubled haploids have a great advantage for diallel mating systems where gene action has to be determined and additive and additive x additive genetic variance has to be estimated (Choo et al., 1979).

Generally speaking, induction of haploids and production of doubled haploids, either by anther culture or interspecific crossing, may be considered as a novel approach to genetic studies which is potentially usable in plant breeding.

References

Baenziger, P. S., Nesenberg, D. M., Schaeffer, G. W., Galun, E., and Feldman, M., 1983. Variation among anther culture derived doubled haploids "Kitt" wheat. Proc. 6th Inter. Wheat Genet. Symp., Kyoto, p. 575–582.

Baenziger, P. S., Hudirka, D. T., Schaeffer, G. W., and Lazar, M. D., 1984. The significance of doubled haploid variation, In: J. P. Gustafson (Editor), Gene Manipulation in Plant Improvement. Plenum Press, New York, p. 385–414.

Bingham, E. T., 1969. Haploids from cultivated alfalfa. *Medicago sativa L.* Nature (UK) 221: 765–866.

Blakeslee, A. F., Belling, F. J., and Farnhman, M. E., 1923. Inheritance in tetraploid Daturas. Bot. Gaz. 76: 329–373.

Bogaert, G. van, 1975. A comparison between colchicine induced tetraploid and diploid cultivars of *Lolium* species. In: B. Nüesch (Editor), Ploidy in Fodder Plants. Report of the EUCARPIA Fodder Crops Section, Zürich, p. 61–78.

Borojević, K., Šesek, S., and Radojević, Lj., 1986. Response of different genotypes on development of callus from anther cultures of wheat. In: Nuclear Techniques and in Vitro Culture for Plant Improvement, International Atomic Energy Agency, Vienna, p. 233–237.

Borojević, S., 1983. (Breeding of high-yielding wheat cultivars in Yugoslavia). Savremena poljoprivreda (Novi Sad), 31 (9–10): 389–403.

Breese, E. L., Stephens, D. E., and Thomas, A. C., 1975. Tetraploid hybrid ryegrass: Allopolyploidy in action. In: B. Nüesch (Editor), Ploidy in Fodder Plants. Report of the EUCARPIA Fodder Crops Section, p. 46–49.

Brown, W. V., 1972. Textbook of Cytogenetics. C. V. Mozby, Saint Louis, 346 pp.

Cauderon, I., 1986. Cytogenetics in breeding programmes dealing with polyploidy, interspecific hybridization and introgression. In: W. Horn, C. J. Jensen, W. Odenbach and O. Schrieder (Editors), Genetic Manipulation in Plant Breeding. Walter de Gruyter, Berlin-New York, p. 83–104.

Chase, S. S., 1969. Monoploids and monoploid-derivatives of maize (*Zea mays L.*). Botanical Review 35: 117–167.

Choo, T. M., Christic, B. R., and Reinberg, E., 1979. Doubled haploids for estimating genetic variances and scheme for population improvement in self-pollinating crops. Theor. Appl. Genet. 54: 267–271.

Collins, G. B., Taylor, N. L., and DeVerna, J. W., 1984. In vitro approaches to interspecific hybridization and chromosome manipulation in crop plants. In: J. P. Gustafson (Editor), Gene Manipulation in Plant Improvement. Plenum Press, New York, p. 323–383.

D'Halluin, K., and Keimer, B., 1986. Production of haploid sugarbeet (*Beta vulgaris L.*) by ovule culture. In: W. Horn, C. J. Jensen, W. Odenbach and O. Schrieder (Editors). Genetic Manipulation in Plant Breeding. Walter de Gruyter, Berlin-New York, p. 307–309.

Dewey, D. R., 1980. Some applications and misapplications of induced polyploidy to plant breeding. In: W. H. Lewis (Editor), Polyploidy: biological relevance. Plenum Press, New York, p. 445–470.

Dorsey, E., 1936. Induced polyploidy in wheat and rye. J. Heredity. Vol. 27, No. 4.

Doussinault, G., Dosba, F., and Jahier, J., 1983. New results on the improvement of the level of resistance to exespot in wheat. Proc. 6th Int. Wheat Genet. Symp. Kyoto, Japan, p. 193–198.

Dunwell, J. M., 1986. Pollen, ovule and embryo culture as tools in plant breeding. In: P. G. Alderson and L. A. Withers (Editors), Plant Tissue Culture and its Agricultural Application. Butterworth, Guildford, U.K. p. 375–404.

Eigsti, O. J., and Dustin, P., 1957. Colchicine – in Agriculture, Medicine, Biology and Chemistry. The Iowa State College Press, Ames (USA), 470 pp.

Enken, V. (Ed.), 1966. Eksperimentalnaia poliploidia v selekcii rastenii. Izd. Nauka, Novosibirsk.

Evans, L. E., and Jenkins, B. C., 1960. Individual *Secale cereale* chromosome additions to *Triticum aestivum*. I. The addition of individual "Dakold" fall rye chromosomes to "Kharkov" winter wheat and their subsequent identification. Canadian J. Genet. Cytol. 2: 205–215.

Griffing, B., 1975. Efficiency changes due to use of doubled haploids in recurrent selection methods. Theor. Appl. Genet. 46: 367–386.

Hagberg, A., and Akerberg, E., 1962. Mutations and poliploidy in plant breeding. Stockholm.

Hendriksen, A. J. T., 1979. Current breeding methods: sugar beet. In: J. Sneep and A. J. T. Hendricksen (Editors), Plant Breeding Perspectives. Pudoc, Wageningen, p. 203–224.

Hougas, R. W., Peloquin, S. J., and Gabert, A. C., 1964. Effect of seed-parent and pollinator on frequency of haploids in *Solanum tuberosum*. Crop Sci. 4: 593–595.

Hu Han, 1978. Advances in anther culture investigations in China. Proc. Symp. Plant Tissue Culture, May 25–30, 1978. Peking. Science Press, p. 3–10.
Hu Han and Shao, Z., 1981. Advances in plant cell and tissue culture in China. Adv. Agron. 34: 1–13.
Janick, J., and Hughes, H. G., 1974. Production of strawberry tetrahaploids from intergeneric crosses. Proc. 1st. Inter. Symp. on Haploids in Higher Plants, Guelph, Canada, p. 137.
Jensen, C. J., 1986. Haploid induction and production in crop plants. In: W. Horn, C. J. Jensen, W. Odenbach and O. Schrieder (Editors), Genetic Manipulation in Plant Breeding. Walter de Gruyter, Berlin-New York, p. 231–256.
Kasha, K. J., 1974. Haploids in higher plants. Advances and potential. II. Methods of producing haploids. Proc. 1st. Inter. Symp. University of Guelph, Canada, p. 11–50.
Kasha, K. J., and Reinbergs, E., 1981. Recent developments in the production and utilization of haploids in barley. Proc. 4th Int. Barley Genet. Symp., Edinburgh, p. 655–665.
Khush, G. S., 1973. Cytogenetics of aneuploids. Academic Press, New York, 300 pp.
Kihara, H., 1951. Triploid watermelons. Proc. Amer. Soc. Hort. Science, Vol. 58.
Kimber, G., 1984. Evolutionary relationships and their influence on plant breeding. In: J. P. Gustafson (Editor), Gene Manipulation in Plant Improvement. Plenum Press, New York, p. 281–294.
Kis, A., 1973. Hexaploid triticale breeding in Hungary. In: Triticale. Studies and Breeding. Proc. Inter. Symp. Leningrad, USSR, p. 38–46.
Lacadena, I. R., 1974. Spontaneous and induced parthenogenesis and androgenesis. Proc. 1st. Inter. Symp., Guelph, p. 13–32.
Law, C. N., Snape, J. W., and Worland, A. J., 1983. Interspecific chromosome manipulation. Phil. Trans. R. Soc. Lond. B., 292: 509–518.
Law, C. N., Snape, J. W., and Worland, A. J., 1987. Aneuploidy in wheat and its uses in genetic analysis. In: F. G. H. Lupton (Editor), Wheat Breeding. Chapman and Hall, London, p. 71–108.
Lazar, M. D., Schaeffer, G. W., and Baenziger, P. S., 1983. Cultivar×environment interaction of callus and polyhaploid plants from anther culture of wheat. Theor. Appl. Genet. 67: 273–277.
Mišić, T., 1987. Potencijal za prinos zrna i druge osobine sorti i linija ozime pšenice u proizvodnji Vojvodine. Semenarstvo, Novi Sad, 99 pp.
Moraes-Fernandes, M. I., Bd. de, 1983. Viability of haploid production by anther culture using Brasilian wheat genotypes. Rev. Brasil. Genet. VI(2): 261–277.
Morris, R., 1973. Summary of wheat gene locations on chromosomes and chromosome arms by aneuploid and other methods. Depart. Agron., Univer. Nebraska, Lincoln, USA.
Morris, R., 1983. Remodeling crop chromosomes. In: D. R. Wood (Editor). Crop Breeding, American Society of Agronomy, Madison, USA, p. 109–128.
Mujeeb-Kazi, A., Bernard, M., Bekele, G. T., and Miranda, J. L., 1983. Incorporation of alien genetic information from *Elymus giganteus* into *Triticum aestivum*. In: S. Sakamoto (Editor), Proc. 6th Inter. Wheat Genet. Symp. Kyoto, Japan, p. 223–231.
Müntzing, A., 1941. Studies on the properties and the ways of production of rye-wheat amphidiploids. Hereditas, 25: 387–430.
Müntzing, A., 1951. Cytogenetic properties and practical value of tetraploid rye. Hereditas, 37: 17–84.
Müntzing, A., 1979. Results and Problems. In: Advances in Plant Breeding. Supplement 10 to Journal of Plant Breeding. Paul Parey, Berlin, 103 pp.
Okamoto, M., 1957. Asynaptic effect of chromosome V. Wheat Inf. Serv. 5:6.
Olsson, G., 1963. Induced poliploids in Brassica. In: A. Akerberg et al. (Editors), Recent plant breeding research. Svalöf 1946–1961, pp. 179–192.
O'Mara, J. G., 1940. Cytogenetic studies on Triticale. I. A method of determining the effect of individual *Secale* chromosomes on *Triticum*. Genetics 25: 401–408.
O'Mara, J. G., 1951. Cytogenetic studies on Triticale. II. The kinds of intergeneric chromosome additions. Cytologia 16: 225–232.
Ouyang Tsuan-wen, Hu Han, Chuang Chia-chun and Tseng Chun-chi, 1973. Induction of pollen plants from anthers of Triticum aestivum L. cultured in vitro. Scientia Sinica (Peking), Vol. XVI. No. 1: 79–90.
Petrović, S., Worland, A. J., and Snape, J. W., 1986. The possibility of doubled haploids wheat

production and their utilization in breeding programme. Savremena poljoprivreda (Novi Sad), 34, (9–10): 431–439.
Pissarev, V., 1966. Different approaches in Triticale breeding. Proc. 2nd Inter. Wheat Genet. Symp. Lund, Sweden, 1963. Hereditas, Suppl. 2, p. 279–290.
Rajaram, S., Maan, Ch. E., Ortiz-Ferrara, G., and Mujeeb-Kazi, A., 1983. Adaptation, stability and high yield potential of certain 1B/1R CIMMYT wheats. In: S. Sakamoto (Editor). Proc. 6th Inter. Wheat Genet. Symp., Kyoto, Japan, p. 613–621.
Rigin, V., and Orlova, I. N., 1977. Pšenično-ržanie amfidiploidi. Izd. Kolos, Leningrad, 277 pp.
Riley, R., and Chapman, V., 1958. Genetic control of the cytologically diploid behaviour of hexaploid wheat. Nature, 182: 713–715.
Riley, R., 1968. The basic and applied genetics of chromosome pairing. Proc. 3rd Int. Wheat Genet. Symp., Canberra, Australia, p. 185–195.
Riley, R., and Law, C. N., 1984. Chromosome manipulation in plant breeding: progress and prospects. In: J. P. Gustafson (Editor), Gene Manipulation in Plant Improvement, Plenum Press, New York, p. 301–322.
Schaeffer, G. W., Baenziger, P. S., Worley, J., 1979. Haploid plant development from anthers and in vitro embryo culture of wheat. Crop Sci. 19: 697–702.
Sears, E. R., 1954. The aneuploids of common wheat. Univ. Missouri Agr. Exp. Stat. Res. Bull. 572: 1–59.
Sears, E. R., 1956. The transfer of leaf-rust resistance from *Aegilops umbellulata* to wheat. Brookhaven Symp. in Biology, 1: 1–22.
Sears, E. R., 1966. Nullisomic-tetrasomic compensations in hexaploid wheat. Chromosome manipulation and Plant Genetics. Heredity 20: 29–45.
Sears, E. R., 1972. Chromosome engineering in wheat. Stadler Symposia 4, Univ. Missouri, Columbia, USA, p. 23–38.
Sears, E. R., 1976. Genetic control of chromosome pairing in wheat. Annu. Rev. Genet. 10: 31–51.
Simpson, E., and Snape, J. W., 1981. The use of doubled haploids in a winter barley programme. Proc. of the Fourth Barley Genet. Symp., Edinburgh, p. 716–720.
Sitch, L. A., and Snape, J. W., 1986. Intrachromosomal mapping of crossability genes in wheat (*Triticum aestivum*). Z. Pflanzenzüchtung, 96: 304–319.
Snape, J. W., de Buyser, J., Henry, J., and Simpson, E., 1986. A comparison of methods of haploid production in a cross of wheat, *Triticum aestivum*. Z. Pflanzenzüchtung 96: 320–330.
Snape, J. W., Simpson, E., Parker, B. B., Friedt, W., and Foroughi-Wehr, B., 1986a. Criteria for the selection and use of doubled haploid systems in cereal breeding programmes. In: W. Horn, C. J. Jensen, W. Odenbach and O. Schrieder (Editors), Genetic Manipulation in Plant Breeding. Walter de Gruyter, Berlin-New York, p. 217–229.
Sneep, J., and Hendriksen, A. J. T. (Eds.), 1979. Current breeding methods. In: Plant Breeding Perspectives. Pudoc, Wageningen, p. 104–233.
Sunderland, N., 1974. Anther culture as a mean of haploid induction. Proc. 1st. Inter. Symp., Guelph, p. 91–122.
Sunderland, N., 1983. On the use of microspores for genetic modification. In: T. Kruge, C. P. Meredith and A. Hollaender (Editors), Genetic Engineering of Plants. Plenum Press, New York, p. 315–332.
Tsunewaki, K., Noda, K., and Fusijawa, T., 1966. Comparative gene analysis of common wheat and its ancestral species. III. Glume hairiness. Genetics 53: 303–311.
Tsunewaki, K., Noda, K., and Fusijawa, T., 1968. Haploid and twin formation in a wheat strain Salmon with alien cytoplasmas. Cytologia 33: 525–538.
Vasil, I. K., Ahuja, M. R., and Vasil, V., 1979. Plant tissue cultures in genetics and plant breeding. Adv. Genetics 20: 127–215.
Varughese, G., Barker, T., and Saari, E., 1987. Triticale CIMMYT, Mexico, D.F., 32 pp.
Wolski, T., and Tymieniecka, F., 1983. The first Polish Triticale variety. Cereal Research Communications 11: 139–141.
Wolski, T., 1988. Progress in winter triticale breeding. In: J. Lekeš (Editor), Kromerišské Obilnárské Dny. II. 9–10th of March 1988 (in press).

Yang, H. J., and Zhou, C., 1982. In vitro induction of haploid plants from unpollinated ovaries and ovules. Theor. Appl. Genet. 63: 97–104.
Zeller, F. J., 1973. 1B/1R wheat-rye chromosome substitutions and translocations. Proc. 4th Inter. Wheat Genet. Symp., Columbia, Missouri, p. 209–221.
Zeller, F. J., and Hsam, S. L. K., 1983. Broadening the genetic variability of cultivated wheat by utilizing rye chromatin. In: S. Sakamoto (Editor), Proc. 6th Inter. Wheat Genet. Symp., Kyoto, Japan, p. 161–174.
Zillinsky, F. J., 1974. The development of triticale. Advan, Agron. 26: 315–348.

19. GENETIC ENGINEERING

Several new methods and techniques have been developed in recent years which, in addition to their importance for fundamental research in the fields of genetics, biochemistry, physiology, etc., may become exceedingly important in plant breeding. These are: 1) a method of gene transfer from one organism to another by a technique of recombinant deoxyribonucleic acid (rDNA) or genetic engineering in a narrow sense, 2) a method of protoplast fusion of various species, and 3) the use of somaclonal variation occurring in tissue and cell cultures. In addition to being directly applicable in breeding some plant species, tissue and cell cultures are a basic implement for various gene manipulations. This is why they are the first to be discussed in this chapter.

Tissue and cell culture

Already at the beginning of this century, Haberlandt (1902) emphasized the totipotency of somatic cells in plants and pointed out the possibility of producing plants by means of tissue culture. However, it took more than half the century to master *in vitro* methods of plant growing (cell suspension, hormone regulation of growth and differentiation of plant organs, and regeneration of whole plants). The first successful uses of tissue culture were in carrot and tobacco hybrids (Skoog and Miller, 1957; Reinert, 1959; Steward, 1963; Vasil and Hildebrandt, 1965; etc.) (Fig. 19.1).

Thanks to the discovery of suitable methods of plant cell and tissue extraction, success was achieved in inducing rapid cell division and obtaining homogeneous callus tissue. The tissue is then cultivated in a nutritive medium containing, among other components, growth hormones – auxine to initiate root growth, and kinetine to initiate vegetative growth – resulting in the development of whole plants. Vasil et al. (1979) made a list of plant species of which whole plants have been obtained by means of tissue culture. More success has so far been achieved with annual plants than with perennial plants as well as with horticultural plants and grasses than with legumes and forest trees.

Clonal propagation by a culture of meristem tissue has become a conventional method of multiplying orchids and other horticultural plants. The decisive advantage of the method is that it renders plant material free of viruses which used to plague the production of these plants.

It would be of premier importance to succeed in the clonal propagation of forest trees by tissue culture methods because of generation time involved in breeding these plant species. Good success has been achieved in poplars and pines (Winton, 1968; Chalupa, 1974; Sommer et al., 1975).

Fig. 19.1 Plant regeneration obtained from a leaf segment of African violet (*Saintpaulia ionatha* Wendl.) after three weeks in culture medium following the process of somatic embryogenesis: embryonic callus → somatic embryo → plantlet. (Courtesy of Dr. Ljiljana Radojević, Institute of Biology, Belgrade)

A method of embryo culture has been used as an effective method of producing those inter-specific and inter-generic hybrids in which the crossing itself is successful but viable seeds cannot be obtained due to underdeveloped endosperm.

Collins et al. (1984) give examples of inter-specific and inter-generic hybridization where, via *in vitro* pollination and fertilization, viable embryos or plants have been obtained. Mostly species from the *Nicotiana* genus are involved, plus two *Gossypium* and two *Zea* species. In regard to *in ovule* embryo culture, some success was achieved with crosses of different *Gossypium*, *Brassica*, *Chrysanthemum* and *Nicotiana* species.

Although tissue culture is aimed primarily at a rapid propagation of superior genotypes which are identical to the source material, on account of advances in the technique achieved in a number of plant species attempts are made to develop *in vitro* selection which would allow the use of positive variants and the elimination of negative ones. Selection is applicable at the level of callus, suspension or plated cells (Meredith, 1984). It has been established that only a limited number of phenotypic traits is detectable in tissue cultures, because most traits come to expression only after the differentiation of plant parts and the development of the whole plant. Prospective uses are evident of *in vitro* selection for resistance to salinity, tolerance to mineral stress, and in some cases resistance to disease with a high probability that regenerated plants will show that trait and that it will be transmitted to their progeny (Chaleff, 1983; Meredith, 1984). It may be stated that tissue and cell culture techniques are not sufficiently advanced for the selection of many crops to rely solely on them even for the traits mentioned above, and therefore they should be combined with other breeding methods.

Tissue and cell culture may serve for the detection of preexisting mutations as well as mutations which took place during the growth in culture medium (Lörz and Brown, 1986).

Protoplast fusion

Studies of cell and tissue culture have provided valuable information on processes which take place in the cell, the function of its parts, and the role of the cell in tissue and in the entire plant. It has also been found that a single isolated cell is incapable of expressing all function which it ordinarily exercises as a part of an organism. One of the reasons for this is that the cell wall represents a barrier to the circulation of substances from and into the cell, especially when it is isolated. It moved researchers to try to remove the cell wall in order to obtain the wall-free plant cell, the so-called *protoplast*. At the beginning, the cell wall used to be removed mechanically (Klercker, 1892). Later on, a chemical procedure, by means of cellulose enzymes, pectinase, etc., was introduced (Cocking, 1960).

Protoplasts are most frequently isolated from leaves, vegetational apexes, roots, root nodules in legumes, and microspores (Vasil, 1976). Whole plants have been successfully regenerated from isolated protoplasts in about 50 plant species. The regeneration has been most successful with species of the genera *Brassica*, *Nicotiana*, *Petunia*, and *Solanum* (Vasil et al., 1979; Collins et al., 1984).

Protoplast culture offers a range of possibilities. For example, two protoplasts may be fused which is not possible with cells which have the cell wall. This has led to the idea of fusing cells of two different species that cannot be crossed sexually in order to obtain synthetic species hybrids. This is an explanation why the method of protoplast fusion is also called *parasexual hybridization.*

The method may be outlined as follows. Segments of leaves or other plant parts are used for the preparation of a suspension of cells, and their cell walls are removed enzymically (Cocking, 1960). Millions of protoplasts of each species must be isolated.

Suspensions of protoplasts coming from two plant species are blended and centrifuged in order to achieve maximum mixing and fusing of protoplasts. The suspension is then placed on a nutritive medium in Petri dishes, taking care that optimum conditions regarding sterility, temperature, light, etc. are provided. After some time, a small number of cells (1 on 1 to 8×10^{-5}) begin to form a callus and develop into a plant (Power et al., 1976). These plants should be examined cytologically for the number of chromosomes in somatic cells to find plants that are the result of the fusion of protoplasts of the two species (Vasil, 1976).

The first to carry out a successful fusion of protoplasts from *Petunia* were Power et al. (1976) in England and Carlson et al. (1972, 1975) in the U.S.A., on a species of *Nicotiana*. A somatic hybrid between *Nicotiana glauca* and *M. langsdorfii* differed from the source material in a number of traits including leaf form and plant height. Besides deviant plants in regard to chromosome number, there was a plant whose seeds had $2n = 42$ chromosomes, i.e., the full set of 24 chromosomes derived from *N. glauca* and 18 chromosomes from *N. langsdorfii*. The hybrid was a proof of the feasibility of somatic hybridization between two species that cannot be crossed sexually.

Somatic hybrid plants of several inter-specific and inter-generic plants have subsequently been obtained via protoplast fusion (Table 19.1).

Table 19.1 Interspecific and intergeneric somatic hybrid plants obtained via protoplast fusion

Parents	References*
Interspecific somatic hybrid plants.	
Brassica oleracea + B. campestris	Schenck and Robbelen (1982)*
Daucus carota + D. capillifolius	Dudits et al. (1977); Kameya et al. (1981); Toshiaki et al. (1981)
Datura innoxia + D. candida	Schieder (1978a)
D. innoxia + D. sanguinea	Schieder (1980)
D. innoxia + D. stramonium	Schieder (1978a)
Medicago sativa + M. falcata	Teoule (1983)
Nicotiana glauca + N. langsdorffii	Carlson et al. (1972); Smith et al. (1976); Chen et al. (1977); Chupeau et al. (1978); Uchimaya et al. (1983)
N. knightiana + N. sylvestris	Maliga et al. (1977); Menczel et al. (1978)
N. tabacum + N. alata	Nagao (1979)
N. tabacum + N. debnevi	Gleba et al. (1979); Kosakovs'ka (1980)
N. tabacum + N. glauca	Evans et al. (1980); Uchimiya (1982)
N. tabacum + N. glutinosa	Nagao (1979); Uchimiya (1982)

Table 19.1 (continued)

Parents	References*
N. tabacum + *N. knightiana*	Maliga et al. (1978); Menczel et al. (1981)
N. tabacum + *N. nesophila*	Evans et al. (1981); Evans et al. (1982)
N. tabacum + *N. otophora*	Evans et al. (1983)
N. tabacum + *N. repanda*	Nagao (1982)
N. tabacum + *N. rustica*	Nagao (1978); Iwai et al. (1980); Douglas et al. (1981); Peitian et al. (1981)
N. tabacum + *N. stocktonii*	Evans et al. (1981)
N. tabacum + *N. sylvestris*	Melchers (1977); Zelcer et al. (1978); Evans (1979); Aviv and Galun (1980); Medgyesy et al. (1980); Hein et al. (1983)
Petunia hybrida + *P. axillaris*	Izhar and Power (1979)
P. hybrida + *P. parodii*	Power et al. (1976, 1977); Cocking et al. (1977); Izhar et al. (1983)
P. parodii + *P. inflata*	Power et al. (1979)
P. parodii + *P. parviflora*	Power et al. (1980)
Solanum nigrum + *S. tuberosum*	Binding et al. (1982)
S. tuberosum + *S. chacoense*	Butenko and Kuchko (1979)
Intergeneric somatic hybrid plants	
Arabidopsis thaliana + *Brassica campestris*	Gleba and Hoffmann (1978, 1979, 1980); Hoffmann and Adachi (1981)
Datura innoxia + *Atropa belladonna*	Krumbiegel and Schieder (1979, 1981)
Daucus carota + *Aegopodium podagraria*	Dudits et al. (1979)
Nicotiana tabacum + *Petunia hybrida*	Binding (1976); Xianghui et al. (1982)
N. tabacum + *Salpiglossis sinuata*	Nagao (1982)
Solanum tuberosum + *Lycopersicon esculentum*	Melchers et al. (1978)

* For the titles of references see literature cited by Collins et al. 1984.
Source: Collins et al. 1984, p. 350–352.

It is an old desire to cross the tomato and the potato and develop a plant bearing both tomato fruits and potato tubers. It would be in fact a plant with doubled yield potential. Attempts to cross the two species sexually have failed because of the total incompatibility that exists between the species. The method of protoplast fusion has finally enabled the crossing of these two species which are both members of the *Solanaceae* family. Melchers (1978) in Germany succeeded in obtained couple of plants possessing chromosomes and cytoplasms of both species Considering a number of shortcomings of the method that remain yet to be solved, it is not surprising that the plants are weak, stunted and deficient in some chromosomes of both species, which creates additional problems. Nevertheless, when the method of protoplast fusion is improved, it will be possible to obtain hybrid plants with valuable traits and perhaps even new synthetic species will be developed.

No matter how promising the method of protoplast fusion may be in producing new synthetic species that cannot be obtained by any other method, it should not be expected to be universally applicable in plant breeding. Certain species and genera are so different that genes become inhibited in the fused cells, or there occur new gene interactions resulting in plant forms which are useless in breeding. It has been confirmed by examples where fusion products were observed to undergo division but no plantlet regeneration was obtained, such as *Vicia hajastana* + *V. villosa*, *Daucus carota* + *Hordeum vulgare*, *Glycine max* + *Zea mays*, *Petunia hybrida* + *Atropa belladonna* and several others (for literature see Potrykus et al., 1983; Collins et al., 1984; Cocking, 1984; Schneider et al., 1986).

The fusion of these distant species should hardly be expected to provide anything of practical value even when the method is much improved, but these products have scientific value because they may show, at the molecular level, what is genetically common and genetically different among various species and genera.

Somatic cybridization

Protoplast culture opens up several other possibilities. One of them is the transfer of cell organelles from one species to another and the production of a hybrid population of organelles, which is not usually achieved by sexual hybridization since most species contain their own organelles obtained from the maternal species through the egg cell. Organisms possessing cytoplasm and organelles from one species and the nucleus from another species are called "cybrids". A cybrid between a normal cultivar of *Brassica napus* and a cms cultivar of *Raphanus* was produced in which only mitochondria from the cms cultivar were present while the chloroplasts and the nucleus came from *B. napus* (Pelletier et al., 1983). A similar intergeneric hybrid was obtained between *Arabidopsis thaliana* and *B. campestris* (Gleba and Hoffman, 1980) (Table 19.1). Attempts were made to develop other cybrids but no plantlet regeneration was obtained (Collins et al., 1984).

Moreover, by protoplast fusion it is possible to obtain new combinations of organellar and nuclear DNA, thus creating new genetic variation (Glimelius et al., 1986).

As the protoplast fusion technology advances it is hoped that some attempts will yield results which can be used in practical plant breeding.

Protoplast culture is a rather convenient medium for gene transfers that will be discussed in the next paragraph.

Gene transfer

The methods and techniques discussed in the previous paragraphs are called genetic engineering in a broad sense, although they do not deal with transfer of individual genes but deal with the manipulation of plant cells where entire nuclear or organellar genomes are either incorporated or fused. The transfer of a particular gene from one organism to another by recombinant DNA techniques is considered to be in fact genetic engineering proper.

Thus, genetic engineering is a direct manipulation of genetic material and it involves transfer of desirable gene(s) from one organism to another as well as transfer of new genes obtained by chemical synthesis.

The discovery of restriction enzymes has allowed for the partitioning of the DNA molecule at certain points, each fragment being mostly an individual gene, and the recombining of these fragments into a new chain of recombinant DNA.

A considerable number of genes extracted from viruses, bacteria, yeasts, fungi, fruit flies, plants, animals and humans have been isolated and cloned (Watson et al., 1983). They are maintained in so-called gene libraries.

Genetic engineering implies the following procedure: selection of a donor of a desirable gene, the selection of a vector or technique of transmitting the gene to the recipient and the selection of a recipient in which the gene will function normally.

Recombinant DNA techniques have been fairly successful on the level of prokaryotes in the production of some medicines, organic syntheses, etc. The largest commercial success has been achieved in the production of insuline and interferone (Gilbert and Villa-Komaroff, 1980; Watson, et al., 1983).

The application of genetic engineering to plants has proved to be quite difficult. Accordingly, modest results have been obtained. One of pioneer attempts was the transfer of nitrification genes (nif-genes) from *Rhizobium* bacteria to plants which do not live in symbiosis with these bacteria, such as maize, wheat and other monocots. However, there are plant species such as soyabean, beans and clover on whose roots the bacteria form nodules which fix atmospheric nitrogen making it available to plants. Attempts have been made to transfer the nif-gene from *Rhizobium*, via a bacteriophage as a vector, into a plant species *Trema canabina* (Dixon and Postgate, 1972; Hardy and Havelka, 1975; Postgate, 1975). However, it turned out that there exist not one but a number of nif-genes, that they cannot be activated unless promotor genes are transferred too, and that there exist specific genes in the host which affect the function of nif-genes. In fact, it was found that it is a complex system which is not easy to transfer (Verma and Long, 1983). Otherwise, the transfer of genes controlling the fixation of atmospheric nitrogen into non-legumes would undoubtedly lower the use of nitrogen fertilizers which at the moment are the largest expense in the nutrition of plants.

Besides viruses used as vectors, there are other methods of transferring a desirable gene into the recipient. Plasmids have been found suitable for that purpose, especially the Ti-plasmid (tumour inducing plasmid) which is isolated from a soil bacterium *Agrobacterium tumefaciens.*

Plasmids are circular molecules of DNA which behave like mini chromosomes. They are present in almost all microorganisms. The plasmid may be cut by means of restriction enzymes and the desirable gene inserted into it. After that, the plasmid is linked by means of other enzymes. This is the so-called chimeric plasmid which has to be incorporated back into the bacterium which is then used to infect the recipient organism. If plasmid DNA is integrated into the genome of the recipient and transferred genes shows its expression, it may be considered that a transformed or transgenic organism has been obtained (Fig. 19.2).

The simplest method of plant transformation by Ti-plasmids is the disk method

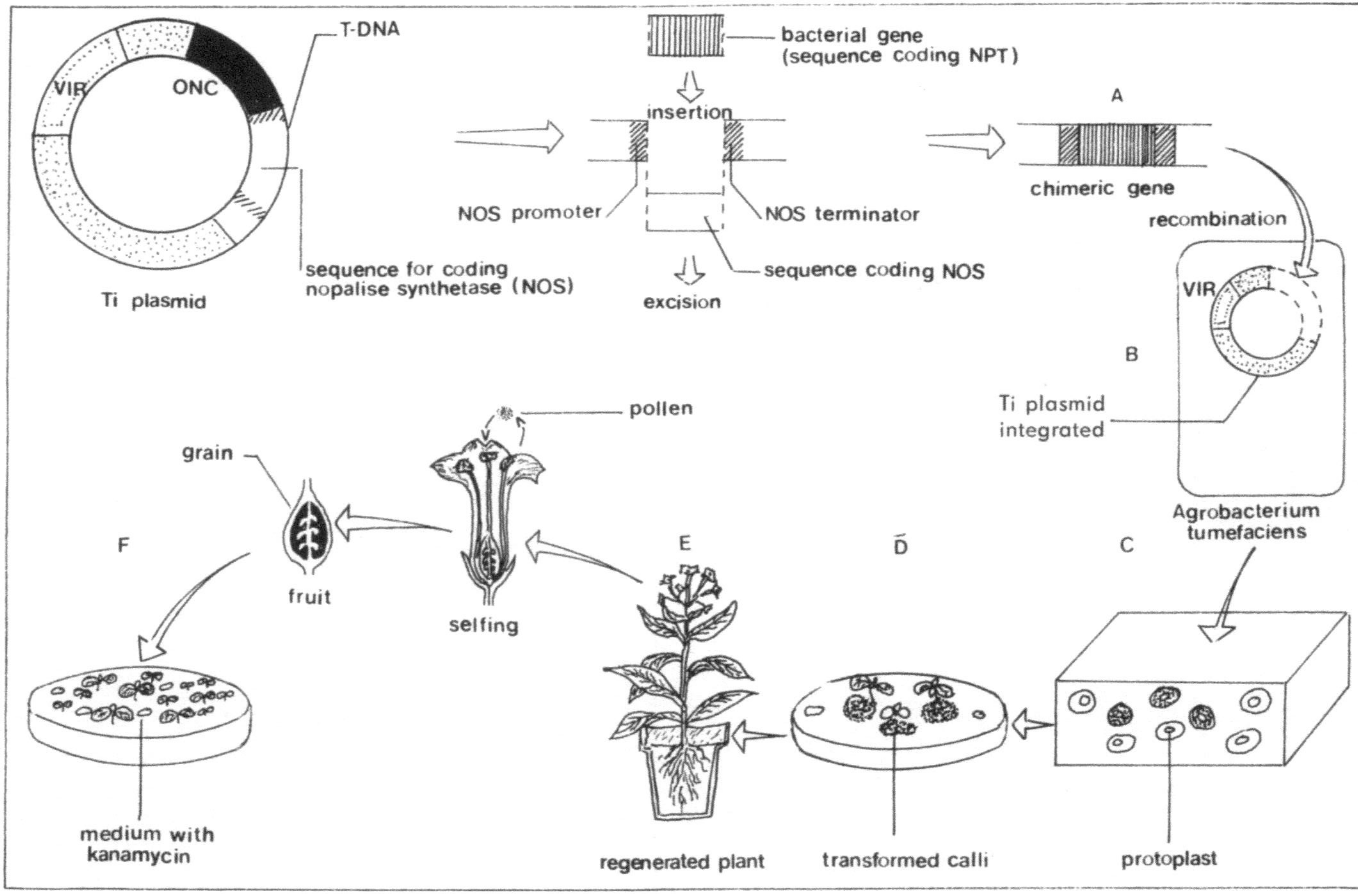

Fig. 19.2 (legend see p. 303)

(Horsch et al., 1985). Leaf disks about 6 mm in diameter are infected by *Agrobacterium tumefaciens* that contains a chimeric plasmid with the desirable gene. So far, genes for resistance to some antibiotics, e.g., kanamycin and carbamycin, have been transferred by this method. Some plants are resistant to these antibiotics and some are not. After about one month in a culture medium, whole plants may be regenerated from the tobacco leaf disks. Only those plants survive which possess the gene for resistance to the antibiotics, i.e., the so-called transformed or transgenic plants. The method has been applied successfully to plant species such as petunia, tomato, soyabean, sugar beet, etc. (Zambryski et al., 1984; Horsch et al., 1985).

Gene transfer by direct incorporation of a Ti-plasmid into the protoplast is another possibility which has been more effective in petunia and other dicots than in monocotyledonous plants (Cocking and Davey, 1987). The first results of direct transfer of genes for resistance to kanamycin were obtained with protoplasts of *Triticum monococcum* and *Lolium multiflorum* (Lörz et al., 1985; Potrykus et al., 1985).

Micro-injection of DNA into the nuclei of protoplasts or into inflorescences is also a new method of gene transfer which has been demonstrated with kanamycin resistance genes incorporated into floral tillers of rye (Schell, 1987).

The transfer of genes for resistance to antibiotics into plants is used only as a marker for other transfers which might be useful in plant breeding. For example, the transfer of genes for resistance to herbicides would be an important achievement in agriculture.

The herbicide glyphosphate is an inhibitor of the enzyme 5-enolpyruvylshikimate-3-phosphatase (EPSP) synthase in plants. Comai et al. (1985) have isolated a mutant gene "aro *A*" for EPSP from the bacterium *Salmonella typhimurium* and transferred it to Ti-plasmid of *A. tumefaciens* by which they have infected tobacco. Transformed plants showed some resistance to herbicides. Shah et al. (1986) have isolated a complementary DNA clone encoding EPSP synthase from a glyphosphate-tolerant petunia hybrid cell line. A chimeric EPSP synthase gene was made with the use of the cauliflower mosaic virus 35S promoter and introduced into petunia cells. Transformed petunia plants showed tolerance to the herbicide.

An experiment performed recently by De Block et al. (1987) seems to be even more successful. Some commercial herbicides contain phosphinotricin (PPT) which is

---

Fig. 19.2 Genetic manipulation with plants requires several steps which in principle include transfer of a foreign gene to be expressed in another organism. The gene used is a chimeric gene constructed in the laboratory consisting of three parts: one bacterial gene (sequence coding for neomycin phosphotransferase or NPT which confers resistance to antibiotic kanamycin, one promotor and one terminator of the T-DNA of Agrobacterium; the last are neccessary to enable the function of the gene in new environment. (A). The gene is introduced into Ti plasmid (which does not carry tumour genes) and then incorporated into Agrobacterium (B). With such cells protoplasts of tobacco plants are infected (C). After some weeks, protoplasts form calli (D) from which whole plants are regenerated (E). To test whether regenerated plants posses resistance to kanamycin, their seeds are planted into medium containing kanamycin (F). Plantlets which are green and grow posses the bacterial gene, i.e. enzyme NPT which is able to detoxify kanamycin, while those without the bacterial gene stay white since kanamycin destroys chloroplasts. There is about one quarter of white plantlets, thus showing Mendelian segregation ratio.
(This drawing is taken from Tempé and Schell (1987) from La Recherche, Vol. 18, No. 188, p. 702, who made it excellently in colour, which was not possible in this book)

an inhibitor of glutamine synthase in plants and is used as a non-selective herbicide. The authors had the idea to use the bar gene which confers resistance of *Streptomyces hygroscopicus* to bialaphos (a tripeptide containing PPT) and encodes phosphinothricin acetyltransferase (PAT) which is able to convert PPT into the non-toxic acetylated form.

Agrobacterium vector was used to transfer the bar gene into tobacco, tomato, and potato plants. Transgenic plants showed an active production of PAT enzyme and were completely resistant to high doses of commercial formulations of phosphinothricin and bialaphos. This is an example of a successful approach to obtaining herbicide-resistant plants by their being able to detoxify herbicides.

Powell-Abel et al. (1986) performed an experiment which indicates that plants can be genetically transferred for resistance to virus diseases. A chimeric gene having a cloned cDNA of the coat protein (CP) gene of tobacco mosaic virus (TMV) was incorporated into tobacco cells on a Ti-plasmid of *A. tumefaciens* without tumor inducing genes. Regenerated plants expressed TMV mRNA and CP as a nuclear trait.

Recently, there have been several experiments attempting to transfer genes for resistance to virus diseases into plants, which seems to be easier than the transfer of genes for resistance to fungi.

Of course, the recombinant DNA technique offers many other possibilities. Jaynes et al. (1986) had the idea to improve plant protein composition, since many crop plants which are used as staple foods (rice, wheat, maize, potatoes, cassava, etc.) are deficient in some or all essential aminoacids. The authors synthetized by machine fragments of DNA which code for high content of essential aminoacids. These fragments called the "high essential aminoacid encoding gene" (HEAAE-gene) were cloned and then expressed in bacteria. The gene was introduced into the potato plant by means of *Agrobacterium rhizogenes* plasmid (Ri plasmid). In the tissue culture of the potato, hairy roots developed from which transformed plants have been regenerated. The authors detected the expression of HEAAE protein, but whether this will be sufficient to change the protein composition of the potato needs further study.

It is interesting to report that Ow et al. (1986) succeeded in transferring the luciferase gene from the firefly (*Photinus pyralis*), under the control of a cauliflower mosaic virus 35 S RNA promotor, into plant protoplast cells of *Daucus carota* by electroporation and into plants of *Nicotiana tabacum* by the use of the Ti-plasmid of *A. tumefaciens*. Light produced by the action of luciferase was detected in transfected plant cells and transgenic plants. This phenomenon may be of interest in horticulture, but even more it may serve as a marker for manipulation of other genes.

Agrobacterium – mediated transformation is still a rather complicated procedure and other easier methods have been tried. One of them is a direct gene transfer in *in vitro* cultured plant cells. This method of DNA injection imposed several problems, among them the difficulties in penetrating the cell wall, damage to the injected cells, existence of chimeric tissue because not all cells in the culture will be transformed, etc. (Reich et al., 1986; Schweiger et al., 1987).

Direct injection into plants, i.e., into seeds, embryos, or pollen has been tried earlier (Ledoux, 1975), but whether the foreign gene is integrated and stably inherited has not been proved yet (Schell, 1987). There is recent evidence for transformation by direct injection of plasmid DNA into young floral tillers of rye (De la Pena et al., 1987) who succeded in obtaining three transgenic rye plants (Lörz et al., 1986, 1988).

There are more examples of the use of genetic engineering in plants which are described in excellent review papers by Tempé and Shell (1987) and Shell (1987). The recombinant DNA technique is undoubtedly a fascinating achievement of human thought but miracles should not be expected to ensue. There is yet much to learn about metabolic processes on the molecular level, processes of cell differentiation and organogenesis, barriers to integration and functioning of transferred alien genes, etc. And even when these discoveries are made, it should not be expected that genetic engineering can be the sole technique of plant breeding or other branches of biotechnology.

A transformed plant with an incorporated gene for resistance to a disease or a herbicide may lose its resistance as a result of the occurrence of new parasitic races or the introduction of new herbicides, just as it happens to genes incorporated by recombination breeding. The transfer of one or a few genes from one organism to another is only a part of the breeding process where gene recombinations for multiple traits are necessary.

Somaclonal variation and plant breeding

In Chapter 3 it was mentioned that somaclonal variation may be counted among new sources of genetic variability, and descriptions were given of how it occurs and what it means genetically. This chapter will dwell on potential uses of somaclonal variation in plant improvement.

Since Lutz's (1966) report of an enormous variation observed among single cells isolated from callus cultures of tobacco plants, so much experimental data and information had been accumulated over the period of 20 years that in 1985 it was possible to organize the world-wide symposium on "Somaclonal Variation and Crop Improvement" (Semal, 1986).

Scowcroft (1977), Scowcroft and Larkin (1982), Larkin and Scowcroft (1981, 1983), Orton (1984), Ahloowalia (1986), Lörz and Brown (1986) made extensive reviews of somaclonal variation which appears in cell and tissue cultures and its possible uses in crop improvement.

The most frequent somaclonal variants belong to chlorophyll deficiencies, such as albinos, xantha, viridis, variegated, etc. In cereals and forage crops, such changes have no value in breeding, but in some horticultural plants they may be of interest.

The second most frequent group of somaclonal variations relates to changes in chromosome number and structure, such as polyploidy, aneuploidy and various chromosome abberations. These changes result in genomic instability and large number of such variants are eliminated in subsequent sexual generations and they also have no practical value in plant breeding. In vegetatively propagated plants, e.g., potato and sugar cane, somaclonal karyotypic variants may be propagated true but so far they have not proved to be better than the parental cultivars.

The next group of somaclonal variations involves changes in quantitative characters, mainly in reduced height and fertility, prolonged maturity, tiller number, spike morphology, etc. A few examples of resistance to some diseases or chemicals were recorded such as resistance in potato to *Phytophthora infestans* and *Fusarium oxysporum* (Behnke, 1979, 1980), of maize to *Helminthosporium maydis* (Gegenbach et al., 1977), and of tobacco to herbicides (Chaleff and Ray, 1984). Even though those changes can be considered as variations, their use in breeding is very limited due to the synergetic effects on other traits, unstable expression in different environments, and unknown genetic transmission in subsequent generations (Ahloowalia, 1986).

Unlike induced mutations in which genetic changes are provoked intentionally by treatments with radiation or chemicals, somaclonal variation is not sought for but is rather a by-product in tissue and cell cultures, resulting from imperfect handling of tissue culture techniques employed to regenerate and propagate certain plant genotypes. Therefore, somaclonal variation is a chance incident, utterly uncontrolable and usually negative.

If some somaclonal variants, which arise either from pre-existing mutations or newly induced mutations, represent something useful, they may be used in plant breeding.

In any case, the breeder cannot wait to see if some genetic variability will or will not occur. He must work on the development of new genetic variability, persistently and on a large scale. This can be primarily achieved by gene recombination after hybridization of selected parents. If some desirable somaclonal variants are observed, they may be used making sure beforehand that they are stable and of known inheritance. Otherwise, many useless plants will be obtained in segregating generations.

* * *

Genetic engineering has produced some good results on the level of prokaryotes (insulin, interferon, etc.) but the results achieved in eukaryotes may be considered only as preliminary. This is understandable if the complexity of the biological system and the dificulties of methodological and technical nature are taken into consideration. It is necessary to keep working on the development of new methods and techniques but they alone should not be expected to be sufficient to produce new plant cultivars. Transformed plants are transformed in only one gene, and this gene has to be combined with other desirable genes which come via conventional breeding methods.

Plant improvement, i.e., development of new cultivars, should be considered as an integral system based on genetic laws, natural and artificial selection, and different methods of creating and utilizing genetic varibility obtained by gene recombinations, mutations, and transfers. New methods and techniques should therefore be taken as an integral part of the breeding system, not as a replacement for the conventional recombination breeding nor in the least as a self-sufficient breeding system.

References

Ahloowalia, B. S., 1986. Limitations to the use of somaclonal variation in crop improvement. In. J. Semal (Editor), Somaclonal Variations and Crop Improvement. Martimus Nijhoff Publ. Dordrecht, p. 14–27.

Behnke, M., 1979. Selection of potato callus for resistance to culture filtrates of Phytophthora infestans and regeneration of resistant plants. Theor. Appl. Genet. 55: 69–71.

Behnke, M., 1980. Selection of dihaploid potato callus for resistance to the culture filtrate of Fusarium oxysporum. Z. Pflanzenzucht. 85: 254–258.

Carlson, P. S., Smith, H. H., and Dearing, R. D., 1972. Parasexual interspecific plant hybridization. Proc. Nat. Acad. Sci. USA 69: 2292–2294.

Carlson, P. S., and Polacco, J. C., 1975. Plant cell cultures: Genetic aspects of crop improvements. Science, 188: 622–625.

Chaleff, R. S., 1983. Considerations of developmental biology for the plant cell geneticist. In: T. Kosuge, C. P. Meredith and A. Hollaender (Editors), Genetic Engineering of Plants, An Agricultural Perspectives. Plenum Press, New York, p. 257–270.

Chaleff, R. S., and Ray, T. B., 1984. Herbicide-resistant mutants from tobacco cell cultures. Science, 223: 1148–1151.

Chalupa, V., 1974. Control of root and shoot formation and production of trees from poplar callus. Biol. Plant. 16: 316–320.

Cocking, E. C., 1960. A method for the isolation of plant protoplasts and vacuoles. Nature (London) 187: 972–929.

Cocking, E. C., 1984. Use of protoplasts: potentials and progress. In: J. P. Gustafson (Editor), Gene Manipulation in Plant Improvement. 16th Stadler Genetics Symposium. Plenum Press. New York, p. 415–426.

Cocking, E. C., and Davey, M. R., 1987. Gene transfer in cereals. Science, 236: 1259–1262.

Collins, G. B., Taylor, N. L., and DeVerna, J. W., 1984. In vitro approaches to interspecific hybridization and chromosome manipulation in crop plants. In: J. P. Gustafson (Editor), Gene Manipulation in Plant Improvement. 16th Stadler Genetics Symp. Plenum Press. New York, p. 323–383.

Comai, L., Sen, L. C., and Stalker, D. M., 1985. An altered aro A gene product confers resistance to the herbicide glyphosate. Science, 221: 370–371.

De Block, M., Botterman, J., Vandewiele, M., Dockx, J., Thoen, C., Gosselé, V., Rao-Movva, N., Thompson, C., Van Montagu M., Leemans, J., 1987. Engineering herbicide resistance in plants by expression of a detoxifying enzyme. EMBO Jour. 6: 2513–2518.

De la Pena, A., Lörz, H., and Shell, J., 1987. Transgenic rye plants obtained by injection of DNA into young floral tillers. Nature, 325: 274–276.

Dixon, R. A., and Postgate, J. R., 1972. Genetic transfer of nitrogen fixation from *Kleibsiella pneumoniae* to *Echerichia coli.* Nature, U.K. 227: 102–103.

Doussinault, G., Dozba, F., and Jahier, J., 1983. New results on the improvement of the level of resistance to eyespot in wheat. Proc. 6th Int. Wheat Genet. Symp. Kyoto, Japan, p. 193–198.

Gegenbach, B. G., Green, C. E., and Donoran, C. M., 1977. Inheritance of selected pathotoxin resistance in maize plants regenerated from cell culture. Proc. Nat. Acad. Sci. (USA), 74: 5113–5117.

Gilbert, W., and Villa-Komaroff, L., 1980. Useful proteins from recombinant bacteria. Sci. American 242 (4): 68–84.

Gleba, S. S., and Hoffmann, F., 1980. "Arabidobrassica": A novel plant obtained by protoplast fusion. Plant 149: 112–117.

Glimelius, K. Fahlesson, Sjödin, C., Sundberg, E., and Djupsjöbacka, 1986. Somatic hybridization and cybridization as potential methods for widening of the gene-pools of crops within Brassicaceae and Solanaceae. In: W. Horn, C. J. Jensen, W. Odenbach and O. Schieder (Editors) Genetic Manipulation in Plant Breeding. Walter de Gruyter, Berlin, p. 663–683.

Haberlandt, C., 1902. Kulturversuche mit isolierten Pflanzenzellen. Sizungser. Math-Naturwiss. Kl. Kaiser Akad. Wiss. Wien (Vienna) 111: 69–92.

Hardy, R. F., and Havelka, V. D., 1975. Nitrogen fixation research: a key to world food? Science, 188: 633–643.

Horsch, R. B., Fry, J. E., Hoffmann, D., Eichholtz, Rogers, S. G., and Fraley, R. T., 1985. A simple and general method for transfering genes into plants. Science 227: 1229–1231.

Jaynes, J. M., Yang, M. S., Espinoza, N., and Dodds, J. H., 1986. Plant protein improvement by genetic engineering: use of synthetic genes. Trends in Biotechnology, p. 314–320.

Klercker, J. A., 1892. Eine Methode zur Isolierung lebender Protoplasten. Oefvers. Vetenskaps. Akad., Stockholm 9: 463–471.

Larkin, P. J., and Scowcroft, N. R., 1981. Somaclonal variation – A novel source of variability from cell cultures for plant improvement. Theor. Appl. Genet. 60: 197–214.

Larkin, P. J., and Scowcroft, W. R., 1983. Somaclonal variation and eyespot toxin tolerance in sugarcane. Plant Cell Tissue Organ Culture, 2: 111–121.

Law, C. N., Snape, J. W., and Worland, A. J., 1983. Interspecific chromosome manipulation. Phil. Trans. R. Soc. Lond. B., 292: 509–518.

Ledoux, L. (Editor), 1975. Genetic Manipulation with Plant Material. Plenum Press, New York.

Lörz, H., Baker, B., and Schell, J., 1985. Gene transfer to cereal cells mediated by protoplast transformation. Mol. Gen. Genet. 199: 178–182.

Lörz, H., and Brown, P. T. M., 1986. Variability in tissue culture derived plants – possible origins, drawbacks and advantages. In: W. Horn, C. J. Jensen, W. Odenbach and O. Schieder (Editors), Genetic Manipulation in Plant Breeding, Walter de Gruyter, Berlin, p. 513–534.

Lörz, H., Göbel, E., and Brown, P., 1988. Advances in tissue culture and progress towards genetic transformation of cereals. Plant Breeding 100: 1–25.

Lutz, A., 1966. Obtention de plantes de tabac á partir de cultures unicellulaires provenant d'une souche anergiée. C.R. Acad. Sc. Paris, 262: 1856–1858.

Melchers, G., 1978. Somatic hybrid plant of potato and tomato regenated from fused protoplasts. Carlsberg Research Communication 43: 203–218.

Meredith, C. P., 1984. Selecting better crops from cultured cells. In: J. P. Gustafson (Editor), Gene Manipulation in Plant Improvement. 16th Stadler Genetics Symp. Plenum Press. New York, p. 503–528.

Orton, T. J., 1984. Somaclonal variation theoretical and practical considerations. In: J. P. Gustafson (Editor), Gene Manipulation in Plant Improvement. 16th Stadler Genetics Symposium. Plenum Press. New York, p. 427–468.

Ow, D. W., Wood, K. V., DeLuca, M., de Wet, J. R., Helinski, D. R., Howell, S. H., 1986. Transient and stable expression of the firefly luciferase gene in plants cells and transgenic plants. Science, 234: 856–859.

Pelletier, G., Primard, C., Vedel, F., Chitrit, P., Remy, R., Rouselle, P., and Renard, M., 1983. Intergeneric cytoplasmic hybridization in cruciferae by protoplast fusion. Mol. Gen. Genet. 186: 453–459.

Postgate, J., 1975. The physiology and genetics of nitrogen fixation. In: L. Ledoux (Editor), Genetic Manipulation with Plant Material. Plenum Press, New York, p. 123–134.

Potrykus, I., Harms, C. T., Hinnen, A., Hutter, R., King, P. J., and Shillito, R. D., 1983. Protoplasts 1983. Lecture Proceedings 6th Inter. Protoplast Symp., Experientia Supplementum, 46, 269 pp.

Potrykus, I., Saul, M. W., Petruska, J., Pazzkowski, J., and Shillito, R., 1985. Direct gene transfer to cells of a graminaceous monocot. Mol. Gen. Genet. 199: 183–188.

Powell-Abel, P., Nelson, R. S., Barun, D., Hoffmann, N., Rogers, S. G., Fraley, R. T., and Beachy, R. N., 1986. Delay of disease development in transgenic plants that express the tobacco mosaic virus coat protein gene. Science, 232: 738–743.

Power, J. B., Frearson, E. M., Hayward, C., George, D., Evans, P. K., Berry, S. F., and Cocking, E. C., 1976. Somatic hybridisation of *Petunia hybrida* and *P. parodii*. Nature (London) 263: 500–502.

Reich, T. J., Iyer, V. N., Miki, B. L., 1986. Efficient transformation of alfalfa protoplasts by the intranuclear microinjection of Ti-plasmid: BioTechnology, 4: 1001–1004.

Reinert, J., 1959. Über die Kontrolle der Morphogeneze und die Induktion von Adventivembryonen an Gewebekulturen aus Karotten. Plant 53: 318–338.

Schell, J. St., 2987. Transgenic plants as tools to study the molecular organization of plant genes. Science 237: 1176–1183.

Schieder, O., Hein, T., and Kohn, H., 1986. Plant cell fusion as a tool for genetic manipulation. In: W. Horn, C. J. Jensen, W. Odenbach und O. Schieder (Editors), Genetic Manipulation in Plant Breeding, Walter de Gruyter, Berlin, p. 641–652.

Schweiger, H. G., Dirk, J., Koop, H. W., Kranz, È., Neuhaus, G., Spangenberg, G., and Wolff, D., 1987. Individual selection, culture and manipulation of higher plant cells. Theor. Appl. Genet. 73: 769–783.

Scowcroft, W. R., 1977. Somatic cell genetics and plant improvement Ad. Agronomy 29, 39–82.

Scowcroft, W. R., and Larkin, P. J., 1982. Somaclonal variation: A new option for plant improvement. In: I. K. Vasil, N. R. Scowcroft and K. J. Frey (Editors), Plant Improvement and Somatic Cell Genetics. Academic Press, New York, p. 159–178.

Semal, J., 1986. Tissue culture and crop improvement: scientific dream or new science. In: J. Semal (Editor), Somaclonal variations and crop improvement. Martimus Nijhoff Publ., Dordrecht, p. 8–13.

Shah, D. M., Horsch, R. B., Klee, H. J., Kishore, G. M., Winter, J. A., Tumer, N. E., Hironaka, C. M., Sanders, P. R., Gasser, C. S., Aykent, S., Siegel, N. R., Rogers, S. G., and Fraley, R. T., 1986. Engineering herbicide tolerance in transgenic plants. Science, 233: 478–481.

Skoog, F., and Miller, C. O., 1957. Chemical regulation of growth and organ formation in plant tissues cultured in vitro. Symp. Soc. Exp. Biol. 11: 118–131.

Sommer, H. E., Brown, C. L., and Kormanik, P. P., 1975. Differentiation of plantlets in longleaf pine (Pinus palustris Mill.) tissue culture in vitro. Bot. Gaz. (Chicago) 136: 196–200.

Steward, F. C., 1963. The control of growth in plant cells. Scientific American, October (167):

Tempé, J., and Schell, J., 1987. La manipulation des plantes. La Recherche 18, 188: 696–709.

Vasil, V., and Hildebrandt, A. C., 1965. Differentiation of tobacco plants from single isolated cells in microculture. Science 150: 889–892.

Vasil, I. K., 1976. The progress, problems and prospects of plant protoplast research. Adv. Agron. 28: 119–160.

Vasil, I. K., Ahuja, M. R., and Vasil., V., 1979. Plant tissue cultures in genetics and plant breeding. Adv. Genetics, 20: 127–215.

Verma, D. P. S., and Long, S. R., 1983. The molecular biology of Rhizobium – legume symbiosis. In: K. Jean (Editor), Intracellular Symbioses. Academic press, New York, p. 211–243.

Watson, J. D., Tooze, J., and Kurtz, D. T., 1983. Recombinant DNA. A short course. Freeman, New York, 260 pp.

Winton, L. L., 1968. Plantlets from aspen tissue cultures. Science 160: 1234–1235.

Zambryski, P., Herrera-Estrella, L., De Block, M., Van Montagu, M., and Schell, J., 1984. The use of the Ti plazmid of *Agrobacterium* to study the transfer and expression of foreign DNA in plant cells. In: A. Hollaender and J. Setlow (Editors), Genetic Engineering. Principles and Methods. Plenum Press, New York, p. 253–278.

20. BREEDING VEGETATIVELY PROPAGATED PLANTS

Many plants species are propagated vegetatively – potato, grapevine, fruit trees, strawberry, hop, cassava, some forest trees, etc. Sexual reproduction is also possible with these species but vegetative propagation is preferred.

Genetic basis of vegetatively propagated plants

A vegetative part taken from a plant, such as a tuber, a root, a stolon, a rhizome, a sprig, a leaf or a stem, may be used for asexual vegetative propagation. Plants arising from a clone are indentical and have the same genetic constitution as the mother plant. If several vegetatively reproducing propagules are taken from a single plant, the plants produced and their subsequent progenies will also be genetically identical. The purpose of clonal propagation is to obtain the largest possible multiplication of a superior genotype resulting in identical and uniform progenies which cannot be obtained by the sexual propagation of that particular variety.

Variability among plants within a single clone results from environmental effects, it is not heritable and may be classified as environmental variability. If there are changes which are transmitted to progenies, they are somatic mutations which are inherited only through the vegetative mode of propagation.

Besides direct vegetative propagation by tubers, roots, rhizomes or sprigs, some horticultural species such as fruits, grapes, roses, etc. are propagated by grafting or transplantation. The recipient plant is called a stock, the vegetative part used for grafting a scion. The purpose of grafting is not the production of new cultivars but the propagation of superior genotypes giving progenies identical to the plants from which the scion had been taken.

After grafting, there may occur larger variability than after propagation without grafting. The resulting changes may be classified into three categories:

- modification-type changes,
- chimera-type changes, and
- mutation-type changes.

Most grafting methods practiced in pomiculture and horticulture bring modification-type changes, such as more or less vigorous growth, larger or smaller fruits, earlier maturation on one stock than on another, etc. These changes are not transferred to progenies by sexual reproduction which means that they have not been fixed in the sexual cells, i.e., they are not heritable. If heritable changes had frequently resulted from grafting, it would have been necessary to find means of preventing them because the traits of cultivars grafted on wild or inferior rootstocks would have been impaired. However, this is not the case, and man extensively uses the positive modifications resulting from grafting.

The changes occurring after grafting in consequence of the mixing of tissues are also classified among non-heritable changes but as they differ from modification-type changes they are called "chimeric changes". Different types of chimeras develop because of different modes of fusing and mixing of the tissues of the stock and the scion (Fig. 20.1). Chimeric changes occurring in the course of ontogenetic plant development are sometimes so striking and unusual that it is difficult to accept them as non-heritable changes. However, whenever it has been tried to retain these changes in progenies by sexual reproduction, the result were invariably only the components of that particular chimera. Chimeras are maintained only by vegetative propagation and this phenomenon is exploited in a number of horticultural species (e.g., *Acer*, *Hlorophytum*, etc.) (Fig. 20.1).

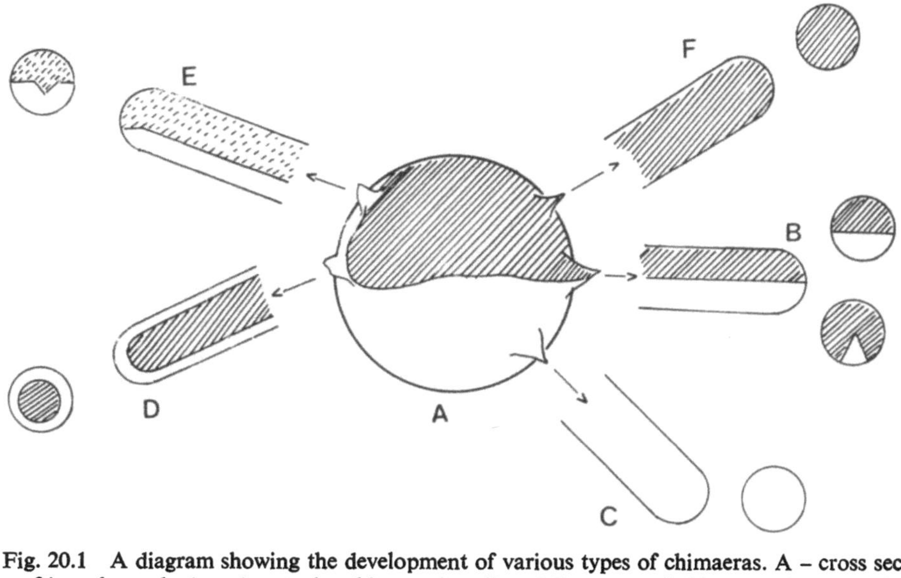

Fig. 20.1 A diagram showing the development of various types of chimaeras. A – cross section of grafting place, shadowed – stock, white – scion, B and E – sectoral chimaeras, D – periclinal chimaera, C – shoot having only scion tissue, F – shoot having only stock tissue.
Source: Tavčar, 1952

Although there are indications that it is possible to induce heritable changes by grafting, it has not been proved experimentally nor has a convincing explanation been offerred. Lisenko (1948) and his followers considered all changes resulting from grafting as heritable changes, although it was proved in a number of instances that the observed changes were in fact mosaic forms, i.e., chimeras, which could not be transferred to progenies by sexual means. Moreover, they claimed that the segregation of traits took place in sexually derived progenies, the segregation ratios being similar to those found in sexual hybridization. Lisenko supplied a theoretical explanation for vegetative hybrids pointing out that each drop of a living body possesses hereditary characteristics, and thus, plastic substances exchanged between stock and scion are bound to transfer its varietal traits.

It is true that various substances come to scion from the stock although the expression "plastic" is absolutely inappropriate. The larger the difference between stock and scion, the more different are the substances exchanged between them. As different organic compounds such as ethylmethansulphonate, ethilimine, etc. do cause heritable changes, a possibility exists for some substances exchanged between stock and scion to be mutagenic and to cause in some cases mutations of a certain gene. This explanation is radically different from that stating that each drop of a living body possesses hereditary characteristics. The fact that the segregation of traits in seed generations was observed in experiments goes in favour of the opinion that the discussed changes are based on mutations. It may happen that substances from stock cause changes in only one allele of a gene locus and not in another allele thus resulting in segregation in seed generations.

Another proof in favour of mutation-type changes is that the frequency of changes after grafting is quite low. If each drop of a living body had possessed hereditary characteristics, changes would have occurred on a large scale. In reality, the rate of changes is expressed in thousandths of one percent.

The methods of grafting or transplantation differ in effects and consequences from irradiation or chemical treatment.

If no changes had been caused in the scion by different stock, the selection of stocks would have been unnecessary. However, these changes are ontogenetic, modificational and non-heritable. Agriculture in general makes use of positive modifications, i.e., the capacity of cultivars to react to better cultural practices and higher doses of mineral fertilizers by a more vigorous growth and increased productivity. However, the more vigorous growth and the increased productivity will not be repeated in the next generation if the environmental factors which had caused them are not provided.

An advantage of grafting is that it allows the propagation of somatic or vegetative mutation, which cannot be achieved by sexual means. Furthermore, grafting allows the stabilization and utilization of heterozygous genotypes, which is not possible with self pollinated plants and with cross pollinated plants only in large populations in equilibrium.

Grafting is a conventional method of improving and growing fruit trees, grapevine and many species of decorative plants.

Initial material for plant improvement includes:

(1) clones or populations of mixed clones usable for the production of stocks and scions, i.e., cultivars, and

(2) populations of seedlings usable for the production of stocks and scions.

Selection of clones

Clones are genetically homogeneous because all plants originate from one ancestral mother plant and they have the same genotype and are identical in all traits, regardless of the degree of eventual heterozygosity. Sometimes, not all members of a given clone have the same genotype, e.g., in the case of the mutation of certain buds or when one branch is diploid and another triploid, etc.

In the old varieties of apple, pear, plum, peach, etc., numerous spontaneous mutation occurred in the course of long-continued exploitation. Clonal selection was applied to positive mutants to produce new commercial varieties which have become popular and are nowadays grown together with the old ones (Table 20.1).

A heterogeneous population of clones consists of individuals which draw origin from different ancestral mother plants, i.e., it contains different genotypes and such clonal populations may be selected for stocks or for scions.

Table 20.1 Some spontaneous mutants used as cultivars in fruit trees

Source cultivars	Mutant cultivar
	Apple
Golden Delicious	Belgolden, Golden B, Smoothee, Auvil Spur, Frazier Spur, Perleberg 3, Yellow Spur
Jonathan	Blackjon, Jonared, All Red Jonathan, Jonathan Cooper selection 2
Delicious	Harrold Red, Red King, Redwin Delicious, Richared, Shotwell Delicious, Stark Delicious, Starking Delicious, Topred Delicious, Vance Delicious, Hi-Early, Hi-Red, Hardispur, MacDonald, Red Chief, Redspur, Starkrimson Delicious, Wellspur
Jonagold	Jonagored, Wilmuta
	Pear
Williams	Max red Bartlett, Stripped Bartlett
Gorham	Grand Champion
Klapp	Starkrimson
	Plum
Agen	Prunier d'Ente GF 707, GF 698, GF 626, GF 303, GF 642, GF 652
Požegača	Bosanka, Korajka, Etscheid, Gunser, Meschenmoser, Schüfer, Wolff
	Peach
Coronet	Early Coronet
Redhaven	Early Redhaven, Garnet Beauty, Sunshine, Pratt's Compact Redhaven
Springcrest	Early crest

Source: Mišić, 1987

Selection of clonal stocks

The selection of clonal stocks is important for a large number of fruits which may be grown only on selected stocks, e.g., apple and quince. For apples, East Malling (EM) stocks, Roemer's clones, etc. are known world-wide. Different stocks are suitable for different cultivars as well as for various climatic and soil conditions. The grafting on the EM stocks ensures better uniformity, earliness and productivity.

At present, apples are mostly grafted onto clonal stocks, although some seedling stocks are still used.

In the case of grapevine, clones are exclusively used for rootstocks. Nowadays, the most extensively used stocks are those resistant to *Phylloxera* species, e.g., *Berlandieri* × *Riparia* and several others.

As success in production does not depend only on the traits of the scion, i.e., the variety, but also on the stock, as stated by Cummins and Aldwinckle (1983), rootstock breeding, either by vegetative or seed multiplication, must satisfy the following requirements:

1) Propagation ability, i.e., propagation requirement of the commercial nursery.
2) Graft compatibility. The stock must be compatible with most commercial cultivars.
3) Yield. The rootstock must induce early flower buds and good fruit set together with a regular and heavy production.
4) Longevity, i.e., good survival under the prevailing conditions.

Since a requirement in modern orchards is for a high uniformity of fruit-bearing trees to allow for simultaneous application of chemicals, mechanized harvest and to meet market demands, rootstocks obtained by clonal selection are preferred to rootstocks from seedlings.

Selection of clonal scions

Material used for selection of clonal scions may be homogeneous and heterogeneous.

Methods of mass and individual selection may be applied in the selection of clones for scions. The procedure of mass selection starts with the positive selection of the plant materials to be used for cloning. After the grafting of scions onto certain stocks, negative selection is conducted to remove weak and diseased nursery plants.

Much progress was made by the application of individual clonal selection. A considerable variability exists in certain cultivars so that individual trees differ in productivity, quality, disease resistance, etc. The individual clonal selection is based on the selection of the best trees or vines, according to important agronomic traits, which are then vegetatively reproduced (i.e., cloned). The individual clonal selection has a long tradition in Western Europe and North America and it was legally regulated there several decades ago (Janick and Moore, 1975; Huglin et al., 1980; cit. lit. Mišić, 1987 and Cindrić, 1987).

To produce new cultivars in the shortest possible time, individual clonal selection is used more and more in species whose domestic populations were considered such a rich material that mass selection was believed to be sufficient to secure a satisfactory progress. An example of a successful and efficient use of individual clonal selection is the development of the walnut cultivar Shampion on the basis of individual trees of a Yugoslav local population (Korać et al., 1988) Fig. 20.2).

Fig. 20.2 Walnut cultivar Shampion obtained by individual clonal selection from a Yugoslav local population (Courtesy of Prof. M. Korać, Fac. of Agric. Novi Sad)

Selection of seedlings

Sexual reproduction of fruit trees produces populations of seedlings which exhibit large genetic variability. The main causes of the variability are cross pollination and mutation. There are three main sources from which to select seedlings:

1) *Natural populations.* Natural populations of walnut, chestnut, plum, wild apple, etc. are particularly variable.
2) *Artificial populations.* As fruit cultivars are mostly cross pollinated, their seeds develop as a result of mating between flowers of the same cultivar as well as other cultivars in orchards. Therefore, fruit seeds produce a heterogeneous population which serves for the selection of seedlings for new stocks or scions, i.e., cultivars.
3) *Known hybrid populations.* The mating of known cultivars in an orchard may serve to establish the base material from which seedlings are then selected.

Selection of seedlings for stocks

Seedlings from natural populations are most frequently used for the production of stocks. Those are sexually produced stocks, to be distinguished from vegetative stocks or clones.

Seedlings for stocks are grown in a nursery and selected for one or more years, depending on the species in question. Seedling populations being heterogeneous, certain types are usually separated into groups (group selection). Within each group, negative selection is practiced in order to eliminate seedlings which are poorly developed, susceptible to diseases, or possess other undesirable traits. The best seedlings are transplanted and multiplied vegetatively.

Such stocks have been used extensively for a large number of fruits such as peach, plum, almond, cherry, black cherry, raspberry, olive, walnut, chestnut, etc. Presently, they are used less because it is difficult to produce uniform and homogeneous stocks and consequently, the orchard differs in growth and productivity which in turn creates problems in plant protection, picking and fruit quality (Pejkić, 1980). Therefore, vegetative stocks or clonal stocks, which are quite uniform, are increasingly used in plantations.

Selection of seedlings for cultivars

Artificial populations are used for the production of scions from which to select new cultivars. Seedlings from artificial populations are grown in a nursery, with individual selection based mostly on visual observations. That artificial populations may be a rich source of new cultivars is demonstrated by the fact that seedlings found accidentally gave rise to some famous cultivars, e.g., the apple cultivars Jonathan, Delicious, Baldvin, etc., the pear cultivars Anjou, Clerjou, etc., the cherry cultivars Bing, Lambert, etc. (Stanković, 1960).

Planned hybridization

Although natural and artificial populations result for most part from natural hybridization and they are good basic material for the selection of seedlings and clones, they do not fit altogether into the concept of a planned development of cultivars. Cultivars possessing certain desirable traits are developed on the basis of planned crossing preceeded by the selection of parental pairs. This is why controlled hybridization aimed at the development of cultivars is used more and more in pomiculture and viticulture, in spite of the length of the process.

Merits of hybridization are that it comes out with new and desirable gene recombinations and that it allows for the vegetative use of the F_1 generation which frequently contains the best recombination of genes of two parents. Furthermore, vegetative reproduction allows desirable genotypes from the F_2 and subsequent generations to be used without having to wait for them to become homozygous.

A number of grapevines, peaches, and other cultivars of self pollinated fruit species have been developed by planned hybridization. In cross-pollinated species too, new cultivars are made by hybridizing known parents (Alston et al., 1987).

As the establishment of modern fruit and grape plantations is a long and expensive undertaking, horticultural science and practice try to introduce more and more novelties in order to make the undertaking more economical.

Particular attention is paid to rootstock breeding since it differs from the breeding of scion cultivars in a number of characteristics. Cummins and Aldwinckle (1983) mention a few important points. The physical environment of the stock system is very different from that of the fruiting part of the tree and variations from tree to tree are far greater in the rhizosphere than above ground regarding temperature, gas exchange, moisture, soil, etc. The biotic environment is also quite different as far as pathogens, insects, symbiotic relationships, etc. are concerned. Furthermore, the physiology of the root is different from that of the leafy portion of the tree and the stock substantially influences vigor, blossom initiation, fruit set, and phenology of the tree. All these facts have to be taken into account when developing a rootstock breeding programme. According to Cummins and Aldwinckle (1983), three approaches offer substantial possibilities for the improvement of seed-propagated rootstocks.

Fig. 20.3 Columnar apple selection SA 162-130 (= Wijcik open pollinated) on MM 106, 3, 5 years from planting (courtesy of Drs. Richard and Kennett Tobuth, East Malling Institute of Horticulture, U.K.)

a) Inbred, self-fertile seed trees. Most peach seed-lines are inbred lines, highly self-fertile, giving relatively uniform seedlings. So production of inbred lines may be used for peach rootstock breeding and this also seems to be efficient for apricots and cherries.
b) The use of F_1 hybrid seed has been observed earlier and it has proved at present to give superior seedlings to ordinary collections of domestic seedlings. In Canada, a series of six crosses of apple were designed to produce seedlings with outstanding winter hardiness, high productivity and long-lived trees (Spangelo, 1971, cit. lit. Cummins and Aldwinckle, 1983).
c) Vegetative propagation by seed or use of seeds of apomictic origin. Rootstocks for citrus plants are commonly produced by the use of seeds from apomixis. In several species of *Malus*, apomixis is found, and seedlings derived from apomictic sources could significantly lower the cost of tree production in the nursery.

More recently, merits of dwarfing stocks have been emphasized since such stocks are considered to be advantageous for spur-type scion cultivars and for mechanical harvesting systems.

Among the novelties, it is worth mentioning the breeding of columnar apple varieties. In an orchard in British Columbia, in a Canadian apple variety McIntosh, a spontaneous mutant was found, with short internodes and axillary buds which grow into spurs rather than extension shoots (Fisher, 1969, cit. lit. Tobuth, 1984). This mutant, known as Wijcik, is now used in several breeding programmes to produce columnar selections for high density orchards. According to Tobuth (1984), columnar types should prove useful for the amateur gardener as easily managed fruit trees, but also for very intensive plantations (meadow orchards) since the trees would not need to be cut back in alternate years as was the case with earlier meadow orchards (Fig. 20.3.).

In some cultivars of apple, plum, blackberry and citrus, a certain number of seeds develops by apogamy. The resulting seedlings possess only the traits of the mother plant and, on account of their high uniformity, are very useful for the production of stocks.

Apomixis

Some species of forage grasses (*Poa pratensis*, *Paspalus notatum*, *Panicum maximum*, *Eragrostis curvula*, etc.) reproduce by apomixis. This mode of reproduction involves the individual parts of the sexual structures, the seeds developing without the union of male and female gametes (Gustafsson, 1947). There are several types of apomixis of which apogamy, apospory, diplospory, and parthenogenesis are the most frequent.

Apogamy results from the fusion of two haploid cells, either synergids or antipodals. The resulting plant is diploid and it develops normally.

Apospory is the case when the embryo develops from a cell in the embryo sac without chromosome reduction.

Diplospory occurs when the embryo develops from the megaspore mother cell without chromosome reduction.

Parthenogenesis is a type of apomixis when the embryo develops from an unfertilized egg. If the chromosome number remains unreduced during meiosis, the embryo and the plant will be diploid. If the egg cell has a reduced chromosome number, but for some reason fertilization does not take place, the embryo and the plants will be haploid.

Some plant species are obligate apomictics; and some are facultative apomictics in which both apomictic and sexual reproduction occur. Apomictic progeny is identical to the mother plant.

Apomixis like other types of vegetative propagation allow clones to be developed. Apomixis is definitely advantageous for the propagation of superior genotypes, especially if they are heterozygous. In the case of obligate apomictics, in which hybridization and gene recombination are precluded, apomixis reduces genetic variability and lowers breeding success. Facultative apomictics are thus preferred because they are open to gene recombinations and the development of new genotypes.

Use of new methods and techniques

Tissue culture methods, used in some ornamental plants for a long time, have started to be used in pomiculture.

Debergh and Maene (1982) mention that meristem culture was used in orchids as far as 75 years ago. It is a merit of Morel (1963) that it became a routine method of orchid production.

Murashige (1978) reports of three types of propagules for multiplication through tissue culture: 1) enhanced bud breaking, 2) production of adventitious buds, and 3) somatic cell embryogenesis.

Tissue culture is used extensively in strawberries for a rapid clonal propagation of new cultivars (Damiano, 1980; cit. lit. Debergh and Maene, 1982). Satisfactory results have been achieved with tissue culture in apples. In the case of high-density orchards (meadow orchards), which should be established on their own roots to accelerate production, tissue culture appears to be a particularly suitable method (Murashige and Huang, 1987). Other fruit species have also been experimented with, e.g., peaches (Hammerschlag et al., 1987; Matsuta et al., 1986), pears (Browning et al., 1987), walnuts (Heile-Sudholt et al., 1986), etc. The work is still at the beginning because laboratory techniques are just being worked out and the propagation of scion cultivars has not yet become a routine operation.

Embryo culture is applied to obtain seedlings from incompletely developed embryos resulting from interspecific hybridization. Because of a low viability of seeds, embryo culture is almost the sole method of developing very early peach and nectarine cultivars (Hammerschlag et al., 1987).

The production of haploids arises interest in pomiculture because it would allow the crossing of two homozygous parents to produce highly uniform seedling rootstocks as well as new scion cultivars (Lespinasse et al., 1983). These authors examined about 4,000 apple seedings and found only one haploid plant resulting from an unfertilized egg-cell (haploid parthenogenesis). However, application of anther culture may increase the frequency of haploids and subsequently the formation of doubled haploids, as demonstrated by Fei Kai-wei and Xue Guang-rong (1981) with the cultivar Delicious and with the crab apple (cit. lit. Lespinasse et al., 1983).

In recent years attempts have been made to transfer genes by Ti-plasmids from *Agrobacterium tumefaciens*. The first successful transformations in strawberries (Jelenković et al., 1986) and in walnuts (Martin and Davdekar, 1986; cit. lit. Mišić, 1987) indicate that some interesting results may soon be expected.

With respect to difficulties associated with the length of the breeding process in woody fruit trees, new methods may help to accelerate the process, especially a rapid clonal propagation of exceptional individual trees and virus-free plants.

References

Alston, F. H., Tobuth, K. P., and Bailey, T. C., 1987. Programme 1/86. Breeding of apples and pears. Report East Malling Research Station (U.K.) for 1986, p. 19–23.

Browning, G., Ognjanov, V., Passey, A. J., and James, D. J., 1987. Multiple shoot and root regeneration from pear embryo cotyledon explants in vitro. Jour. Hort. Sci. 62 (3): 305–311.

Cindrić, P., 1987. (Contemporary trends in grape-vine breeding). In: M. Osmanagić (Editor), Food and Development, Jugoslovenska Naučna Tribina, Beograd, p. 213–218.

Cummins, J. N., and Aldwinckle, H. S., 1983. Rootstock breeding. In: J. N. Moore and J. Janick (Editors), Methods in Fruit Breeding. Purdue Univ. Press, West Lafayette (USA), p. 294–327.

Damiano, C., 1980. Strawberry micropropagation. In: Anderson, W. C., 1980, 11–12.

Debergh, P. C., and Maene, L. J., 1982. Contribution of tissue culture techniques to horticultural research and production. Proc. XXI st. International Horticultural Congress, Hamburg, 29th August to 4th Sept. 1982, p. 787–798.

Fei Kai-wei and Xue Guang-rong, 1981. Induction of haploid plantlets by anther culture in vitro in apple cv. Delicious. Acta Hort. Sinica, 8 (1): 44.

Fischer, D. W., 1969. Spur-type strains of McIntosh for high density planting. British Columbia Fruit Growers Assoc. Quality Report 14 (2): 3–10.

Gustafsson, A., 1947. Apomixis in higher plants. C. W. K. Gleerup. Lund, 370 pp.

Hammerschlag, F. A., Bauchan, G. R., and Scorsa, R., 1987. Factors influencing *in vitro* multiplication and rooting of peach cultivars. Plant Cell, Tissue and Organ Culture 8: 235–242.

Heile-Sundholt, C., Huetteman, C. A., Preece, J. E., Cambeck van, J. W., and Gaffney, G. R., 1986. In vitro embrionic axis and seedling shoot tip culture of *Juglans nigra* L. Plant Cell, Tissue and Organ Culture 6: 189–197.

Huglin, P., Guillot, R., Valat, C., and Vuittenez, A., 1980. L'evaluation genetique et sanitaire du material clonale de la vigne. Bull. de I.O.IV. 53: 857–882.

Janick, J., and Moore, J. N., 1975. Advances in Fruit Breeding. Purdue Univ. Press, West Lafayette, Indiana, USA. 623 pp.

Jelenković, G., Chin, C., and Billings, S., 1986. Transformations studies of Fragaria ananassa Duch. by Ti plazmids of *Agrobacterium tumefaciens*. XXII. Inter. Hort. Congress, Davis (USA), p. 695.

Korać, M., Slović, D., Cerović, S., and Golušin, 1988. Characteristics of new walnut cultivars – Shampion, Srem, Tisa, Baćka and Mire. Proc. IX. Congress of fruit Growers of Yugoslavia, Novi Sad, (In press).

Lisenko, T. D., 1948. O položenii v biologičeskoi nauke. Seljhozgiz. Moskva.

Lespinasse, S., Godicheau, M., and Duron, M., 1983. Potential value and method of producing haplods in the apple tree, Malus pumila Mill. Acta Horticulturae, 131: 223–230.

Martin, L. A., and Davdekar, A. M., 1986. Infection of walnut (Juglans spp) with recombinant *Agrobacterium tumefaciens*. XXII. Inter. Hort. Congress, Davis (USA), p. 969.

Matsuta, N., Hirabayashi, T., Akihama, T., and Kosaki, I., 1986. Callus formation protoplast of peach cell suspension culture. Scientia Horticulturae, 28: 59–64.

Mišić, P., 1987. (New trends in fruit breeding). In: M. Osmanagić (Editor), Food and Development, Jugoslovenska Tribina, Beograd, p. 205–211.
Morel, G., 1963. La culture in vitro du méristéme apical de certaines orchidées. C.R. Acad. Sci. Paris 256: 4955–4957.
Murashige, T., 1978. The impact of plant tissue culture on agriculture. In: T. A. Thorpe (Editor), Frontiers of Plant Tissue Culture, 1978, *IAPTC*, Calgary, pp. 518–524.
Murashige, T., and Huang, L. C., 1987. Cloning plants by tissue culture: early years, current status and future prospects. Acta Horticulturae 212: 35–42.
Pejkić, B., 1980. Oplemenjivanje voćaka i vinove loze. Naučna knjiga, Beograd, 486 pp.
Spangelo, L. P. S., 1971. Hybrid seedling rootstocks for apple. Canadian Dept. Agr. Pub. 1431.
Stanković, D., 1960. Oplemenjivanje voćaka i vinove loze. Zavod za izdavanje udžbenika, Beograd. 211 pp.
Tavčar, A., 1952. Osnove genetike. Školska knjiga, Zagreb, 493 pp.
Tobuth, K. R., 1984. Breeding columnar apple varieties at East Malling. Scientific Horticulture, 35: 72–77.

21. GENETIC COMPOSITION AND ADAPTABILITY OF THE CULTIVAR

Before discussing the genetic structure or composition of the cultivar, it is necessary to clarify the meaning of the term *cultivar* itself. The term has frequently been used in previous chapters assuming that the reader understands its meaning. Nevertheless, the term should be defined precisely since it is constantly in operation.

What is the cultivar?

Several definitions of the cultivar are in use. However, the one given in the International Code of Nomenclature for Cultivated Plants – 1980 is most appropriate. In Article 10 of the Code it is stated:

"The international term cultivar denotes an assemblage of cultivated plants which is clearly distinguished by any character (morphological, physiological, cytological, chemical, or other), and which, when reproduced (sexually or asexually), retains its distinguishing characters.

The cultivar is the lowest category under which names are recognized in this Code. This term is derived from cultivated variety, or its etymological equivalents in other languages.

Note 1. Mode of origin is irrelevant when considering whether two populations belong to the same or to different cultivars.

Note 2. The concept of cultivar is essentially different from the concept of botanical variety, varietas. The latter is a category below that of species. Names of botanical varieties are always in Latin form and are governed by the Botanical Code. Rules for the formation of cultivar names are set out in the preseent Code (see Arts. 27–32).

Note 3. The term cultivar is equivalent to variety in English, variété in French, variedad in Spanish, variedade in Portugese, varieta or razza in Italian, varieteit or ras in Dutch, Sorte in German, sort in Scandinavian language and Russian, pinzhong (p'inchung) in Chinese, and hinshu in Japanese, whenever these words are used to denote a cultivated variety.

Note 4. The terms cultivar and variety (in the sense of cultivated variety) are exact equivalents. In translations or adaptations of the Code for special purposes either cultivar or variety (or its equivalent in other languages) may be used in the text.

Note 5. Usually a cultivar will comprise a part only of the species, botanical variety or other botanical category under which it is classified. A cultivar may however be co-extensive with any of these.

Note 6. When a forestry provenance is clearly distinguished by one or more characters and, when reproduced, retains its distinguished characters, it may be treated as a cultivar." (p. 12–13).

Genetic composition of the cultivar

As any given cultivar exists in a certain environment, it should establish a relationship with the environment to ensure its optimum prosperity. This relationship is reflected in the adaptive value of the cultivar itself. It is the outcome of the genetic constitution of the genotype(s) comprising the cultivar. According to its genetic composition, the cultivar may be either a:

1) pure line cultivar,
2) pure cultivar,
3) multiline cultivar,
4) mixture of cultivars,
5) population cultivar,
6) synthetic cultivar,
7) hybrid cultivar,
8) polyploid cultivar,
9) clonal cultivar.

Frey (1983) considered only four types of plant cultivars: (1) pure lines, (2) hybrids, (3) multilines, and (4) synthetics. The types of cultivars we have cited refer more to the origin of cultivar development which may be more helpful in breeding as well as in commercial production.

The breeder's decision which of the above genetic compositions to create depends on several factors: whether he deals with self-pollinated or cross-pollinated plants, annuals or perennials, vegetatively or generatively propagated plants, agroecological conditions for which the cultivar is made, and requirements of the market. All these factors must be considered so that the genotype of the cultivar to be developed is clearly defined already at the beginning of the breeding programme.

Pure line cultivar

Pure line cultivars comprise progenies of a single homozygous self-pollinated plant. When multiplied, a cultivar is obtained whose individuals, regardless of their number, have the same genotype. It has already been said (Chapter 4) that this state cannot be maintained for a long time due to mutations, natural hybridization and mechanical admixtures occurring in the course of reproduction.

If a decision is made for a new cultivar to be a pure line, it is most convenient to apply the pedigree breeding method. In the case of self-pollinated plants, experimental materials should not be considered to be a pure line before the F_6 generation. Individual selection in subsequent generations (F_7 to F_{10}) leads to full homozygosity for all traits.

Inbred lines of cross-pollinated species after 6 generations of inbreedings are genetically identical to pure lines of self-pollinated species. In view of inbreeding depression, inbred lines are not grown for yield production themselves but serve almost exclusively as components of heterotic hybrids.

Pure lines cultivars are mostly used in self-pollinated species of vegetables and decorative plants which are predominantly grown in controlled environments (sowing of transplants, in greenhouses and cloches, etc.) and which should be phenotypically uniform to meet the market demand. Pure line cultivars are also used in self-pollinated cereals (wheat, barley, rice, etc.), legumes (soyabean, pea, vetch, etc.) and other intensively produced crops.

Pure cultivar

The pure cultivar is a group of very similar genotypes or lines that is phenotypically uniform. With self-pollinated plants, it is obtained from progenies of a single line which have not become fully homozygous, which happens when a phenotypically uniform line is selected from the F_5 or F_6 generation. Therefore, the pure cultivar is a croup of several pure lines which do not differ much genotypically and which are very uniform phenotypically.

As the pure cultivar is not a pure line, the process of its development is somewhat faster. Already in the F_4 generation it is possible to apply the pedigree breeding method and isolate lines from which a pure cultivar will be selected.

Most widely grown commercial cultivars of wheat, barley, oat, rice, soyabean, vetch, and other self-pollinated plants are in fact pure cultivars.

Whether a pure cultivar will be "more" or "less pure" depends on legal regulations on varietal purity, market requirements, customers' preferences, etc. In countries with intensive production, genetically very homogeneous materials are favoured, in order to diminish competition among individuals in dense stands and to obtain maximum yields. Consequently, a pure cultivar is very uniform because it includes a small number of similar genotypes (lines). In countries with less intensive production and not so well organized seed production, the pure cultivar is "less pure", i.e., it is more heterogeneous on account of a larger number of similar lines constituting it.

Multiline cultivar

The intensification of agricultural production in the last 50 years has increased the phenotypic uniformity and genetic homogeneity of the cultivar, i.e., it has led to the system of monoculture. Increased yields per unit area did result, but costs of crop protection increased also because dense crop populations abundantly fertilized with nitrogen become a favourable environment for the development of diseases and pests. When the problem was recognized, solutions were sought of how to control diseases biologically, i.e., genetically.

Jensen (1952) came out with a concept of the multiline cultivar, Borlaug (1959) worked out the concept in more detail for wheat, and Browning and Frey (1969) supplied the theoretical and practical explanations of the development and growing of multiline cultivars.

The multiline cultivar consists of 6 to 10 isogenic lines which are phenotypically rather uniform but different in genes for resistance to parasites, e.g., rusts and powdery mildew.

The multiline cultivar is developed from a good cultivar serving as the recurrent parent. It is crossed to several cultivars which serve as donors of various genes for resistance to pathogens. Later generations are back-crossed to the recurrent parent and selected for the resistance of the donors. After 5–6 generations of back-crossing, almost isogenic lines in the type of the recurrent parent are produced which differ in resistance to pathogens.

Although the concept of the multiline cultivar has been known for quite some time, its practical application has not been very wide. The best studied examples are the growing of several multiline oat cultivars in Iowa, U.S.A. (Frey and Browning, 1976). One of them will be given here for illustration.

Table 21.1 Pedigree and response to prevalent races of Puccinia *coronata* f.sp. *avenae* of nine isolines of oats and their percentage in foundation seed of the multiline cultivar "Multiline E 74"

CI No.	Pedigree	Response to races			
		290	326	264B	%
9169	$CI8044^4 2\times CI17555^4 \times$ Ceirch du Bach	MR	S	MS	18
9170	$CI8044^5 3\times$ Cliston $\times$ Garry $2\times CI8079$*	R	R	R	22
9172	$CI8044^6 \times CI8001$	MR	MS	MS	4
9176	$CI8044^6 \times$ Acencao	R	MR	S	4
9177	$CI8044^4 3\times$ Clintland $2\times$ Chapman $178\times CI7235$	MR	MS	MS	5
9178	$CI8044^4 3\times$ Bonkee $2\times CI7154\times CI7171$	R	MS	MR	18
9179	$CI8044^4 2\times CI7555\times CI6665$	MS	S	MS	4
9180	$CI8044^4 2\times CI7555\times CI7654$	MR	MR	MS	4
9181	$CI8044^5 2\times$ Clinton $\times CI8081$*	MR	R	R	22
Percentage (R + MR + MS)		100	78	96	

* *Avena sterilis*

Source: Frey and Browning, 1976

The multiline cultivar Multiline E74 has been composed of 9 isolines represented by 4% to 22% from each isoline (Table 21.1). The lines differ in resistance to three prevalent races of *Puccinia coronata* f.sp. *avenae* as well as in some other traits. Some of them had undergone 6 and some only 4 or 5 back-crosses which was decisive in the determination of their percentages in the cultivar.

Only 78% of plants of Multiline E74 are resistant to race 326 but 100% and 96% of plants are resistant to the other two races, 290 and 264B, respectively (Table 21.1). Frey et al. (1979) are of the opinion that 60% of resistance to each pathogen is sufficient to prevent economically pertinent losses.

Each isoline is multiplied separately in amounts needed to compose the multiline cultivar. In the course of commercial exploitation of the cultivar, the pathogen is screened for distribution and for the occurrence of new races so as to be able to counteract it by replacing certain isolines in the cultivar.

Since 1968, 13 multiline oat cultivars have been included in commercial production in the U.S.A. They have been grown annually on about 400,000 ha without suffering economic losses from stem rust (Browning et al., 1979).

The CIMMYT programme in Mexico has not succeeded in producing commercial multiline cultivars of wheat. The principal problem resulted from the decision to try to incorporate resistances to several pathogens (leaf rust, stem rust, powdery mildew, *Septoria*). Another problem is associated with the fact that it is difficult to completely restitute the recurrent parent (Siete Cerros) when quite different donors are used.

In Punjab, India, Dr. K. S. Gill and his associates have succeeded in introducing into commercial production several multiline wheat cultivars derived from the cultivars Kalyansona and PV18 (Gill et al., 1975, 1979). These multiline cultivars, resistant to yellow rust and leaf rust, have outyielded on a two-year average the original pure line cultivars, Kalyansona and PV18. Moreover, they exhibited a higher yield stability, especially in 1975/76 (Table 21.2).

Table 21.2 Yield grain, coefficient of regression and deviation of regression in some multilines obtained on the basis of wheat cultivars Kalyansona and PV 18

Multiline and cultivar	Yield* t/ha	Coefficient of regression 1974/75	1975/76	Deviation of regression 1974/75	1975/76
LSML 1	4.73	1.03	0.98	6.34	4.17
KSML 2	4.61	0.93	0.81	4.68	4.22
KSML 3	4.84	1.02	1.30	4.64	6.22
KSML 4	4.89	1.04	1.28	28.71	1.87
Kalyansona	4.59	0.96	0.69	4.77	27.31
PVML 1	4.67	0.89	0.97	23.75	15.48
PVML 2	4.62	1.09	1.09	10.36	1.97
PVML 3	4.79	0.99	1.13	8.00	7.89
PV 18	4.69	0.98	0.76	5.66	47.97

* 2 year average

Source: Gill et al., 1979

However, controversial opinions are still voiced as to the advantage of multiline cultivars over conventional ones in yield level and/or stability. The solving of the dilemma was greatly facilitated by the results of a long-term study conducted by Allan (1988) in Washington State (U.S.A.). The study, conducted from 1979 to 1986 in 10 to 48 locations annually, included 16 multiline cultivars of soft white club bread wheat which were compared against four compact-type cultivars and three white bread wheat cultivars. The majority of the multiline cultivars consisted of 10 components, and one cultivar even 23 components, which had been mixed in equal weight proportions. The components had been developed by one to four back-crossings of nine American compact-type cultivars and 18 lines resistant to yellow rust.

In relation to the yields of the cultivars, several of the multiline cultivars were superior, most of them had yields on the level of the cultivars and some were inferior (Table 21.3). It should be pointed out that the three cultivars (Tres, Nugaines and Daws) had higher yields than most multiline cultivars while the four compact-type cultivars had lower yields (Table 21.3).

Table 21.3 Mean grain yields of 1 t wheat multilines and their mean yields differences vs. 7 pureline cultivars

				Cultivar							
Multiline	Components	Trials	Multiline kg/ha	Barbee	Paha	Tyee	Tres	Nugaines	Daws	Stephens	LSD
				Difference from multiline (kg/ha)							
Crew 78	10	32	5,863	−457*	−523*	−30	131	−473*	77	−84	213
Crew 79	10	48	5,447	−444*	−541*	−47	144	−306*	110	−78	158
Crew 80	10	46	5,564	−413*	−487*	−30	306*	−487*	212*	128	164
Crew	10	36	6,014	−527*	−642*	−181	84	−578*	−90	−215*	192
Crew 83	10	18	6,253	−830*	−782*	−407*	−242	−252	−3	−239	285
Paha ML	11	30	5,793	−313*	−447*	84	369*	−195	238*	97	217
Barbee ML	8	30	5,571	−91	−225*	306*	591*	27	460*	319*	217
Moro ML	5	28	5,638	−134	−228*	290*	615*	30	390*	309*	227
Faro sib ML	10	28	5,853	−349*	−447*	71	400*	−185	175	−94	227
78 MLA	10	14	4,140	−158	−202	228	393*	396*	547*	228	240
78 MLB	10	14	4,277	−295*	−339*	91	256*	259*	410*	91	240
79 ML	10	14	4,199	−137	−181	249*	414*	417*	568*	249*	240
Tres CX	11	18	5,783	−404*	−420*	57	302*	137	134	16	294
84 ML	23	16	5,991	−699*	−642*	−276	10	−91	91	185	303
85 MLA	11	12	6,156	−639*	−736*	−471*·	−88	−37	−162	−353*	352
85 MLC	11	10	6,391	−820*	−847*	−645*	−199	−242	−313	−545*	384

* Multiline and cultivar differ significantly based on the LSD of $P < 0.05$.

Source: Allan, 1988

On the whole, the multiline cultivars exhibited a yield stability similar to that exhibited by the cultivars. Some of them were better and some poorer than the latter cultivars.

Furthermore, the multiline cultivars did not outyield the cultivars on account of a higher level of resistance to yellow rust, because the latter also had a good resistance to yellow and leaf rust.

It may be concluded on the basis of the results of Allan (1988) that multiline cultivars may be superior to pure line cultivars but not necessarily. The success of a multiline cultivar depends on its components, the success of a pure cultivar on its gene recombination pattern.

The multiline cultivar has been conceived almost exclusively as a means of genetic control of parasites. In a crop comprised of genotypes differing in resistance, the initial inoculum (x_0) and degree of infection (r) are inevitably reduced (Van der Plank, 1963; Browning and Frey, 1969). This, on one side, lowers yield reductions brought about by disease attack and, on the other side, prolongs the effective resistance to the pathogen in the cultivar (see Chapter 14).

Nevertheless, multiline cultivars have never become widely accepted although the concept they stand for seems to be the most efficient means of controlling diseases which take the largest part of breeding effort and are one of the most important single factors of yield reduction in cultivated plants. The chief reasons for this are difficulties encountered in composing lines for the multiline cultivar, their maintenance and substitution, and especially the fact that the multiline cultivar keeps the yield stagnant at the level of the recurrent parent while superior new cultivars may be developed with the same effort. Seed production presents another set of problems. In countries with a long tradition in the production of seed of pure line cultivars, the shift to the production of seed of "impure cultivars" is slow because the control of multiline production is difficult and its abuse easily managed.

Mixture of cultivars

Mixtures of cultivars are mechanical mixtures of two or more cultivars composed at a certain ratio and used for sowing. Therefore, mixtures of cultivars consist of different genotypes which are more different than the components of the multiline cultivar.

When there are several limiting environmental factors and no available cultivars to counteract all of them, a solution is sought by making a mixture of cultivars. This happened in Yugoslavia after the introduction of Italian wheat cultivars in the 1960s which were not sufficiently winter-hardy. Double and triple mixtures of cultivars differing in resistance to low temperatures, lodging, rusts and powdery mildew had to be tested (Borojević and Mišić, 1962).

The cultivars which were more winter-hardy (e.g., San Pastore, Elia) produced larger numbers of plants and spikes per m^2 and therefore contributed more to the total yield than did the less winter-hardy cultivars (Fortunato and Mara) although they were sown in equal proportions (Table 21.4). The mixture of cultivars equalled the average yield of the pure cultivars. Other investigations also indicate that mixtures of cultivars yield to the level of the better cultivar or equal the average for pure cultivars (Orlyanskaya and Polyakov, 1950; Nasypayko, 1959). In Hungary Kovacs and Arbanyi (1985) examined 10 combinations of cultivar mixtures of wheat in two locations and found that some mixtures outyielded the better cultivar in one location but not in the other. The authors concluded that the cultivar mixtures were superior to the purely sown cultivars only in yielding stability.

A classical example of differential survival in mixtures was reported by Suneson (1949) as already discussed in Chapter 12. Four adapted barley cultivars were mixed in equal proportion. After 16 years, the cultivar Atlas comprised 88% of the mixture while the cultivars Hero and Vaughn only 1% and 0%, respectively (Table 21.5).

Table 21.4 A study of wheat cultivar mixtures in Novi Sad in 1959/60 and 1960/61

Cultivar and mixture	Planted grains/m²	No. of plants/m²	Productive tillers	Total	Per component	% component	Grain yield t/ha	Yield of mixture in % in relation to components average
San Pastore RA*	600	432	1.45	627			6.77	
San Pastore + Fortunato	350 + 350	485	1.16	562	344 + 218	61.2 38.8	6.03	92.4
Fortunato *rA*	700	446	1.26	561			6.28	
San Pastore + Mara	350 + 350	505	1.18	598	372 + 226	62.2 37.8	6.39	103.2
Mara *rA*	700	433	1.15	497			5.62	
Elia + Mara	300 + 350	448	1.26	567	380 + 187	67.0 33.0	6.15	107.0
Elia *RA*	600	489	1.19	584			5.87	
Leone *Ra*	700	509	1.29	657			5.93	
Leone + Fortunato	350 + 350	444	1.34	595	327 + 268	54.9 45.1	6.00	98.5
San Pastore + Leone + Fortunato	233 + 233 + 233	535	1.17	626	225 + 237 + 164	35.9 37.9 26.3	6.49	102.7
San Pastore + Leone + Mara	233 + 233 + 233	494	1.26	623	246 + 207 + 170	39.5 33.2 27.3	6.17	101.0
				LSD	5% 1%		0.677 0.902	

* Spike colour and awnedness were used as markers for counting the number of spikes/m²: *RA* – red awnless, *rA* – white awnless, *Ra* – red awned.

Source: Borojević and Mišić, 1962

Table 21.5 Analysis of barley cultivars remaining in a mixture propagated for 16 consecutive years

Cultivar	Year of analysis 1933	1940	1948
	percentages		
Atlas	25	63	88
Club Mariont	25	17	11
Hero	25	8	1
Vaughn	25	12	0

Source: Suneson, 1949

When tested in pure stands, Vaughn gave higher yields than Atlas. The competitive advantage of Atlas was attributed to the better root development. In general when grasses are concerned, genotypes with better tillering ability and higher plant height proved to be better competitors in mixtures.

Mixtures of cultivars are seldom used in intensive production if an array of cultivars is available. They are more frequently used with forage grasses and legumes, where even different species are mixed in order to use the variation in the production of biomass in the first and subsequent years and in different cuttings.

Population cultivar

A population cultivar is a cultivar consisting of a large number of different genotypes, mostly those found in natural or local populations, but it has one or more heritable characteristics by which it can be differentiated from other cultivars.

When self-pollinated crops are concerned, each genotype represented in the population is homozygous at a certain frequency. When cross-pollinated crops are concerned, a portion of genotypes is homozygous and the other part is heterozygous at a certain frequency which, in the system of random mating, may come into the state of equilibrium.

Population cultivars of cross-pollinated crops are grown when it is either biologically impossible or economically unprofitable to produce a hybrid cultivar or any other cultivar type. Some grasses and forage legumes are mostly grown as population cultivars.

Synthetic cultivar

The synthetic cultivar consists of several different genotypes made artificially by certain methods, but again having a distinctive heritable characteristics. Those are mostly cross-pollinated plants such as maize, lucerne, etc.

Using the recurrent selection method, many synthetic populations of maize have been made: Iowa State Stiff Stalk (SSS_{III}) (Lonnquist and Williams, 1967), Iowa Long Ear Synthetic (BSLE) (Hallauer, 1972), Alph (Russell et al., 1973), BsCB1, BSSS(HT) (Eberhart et al., 1973).

Maize synthetics have mostly been used for the production of superior inbred lines for hybrids although many of them were good enough performers to be grown themselves in commercial production. Whenever maize hybrid production encounters a problem, due to susceptibility to diseases or pests, an old argument comes out whether synthetic maize cultivars *per se* would have been as good as maize hybrids if the same efforts and funds had been invested in them. A number of trials have been conducted in that context which, according to Duvick and Brown (1979), proved beyond doubt the superiority of the hybrids over the synthetics (Table 21.6). These authors concluded that the synthetics are lower yielding because they have been selected mostly for yield while the hybrids are selected also for other traits, e.g., grain quality, resistance to diseases and pests, etc.

Table 21.6 Yields of synthetic varieties per se compared with control hybrids

Variety, hybrid	Yield in t/ha		Yield of syn. in % of hybrid	Authors
	syn.	hybrid		
Syn-3 High	5.9		93.6	Lonnquis (1949)
US 13 (double cross)		6.3		
Mean of nine synthetics	5.3		62.3	Eberhart (1971)
Mean of four hybrids		8.5		
Mean of six synthetics	4.2		65.6	Eberhart (1971)
Mean of four hybrids		6.4		
Mean of nine synthetics	6.2		75.6	Hallauer (1972)
Mean of four hybrids		8.2		
Mean of two synthetics	6.6		97.0	Hallauer (1973)
Mean of six hybrids		6.8		
Mean of two synthetics	4.9		65.3	Russell et al. (1973)
Mean of three hybrids		7.5		
Mean of four synthetics	5.8		79.5	Eberhart et al. (1973)
Mean of four hybrids		7.3		
Mean of synthetics	5.6		79.7	
Mean of control hybrids		7.3		

Source: Duvick and Brown, 1979

Hybrid cultivar

The hybrid cultivar is the F_1 generation obtained by crossing two $(A \times B)$, three $(A \times B) \times C$ or four parents $(A \times B) \times (C \times D)$. The hybrid cultivar exploits hybrid vigour (heterosis) which is frequently manifested in crosses of genetically different inbred lines or varieties.

If the F_1 generation of single-cross hybrids is used, the hybrid cultivar is heterozygous, but all individuals in the populations have the same genotype. In other cases the hybrid cultivar may consist of different genotypes.

In cross-pollinated species such as rye and some grasses, for which it is difficult to produce inbred lines or manual production of hybrid seeds is not economical, the hybrid cultivar is made by means of random mating of different varieties grown in bulk. Such a hybrid cultivar is not uniform genotypically but it represents a mixture of various heterozygous and homozygous genotypes.

As in open-pollinated plants each new reproduction differs from the preceeding one, F_1 hybrids are made wherever it is possible. This brings a double advantage – on one side, hybrid vigour is used better than in synthetic cultivars, and on the other, seed production if fully controlled (Table 21.6).

The maximum state of heterozygosity present in the hybrid variety makes it highly adaptable to various environmental conditions. This is yet another advantage of the hybrid cultivar which gains popularity in many agricultural plants, as mentioned in Chapter 16.

Polyploid cultivar

The polyploid cultivar may vary genetically in dependence of ploidy level and the genetic composition of the parental materials constituting it. In the case of rye, buckwheat, red clover, etc., tetraploid cultivars are mostly heterozygous because the diploid cultivars which compose them are predominantly heterozygous too. Regarding triploids such as sugar beet, watermelon, etc., they are heterozygous in most gene loci. This is why the effects of ploidy as well as hybrid vigour are expressed in them. Because of this double advantage, the production of polyploids at different levels by chromosome engineering gains importance in plant breeding (see Chapter 18).

Clonal cultivar

In many plant species it is possible to develop a cultivar vegetatively, i.e., by cloning. If all clones originate from the same mother plant, the cultivar will be homogeneous and all plants will have the same genotype, unless a mutation occurred in reproductive parts. If clones come from different mother plants, the cultivar will be heterogeneous, i.e., it will consist of different genotypes (see Chapter 20), with some distinct heritable feature.

Depending on differences in genetic composition, different clonal cultivars will react differentially to environmental factors.

Adaptability and yield stability of the cultivar

The adaptability of the cultivar is generally defined as a genetic capacity to realize a high and stable yield in a range of environments (Finlay and Wilkinson, 1963; Allard and Bradshaw, 1964; Oka, 1967; Matsuo, 1975; etc.). As cultivars definitely differ in adaptability to the environment, general or broad adaptability and specific or narrow adaptability may be distinguished (Oka, 1967). There seems to be a regularity for the broadly adaptable cultivars to have stable yields in a variety of environments but at a lower level than the narrowly adaptable cultivars. The latter have a high genetic yield potential and perform exceedingly well in favourable conditions but poorly in unfavourable ones (Finlay and Wilkinson, 1963).

Understandably, growers and breeders are primarily concerned with yield level and stability of the cultivar. Yield level depends on the genetic yield potential, i.e., on positive genes incorporated into the cultivar in the breeding process. Yield stability depends on the cultivar's capacity to react to environmental conditions. A synonymous term for it is phenotypic plasticity (Bradshaw, 1965; Frey, 1983). It is determined by the genetic composition of the cultivar, i.e., by the reaction of its genotypes as individuals and the population on the whole.

In fact it is a matter of interaction between the cultivar and environmental factors, the outcome of which depends on the genetic composition of the cultivar and the intensity of environmental factors, particularly the limiting ones. The result of the cultivar/environment interaction is expressed as the adaptability and stability of the cultivar.

Let us consider the dependence of adaptability and yield stability of cultivars on their genetic composition.

Yield level and stability of the pure line cultivar are based on individual buffering to limiting factors of the environment, i.e., on the capacity to make use of favourable factors of the environment (Allard and Bradshaw, 1964). As all individuals have the same genotype, the adaptability ensues from the reaction norm of this genotype to different environmental conditions, i.e., its phenotypic plasticity. A degree of adaptation depends on the development of individual plants which is determined by seed size as the initial capital, sowing date, coincidence of critical stages of plant development and adverse conditions, etc. It means that yield stability rests on the developmental or physiological homeostasis. As the pure line practically does not possess genetic variability, its adaptability is specific and suitable for narrow agroecological areas. Therefore, pure lines with high genetic yield potentials are grown in favourable environments in which most factors are under control.

Regarding the pure cultivar whose genotypes are highly similar, its adaptability and yield stability also rest on individual buffering, i.e., physiological homeostasis. The genotypes, being very similar, react identically to environmental factors such as temperature, pathogens, etc. If the genotypes are more different and there is a larger number of them, genetic homeostasis, i.e., population buffering is also present.

In any case, pure cultivars are more adaptable than pure line cultivars which explains why the former are regularly grown over a larger area than the latter.

The multiline cultivar, owing to its genetic variability, is more adaptable than the pure cultivar, especially to parasite attack which also makes it a more stable performer. The stability of multiline cultivars is based on population buffering (Allard and Bradshaw, 1964), i.e., on the genetic homeostasis. In other words, some genotypes are resistant to one group of pathogenic races, the other genotypes to another group, while the population on the whole reduces the initial inoculum and infection intensity. This effect is called "population resistance" (Frey, 1983).

Regarding their productivity, multiline cultivars may be as productive as pure cultivars because they are made from the best pure cultivars, as illustrated by examples of wheat and oats.

The mixture of cultivars and especially the population cultivar display a broad adaptability to various environmental conditions. One set of genotypes may successfully counteract the limiting factors of the environment (low temperatures, rust attack, etc.) in one location or one year, another set will counteract the limiting factors (high temperatures, powdery mildew attack, etc.) in another location or another year. This population buffering and genetic homeostasis will thus ensure a high yield stability of the mixture of cultivars or the population cultivar, although some reports indicate the contrary, as in the case of the barley mixture (Table 21.4). However, the yield level of these cultivar types is lower than that of the pure cultivar which explains why they have been virtually abandoned in intensive agricultural production.

With cross-pollinated plants, population cultivars as well as synthetic cultivars contain homozygous and heterozygous genotypes. Thus, their adaptability rests on the genetic homeostasis including not only heterogeneity but also heterozygosity which gives them a larger reaction range than that in population cultivars, cultivar mixtures, or multiline cultivars of self-pollinated plants.

According to the investigations of Allard and Workman (1963) on *Phaseolus lunatus*, and Griffing and Langridge (1963) on *Arabidopsis thaliana*, homozygous and heterozygous plants did not differ in their adaptability in optimum conditions. In adverse conditions, however, the heterozygotes were better performers.

It is a logical conclusion, which has already been confirmed in practice, that genetically homogeneous cultivars such as the pure line or pure cultivar should be grown in favourable or optimum conditions in order to fully express their genetic yield potential. Such conditions may be provided in relatively limited areas. In larger agro-ecological areas and less favourable conditions, priority should be given to heterogeneous and heterozygous cultivars.

The hybrid cultivar has a specific status in respect to adaptability and yield stability. The hybrid cultivar is genetically homogeneous, because all F_1 plants are uniform but heterozygous. It practically does not possess genetic variability but, unlike the pure line or the pure cultivar, it does possess full heterozygosity as a basis for full expression of the physiological homeostasis. This is why hybrid cultivars have a high genetic yield potential as well as high adaptability to and yield stability in different environmental conditions.

Double-cross maize hybrids used to be grown intensively but they have been found to have a somewhat lower yield but higher adaptability and yield stability than do the single-cross hybrids. Conversely, some investigations indicate that certain single-crosses have very good adaptability and yield stability and are quite responsive to improved environments (Eberhart and Russell, 1969; Russell, 1974). This information is certainly important for the choice of maize hybrids for certain environments.

Analysis of adaptability, stability, and productivity of the cultivar

The productivity of the cultivar expresses the yield level achieved in different environmental and agricultural conditions over a number of localities and years. Genetic yield potential of the cultivar is indicated by the maximum or record yield as already discussed in Chapter 15.

According to Finlay and Wilkinson (1963) who studied over 400 different barley cultivars in widely diverse agroecological conditions of Australia, the adaptability and stability of the cultivar can be measured by a yield regression coefficient of each cultivar in relation to the average yield of all cultivars grown in a certain location. Cultivars with the regression coefficient (b_1) around 1.0 exhibit average stability, those with the coefficient above 1.0 inferior stability, and those with the coefficient under 1.0 superior stability. However, the regression coefficient has been found insufficient to provide a real picture of yield stability and other methods are more in use.

Eberhart and Russell (1966), in addition to the mean value of yield and regression coefficient, introduced a parameter of deviation from regression (environmental index and yield) for the estimate of yield stability. They maintained that a stable cultivar should have a high average yield, the regression coefficient around 1.0, and the smallest possible deviation from regression.

These parameters may be calculated by the following equation:

$$Y_{ij} = \mu_i + \beta_i I_j + \sigma_{ij},$$

where Y_{ij} is the average yield of cultivar i in the location j, μ_i is the average yield of cultivar i in all locations, β_i is the regression coefficient showing the cultivar's reaction to different location, I_i is the environmental index obtained by subtracting the average for all cultivars in location j from the average for all cultivars and all locations, and σ_{ij} is the deviation from regression of cultivar i in location j.

Stroike and Johnson (1972) analysed the parameters of yield stability for the International Winter Wheat Performance Nursery (IWWPN) and concluded that deviation from the sum of squares gives a more realistic picture on cultivar behaviour than deviation from regression.

Genotypic stability may be estimated by other analytical procedures too. Becker (1981) states that according to "biological concept" a stable genotype should have a minimal variance under different environmental conditions, while according to "agronomic concept", a stable genotype should show minimal interactions with environments as measured by the ecovalence (Wricke, 1964). Analysing the data for different crops, Becker came to the conclusion "that the widely used method of regressing the yield of a given genotype in the various environments on the respective means of all genotypes under test may be regarded as a combination of these two concepts, for coefficients of regression are almost perfectly correlated with variances and mean squares for deviation from regression are almost perfectly correlated with ecovalence" (p. 835).

In reviewing present-day stability statistics, Lin et al. (1986) emphasize that they are related to three concepts, according to which a genotype may be considered stable (1) if its among-environment variance is small, (2) if its response to environments is parallel to the mean response of all genotypes in trial or (3) if the residual mean square from a regression model on the environmental index is small.

Which of the methods of stability statistics a scientist will use depends on whether he is more concerned with the stability over a wide range of environments or with the relative stability of a group of cultivars included in the experiment. As Becker

(1981) has shown, differences between either method of stability statistics applied are very small. Moreover, as we shall see from the analysis given below, statistical data are very helpful in making general conclusions, but statistics on the stability of yield are seldom sufficient for making specific decisions and therefore other parameters should also be taken into consideration.

International trials on rice, wheat, barley and other crops offer excellent opportunities to estimate the adaptability, productivity and yield stability of various cultivars in large and small areas and in different growing conditions. The wheat trials include over 30 countries and more than 50 locations per year. Table 21.7 shows

Table 21.7 Yield stability parameters for some wheat cultivars tested in the International Winter Wheat Performance Nursery (IWWPN), for three years in 96 locations

Cultivar	Test year	Average yield t/ha	Regression coefficient	Deviation from mean of squares
Sava	1971	4.05	1.04	107.9
(Yugoslavia)	1972	4.17	1.26	56.9
	1973	4.48	1.41	52.7
	x	4.23		
Bezostaia-1	1971	4.02	0.94	22.3
(USSR)	1972	4.33	0.90	53.6
	1973	3.85	0.88	20.4
	x	4.07		
Centurk	1971	4.01	1.20	36.2
(USA)	1972	3.92	0.91	81.7
	1973	4.07	0.97	46.8
	x	4.00		
Strampelli	1971	3.78	1.14	121.1
(Italy)	1972	3.16	0.99	179.5
	1973	3.78	1.05	95.5
	x	3.57		
Starke	1971	2.88	1.34	171.0
(Sweden)	1972	2.77	1.01	141.4
	1973	2.40	1.03	118.8
	x	2.69		

Source: Borojević, 1975

the performances of five typical cultivars out of the 30 tested. Judging on the basis of regression coefficient and deviation from mean of squares, Bezostaia-1 and Centurk displayed the highest stability, then Sava, Strampelli, and Starke (Fig. 21.1). Regarding productivity, Sava was first place, Bezostaia-1 the second, and Starke the last (Table 21.7). According to the distribution of rank in the 96 locations, Sava ranked first in 16 locations and second to fifth in 30 locations, etc., which was much better than the other cultivars. The cultivars with a high genetic yield potential performed better in favourable conditions, e.g., in locations with the average yield above 5.0 t/ha, while cultivars, with a lower regression coefficient performed better in locations with the average yield under 3.0 t/ha (Table 21.8).

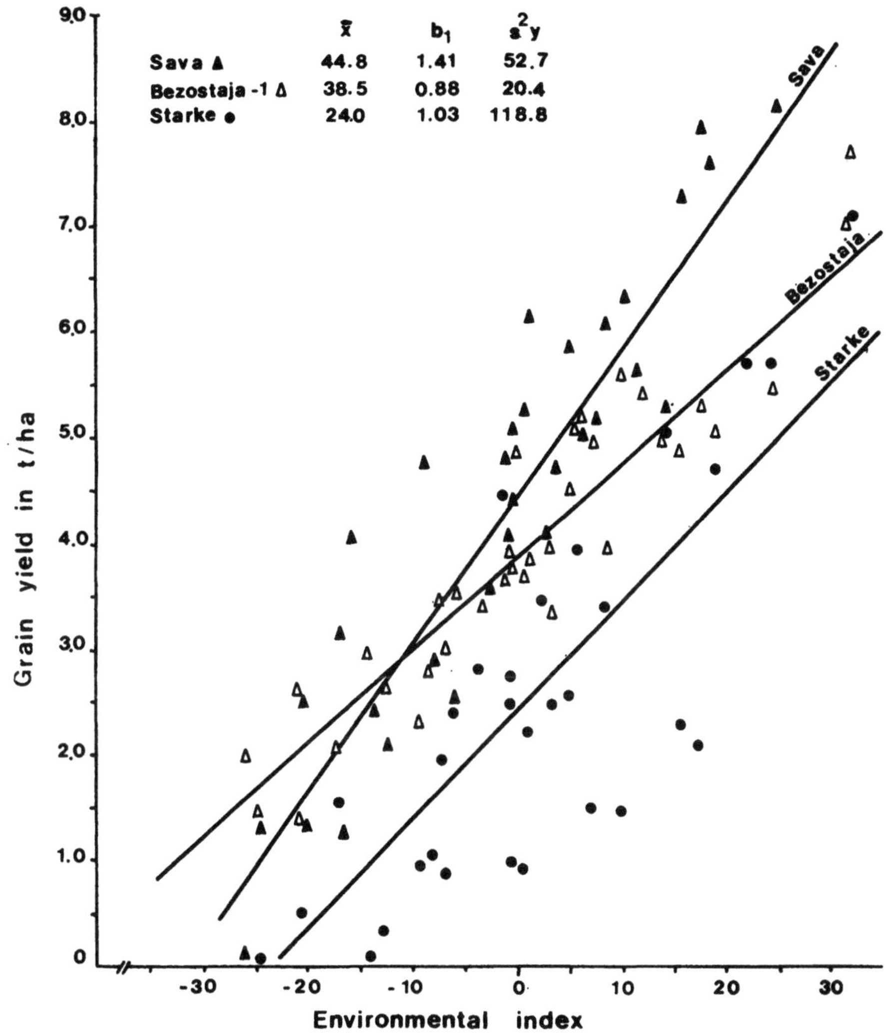

Fig. 21.1 Regression of grain yield on environmental index for three wheat cultivars grown in IWWPN in 1973 (after Borojević, 1975)

The studies of adaptability and yield stability for rice (Tanigushi and Yokayama, 1975) and wheat (Stroike and Johnson, 1972) indicated no correlation between average yield of the cultivar and regression coefficient. Matsuo (1975) tried to explain it by stating that genes controlling productivity are different from those controlling yield stability. It ensues that it is possible to produce cultivars with a high genetic yield potential and a high yield stability. This was confirmed by several new rice cultivars (Matsuo, 1975) and several high-yielding wheat cultivars which took first places in several years (Bezostaia-1, Sava, Talent, Jubiley, Partizanka, etc. (Wilhelmi et al., 1979).

Table 21.8 The rating of the yields of the wheat cultivars from Table 21.7 for the 3-year period (1971–1973), among 30 cultivars

Cultivar	Rating of yield 1	2–5	6–10	11–15	16–20	21–25	26–30
	No. of locations (96 in total)						
Sava	16	30	15	12	13	7	7
Bezostaia-1	4	21	34	18	13	4	2
Centurk	5	22	19	17	23	8	2
Strampelli	7	18	13	11	10	12	18
Starke	2	7	5	6	7	17	51
	No. of locations with yield above 5 t/ha (13)						
Sava	4	6	1	1	1		
Betostaia-1			3	4	4	2	
Centurk	1	6	1	1	3	1	
Strampelli		3	1	3	2	3	1
Starke	1		2	1	1	3	5
	No. of locations with yield under 3 t/ha (36)						
Sava	2	9	5	5	7	6	2
Bezostaia-1	2	12	10	4	5	2	1
Centurk	1	8	10	5	8	2	1
Strampelli	1	6	4	1	3	10	8
Starke	1	1	1	1	2	9	19

Source: Borojević, 1975

Most wheat cultivars tested in the international trials are pure cultivars according to their genetic composition. However, they differ in the degree of "purity": some of them are composed of a small number of different genotypes and some include several different genotypes. Hence the differential reactions in various locations, which bear effect on the regression coefficient and deviation from regression. Still, the largest differences in the cultivars' reactions are due to genetic differences in those traits which allow the cultivar to counteract the limiting factors of the environment.

The cultivars Starke and Sava have equal genetic yield potentials but Starke is superior in performance to Sava when grown in Sweden and the reverse situation occurs in Yugoslavia. Sava is too early and not sufficiently winter hardy for Swedish conditions, so that it may suffer damage from late frosts. Starke is too late for Yugoslav conditions and it suffers from forced maturation which causes large yield reductions. Considered globally, Sava demonstrated a higher adaptability than Starke in a large number of locations.

The farmer is interested in the cultivar that will bring high yield and display good yield stability over a period of several years, until it is replaced by a superior cultivar.

That may be a cultivar which has a narrow or specific adaptability, as was the case with Starke which was the leading cultivar in southern Sweden for a number of years.

Therefore, there is no need to contrast the broad or general adaptability against the narrow or specific adaptability. The point is to develop as many cultivars with either type of adaptability as possible, in order to effectively utilize different environmental and agricultural conditions and to efficiently exploit the positive cultivar/environment interaction and counteract the negative interaction.

Interaction cultivar/environment

It is evident that the adaptability and stability of the cultivar depend on its genetic composition, i.e., on the number of different genotypes that comprise it and on the homozygosity or heterozygosity of the genotypes. The adaptability and stability are even more dependent on the traits which allow the cultivar to counteract then limiting factors and use favourable environmental factors.

Taking into consideration the above as well as the fact that environmental factors are perpetually variable in time and space, there exists a continuous interaction between the cultivar and the environment. It makes it difficult for breeders to distinguish among the large numbers of new varieties that are developed each year, which are truly superior and should be recommended for commercial growing. To introduce more reliability and certainty into the process of decision making as to which varieties should be registered as new cultivars, it is necessary to establish a network of comparative yield trials. In these prospective varieties are compared against standard cultivars in different agroecological regions (in a large number of locations), for several years.

According to Mather (1949), Falconer (1964), and Mather and Jinks (1971), the general premise for such trials may be expressed by a linear equation which includes the phenotype, i.e., yield or some other quantitative trait:

$$X = \mu + g + e + ge,$$

where μ is the mean value of the population (trial), g is the value of the genotype (cultivar), e is the value of the environmental effect, and ge is the interaction between the genotype and the environment. This formula is evidently derived from the general formula $P = G + E$ (see Chapter 6).

To obtain reliable data on the share of the genotype and environmental factors in the formation of yield or other traits, the environmental variability should be further separated. This is done by inclusing in the trials replications arranged in randomized blocks, in several locations (to cover the effect of various soils, cultural practices, etc.), and over several years (to cover environmental variability in rainfall, temperature, and other factors).

Therefore, the equation for the estimation of genotypic value and the share of environmental factors in the realization of yield (or some other quantitative trait) is as follows:

$$X = \mu + \underbrace{g}_{\substack{\text{genetic}\\ \text{effect}}} + r + \underbrace{l + y + ly}_{\substack{\text{eviron-}\\ \text{mental}\\ \text{effect}}} + \underbrace{gl + gy + gly}_{\substack{\text{interaction}\\ \text{cultivar/}\\ \text{environment}}} + \underbrace{e}_{\substack{\text{random}\\ \text{effects}}},$$

where g is the effect of the genotype, r is the effect of replication, l is the effect of location, y is the effect of year, gl is the interaction genotype/location, gy is the interaction genotype/year, gly is the interaction genotype/ location/year, and e is the effect of random and unidentifiable environmental factors.

Trials which cover all these parameters may be calculated by the analysis of variance if all treatments are chosen at random. As this is not always possible in agricultural experiments, one should be satisfied with what one can get. For example, the factor of year may be taken as randomly chosen. Although locations are not selected at random but according to available experimental facilities, they may be considered as randomly chosen if there are many locations spread over a large territory Regarding the factor of genotypes, they are never chosen at random. They are regularly new cultivars that are due to be tested but since a trial includes several scores of cultivars coming from different breeding institutes, it may be taken as a random choice.

Therefore, a model of the analysis of variance for varieties to be tested in several locations over several years in order to choose those that are most suitable for commercial production is as follows in the Table 21.9.

Table 21.9 A model of the analysis of variance for genotypes tested for several years in a number of locations

Source of variation	Degrees of freedom	Mean of squares	Expected mean of squares
Replications – R	$ly(r-1)$	not relevant	
Years – Y	$y-1$		
Locations – L	$l-1$		
$Y\times L$	$(y-1)(e-1)$		
Genotypes – G	$g-1$	M_1	$\sigma_e^2 + r\sigma_{gly}^2 + rl\sigma_{gy}^2 + ry\sigma_{gl}^2 + rly\sigma_g^2$
$G\times Y$	$(g-1)(y-1)$	M_2	$\sigma_e^2 + r\sigma_{gly}^2 + rl\sigma_{gy}^2$
$G\times L$	$(g-1)(l-1)$	M_3	$\sigma_e^2 + r\sigma_{gly}^2 + ry\sigma_{gl}^2$
$G\times Y\times L$	$(g-1)(y-1)(l-1)$	M_4	$\sigma_e^2 + r\sigma_{gly}^2$
Error ($G\times R$)	$(g-1)(r-1)\quad ly$	M_5	σ_e^2

This model shows the values of the components of variance, i.e., the shares of genetic differences among varieties, interactions between varieties, years and locations, and random effects in the total variance. The values of the components of variance are calculated from the following formulae:

$$\bar{\sigma}_e^2 = M_s \qquad \bar{\sigma}_{gly}^2 = \frac{M_4 - M_s}{r} \qquad \bar{\sigma}_{gl}^2 = \frac{M_3 - M_4}{ry}$$

$$\bar{\sigma}_{gy}^2 = \frac{M_2 - M_4}{rl} \qquad \bar{\sigma}_g^2 = \frac{M_1 - M_2 - M_3 + M_4}{rly}$$

If it is found that there exists an interaction between varieties and years, it means, for example, that variety *G*-1 outyielded variety *G*-2 in one year but not in another year. If interaction is found between varieties and locations, it means, that varieties *G*-1, *G*-3, *G*-7 outyielded cultivars *G*-2 and *G*-5 in locations *L*-1 and *L*-4 but not in locations *L*-2 and *L*-5. If the analysis of variance indicates that the interaction of the second order, i.e., variety/year/location, was significant too, it means that variety *G*-1 outyielded cultivar *G*-5 in *Y*-1 (the first year) and *L*-3 but the reverse occurred in *Y*-2 and *L*-8, etc.

The interaction between genotype/environment and components of variance will be explained on the basis of data for the wheat trials of the Yugoslav Federal Commission for Variety Approval.

To be released into commercial production, a new variety must pass trials conducted for three years in several locations. Presently, wheat varieties are tested in about 15 locations distributed in all wheat-growing regions of the country. However, most experimental locations are situated in the regions with the most intensive wheat production. About 60 prospective varieties are tested against two standard cultivars (3 trials) in each location. The trials are established in a randomized block design in five replications, with the basic size of plot being 5 m^2.

In this analysis, experimental data from 1967, 1968, and 1969 will be used. The data for 7 varieties which were tested in 12 locations for three years are analysed (Table 21.10). The analysis also includes the variances resulting from the variability of the years, the locations, and their interactions. It may be seen that these variances were higher than the others but, since these variances are separated from the variances of cultivars and their interactions, it is not relevant for further considerations here.

The variance of the varieties was higher than the other variances, as well as the *F* tests, which indicates that the tested varieties differed significantly with respect to grain yield. The highest grain yield was produced by variety 6, then by varieties 4, 3, etc. (Table 21.11).

The analysis of variance showed that the interaction between the varieties and the years was significant. 1967 was the most favourable year when the highest average yield was obtained. 1968 and 1969 were considerably less favourable for wheat growing. Varieties 1, 2, 4, and 7 followed this trend, but varieties 3, 5, and 6 performed better in 1969 than in 1968 indicating the interaction between the varieties and the years (Table 21.10).

The interaction variety/location was higher than the interaction variety/year in consequence of a high variability in yield per location. The highest average yields were obtained in locations *L*-5 and *L*-11, the lowest in locations *L*-10 and *L*-7. The

Table 21.10 Analysis of variance for 7 wheat varieties tested in the trials of the Yugoslav Federal Commission for Variety Approval for three years (1967–1969) in 12 locations

Source of variation	Degrees of freedom	Mean squares	F test	Components of variance $\bar{\sigma}^2$	%
Replications – *R*	4	0.080			
Years – *Y*	2	8.872	124.711**		
Locations – *L*	11	32.683	459.404**		
$Y\times L$	22	14.218	199.855**		
Varieties – *G*	6	3.587	50.415**	0.018	13.8
$G\times S$	12	0.236	3.324**	0.000[1]	0.0
$G\times L$	66	0.290	4.078**	0.001	0.9
$G\times Y\times L$	132	0.272	3.823**	0.040	30.8
Error – *E*	1004	0.071		0.071	54.6
Total	1259			0.130	100.0

[1] The value was negative
** significant for 1%

Table 21.11 Interaction cultivar (*C*) × year (*Y*) for grain yield (t/ha) in the 1967–1969 trials in 12 locations (see Table 21.10)

Cultivar	1967	1968	1969	Average
1. Bezostaia-1 (st)	5.19	4.86	4.64	4.90
2. Libellula (st)	5.46	5.11	5.04	5.20
3. OSK-6.9/11	5.55	5.10	5.23	5.29
4. KG-1081	5.71	5.12	5.15	5.33
5. ZG-3815/65	4.90	4.24	4.69	4.61
6. NS-611	5.88	5.15	5.22	5.41
7. NS-440	5.57	5.01	4.96	5.18
Average	5.47	4.94	4.99	

LSD at 5% for cultivars: 0.110, for years: 0.072, for *C*/*Y*: 0.191
LSD at 1% for cultivars: 0.145, for years: 0.095, for *C*/*Y*: 0.251

highest average yields were produced by variety 4 in *L*-5 and variety 3 in *L*-11. Variety 6 was the best performer in the poorest location (*L*-10), variety 2 in the similarly poor location *L*-7 (Table 21.11). In general, varieties 6, 4, and 7 were the best performers in 8 out of 12 locations, but variety 6 was less successful in *L*-8 and *L*-11, variety 7 in *L*-3, *L*-4, *L*-6, and *L*-7, which cause the interaction. Varieties 5 and 1 were the poorest performers in all locations which is an indication of a lack of interaction between them and the locations (Table 21.12).

Regarding the interaction of the second order, i.e., variety/year/location, it was significant too (Table 21.13). On the average for all varieties and locations, the highest yield was obtained in 1967. However, the 1967 average yields for locations *L*-1, *L*-2, and *L*-12 were lower than the 1968 and 1969 average yield in the respective locations, although varieties 4 and 6 had higher yields in 1967 than in 1968 (Table 21.13). In location *L*-5, varieties 6 and 7 had lower yields in 1968 than in 1969, in

Table 21.12 Interaction cultivar (C) × location (L) for grain yield (t/ha) in the 1967–1969 trials (see Table 21.10)

Location	Cultivar 1	2	3	4	5	6	7	$\bar{x}$
1. Novi Sad	4.31	4.38	4.36	5.10	3.91	4.77	4.42	4.46
2. Srbobran	5.12	5.75	5.15	5.25	4.97	5.86	5.73	5.40
3. Osijek	5.20	5.85	6.05	5.58	5.16	6.34	5.39	5.65
4. Bijeljina	5.03	5.82	6.12	5.24	4.55	5.59	5.04	5.34
5. Valjevo	6.11	6.53	6.20	7.21	6.27	6.64	6.74	6.53
6. Kragujevac	5.08	5.49	5.85	5.55	4.71	5.73	5.44	5.41
7. Zaječar	3.44	3.92	3.70	3.89	3.34	3.76	3.87	3.70
8. Peć	5.92	5.22	5.70	5.70	4.91	5.22	5.81	5.50
9. Skopje	5.63	5.72	5.74	6.02	5.52	6.23	5.94	5.83
10. Titograd	2.56	2.77	2.76	2.74	2.33	3.13	2.65	2.71
11. Banja Luka	6.04	6.66	7.16	6.91	5.71	6.43	6.75	6.52
12. Starše	4.33	4.33	4.73	4.72	3.93	5.28	4.39	4.53
Average	4.90	5.20	5.29	5.33	4.61	5.41	5.18	

LSD at 5% for cultivars: 0.144, for locations: 0.110, for C/L: 0.381
LSD at 1% for cultivars: 0.990, for locations: 0.145, for C/L: 0.502

contrast to the other varieties in the trial. In location L-7, all varieties produced low yields in 1969 but variety 1 was in the first place and variety 6 among the last, although a completely reverse picture is obtained when the two varieties are considered on the average for the other locations and years (Tables 21.11, 21,12, 21.13).

If the behaviour of individual varieties is analysed for locations and years, specific reactions in some locations and years will be observed in addition to similar reactions. All this indicates that there existed a fairly high interaction among the varieties, years, and locations. This should receive more attention in future work.

When the entire territory of Yugoslavia is considered, it might be expected that the interaction between variety and location would be higher than the interaction between variety and year, especially if the fact that the trials were conducted for only three years is taken into account. It seems, however, that the annual variation was higher than the variation for the location, at least in one region.

There are numerous data in the literature indicating that interaction variety/year exceeds that of variety/location (Allard and Bradshaw, 1964). This of course depends on how large and climatically diverse is the region represented by the experimental locations. In Yugoslavia, Andjelković (1974) made an analysis of variance of wheat yields for the production regions in the period 1969–1971 and found that the interaction variety/year was higher in two regions and that the interaction variety/location was higher in the other three regions.

These results also demonstrate that, on one side, some locations grouped in one region may be more diverse environmentally than are three consecutive years taken as a whole. On the other side, one period of three consecutive years may differ considerably from another 3-year period. Furthermore, a trial conducted for three years includes genetically different varieties which results in different interaction with environmental factors.

To run a check on the above results, another analysis was made for the 1976–1978 trials which involved 12 prospective wheat varieties and 15 locations. The analysis of variance showed that the interaction variety/year was higher than that for variety/location in the entire country (Table 21.14). which is opposite to the results obtained for the period 1967–1969 (Table 21.10).

Table 21.13 Interaction cultivar/year/location (*C*/*Y*/*L*) for grain yield (t/ha) in the 1967–1969 trials (see Table 21.10)

Cul-tivar	Year	Location 1	2	3	4	5	6	7	8	9	10	11	12
	1967	3.17	3.48	5.30	6.33	7.34	5.79	3.17	6.80	5.88	6.64	6.43	3.90
1	1968	4.32	6.13	5.59	4.36	5.74	4.67	5.08	5.26	5.53	1.00	5.23	5.45
	1969	5.45	5.75	4.69	4.41	5.24	4.78	2.07	5.70	5.48	2.04	6.44	3.63
	1967	3.20	4.06	5.87	6.91	7.66	6.72	3.54	5.84	5.44	5.36	7.54	3.40
2	1968	4.14	6.60	6.35	5.05	6.28	4.78	6.22	4.39	5.08	1.04	5.90	5.45
	1969	5.80	6.57	5.32	5.49	5.64	4.96	2.01	5.42	6.64	1.88	6.53	4.13
	1967	3.40	2.49	5.75	7.42	7.44	7.80	3.84	6.18	6.40	5.36	7.65	3.40
3	1968	3.46	6.08	6.92	4.72	5.72	4.74	5.24	5.03	4.89	1.52	6.85	6.02
	1969	6.21	6.87	5.48	6.22	5.42	5.01	2.00	5.90	6.00	1.90	6.98	4.78
	1967	5.08	3.18	4.86	6.09	8.90	6.33	3.32	7.60	6.40	5.00	7.84	3.91
4	1968	4.37	6.05	6.38	4.48	7.11	5.08	6.27	4.13	4.90	1.22	5.98	5.48
	1969	5.84	6.53	5.49	5.15	5.60	5.24	2.07	5.38	6.75	2.02	6.91	4.76
	1967	3.28	2.99	5.13	5.58	7.02	5.64	3.36	6.68	5.76	4.12	5.92	3.31
5	1968	3.35	5.60	5.58	4.10	5.79	3.94	5.00	2.96	4.69	0.74	5.07	4.06
	1969	5.15	6.32	4.78	3.98	6.00	4.55	1.65	5.10	6.12	2.12	6.13	4.43
	1967	4.66	3.26	6.18	6.42	8.84	7.94	3.82	6.68	6.68	5.32	6.79	3.98
6	1968	3.77	7.02	6.90	4.83	5.40	4.58	5.76	3.30	5.20	1.84	6.29	6.85
	1969	5.86	7.30	5.96	5.52	5.67	4.67	1.70	5.68	6.80	2.22	6.22	5.02
	1967	2.92	4.00	5.82	5.76	9.04	7.26	3.66	6.65	5.92	4.48	7.70	3.60
7	1968	4.20	6.58	6.08	4.35	5.03	4.40	6.14	4.83	5.84	1.46	6.12	5.10
	1969	6.14	6.59	4.25	5.00	6.14	4.66	1.82	5.94	6.06	2.00	6.44	4.47

LSD at 5% 0.661
1% 0.869

Table 21.14 Analysis of variance for 12 wheat varieties tested in the trials of the Yugoslav Federal Commission for Variety Approval for three years (1976–1978) in 15 locations

Source of variation	Degrees of freedom	Mean squares	*F* test	Components of variance $\bar{\sigma}^2$	%
Replications – *R*	4	0.270			
Years – *Y*	2	62.318	869.788**		
Locations – *L*	14	43.887	612.542**		
$Y \times L$	28	8.258	115.260**		
Varieties – *G*	11	3.763	52.528**	0.013	8.4
$G \times Y$	22	0.671	9.368**	0.004	2.6
$G \times L$	154	0.486	6.788	0.008	5.2
$G \times Y \times L$	308	0.364	5.082	0.059	38.1
Error – *E*	2166	0.071		0.071	45.8
Total	2699			0.155	100.0

Analysis of values of components of variance

To further complete a picture of the contribution of genotypic and environmental factors to the realization of the phenotype, values of the components of variance calculated by the formulae given in Table 21.9 will be reviewed.

For the entire 1967–1969 trial with 7 varieties and 12 locations, 13.8% of the total variability for grain yield is attributed to the pure genetic variability, i.e., genetic differences among the varieties, 31.7% to the interaction between the varieties and the environment, and 54.6% to the random environmental factors and experimental error (Table 21.10).

Although the F test indicated that the interactions of the first order, $G \times Y$ and $G \times L$, were about the same as the interactions of the second order, $G \times Y \times L$, the values of the variances $G \times Y$ and $G \times L$ were insignificant when compared with the value of the variance $G \times Y \times L$ (Table 21.10). The variability among the years and among the locations was high on account of the environmental diversity of different regions which brought about a high interaction between the years and the locations ($Y \times L$) which in its turn affected the interaction $G \times Y \times L$ (Tables 21.10 and 21.14).

It should also be mentioned that the value of the components of variance, due to genetic differences among the varieties ($o^2{}_S$), was higher in the 1967–1969 trials than in the 1976–1978 trials (Tables 21.10 and 21.14). This is in good agreement with all other data because the former trials included high-yielding varieties which were registered as cultivars (NS-611, OSK-6.9/11, KG-1081) compared against the standard cultivars Libellula and Bezostaia-1 (Table 21.10), while the latter trials included new high-yielding varieties as well as new standards, Zlatna dolina and Partizanka, which were higher yielding that the previous standards (Table 21.15).

Table 21.15 Interaction cultivar (C) × year (Y) for grain yield (t/ha) in the 1976–1978 trials in 15 locations (see Table 21.14)

Cultivar	1976	1977	1978	Average
1. NS-18-30	6.76	6.94	5.80	6.50
2. NS-12-01	6.80	6.72	5.74	6.42
3. ZG-4364/73	6.38	6.76	5.88	6.34
4. KG-3/270-6-63	6.12	6.66	5.98	6.24
5. L-408/73	6.84	6.68	5.94	6.48
6. NS-2630/1	7.12	7.18	6.28	6.86
7. Ze-1/10	6.14	6.70	5.46	6.10
8. ZG-4387/73	6.82	7.40	6.12	6.78
9. B-336	6.80	6.60	5.48	6.30
10. NS-2568	7.20	7.20	6.30	6.90
11. "Partizanka" (st.)	6.68	6.54	5.84	6.36
12. "Zlatna dolina" (st.)	6.96	7.20	6.00	6.72
Average	6.72	6.88	5.90	

LSD at 5% for cultivars: 0.111, for years: 0.049, for C/Y: 0.171
1% for cultivars: 0.145, for years: 0.065, for C/Y: 0.225

Conclusions on interaction cultivar-environment

The analysis of variance of grain yield of wheat showed that the yield variation was predominantly governed by the interaction between the varieties and the environmental factors (location, year), then by the random and undefined environmental factors (objective and subjective errors in cultural practices and experimental procedure) and finally by differences in the genetic yield potentials of the tested varieties (13.8% in Table 21.10 and 8.4% in Table 21.14). This order may seem strange but one should not forget that wheat varieties, and for that matter varieties of any crop tested in comparative trials for registration, have similar genetic yield potentials, differing genetically by not more than 5 to 20% if the standard cultivars are also at a high level (and only such standard cultivars should be used). When breeding reaches a high level, differences in genetic yield potential of 1% are very important and they may bring large economic gains.

Because of relatively small differences among varieties and the high effect of the environment on them it becomes difficult to choose truly superior yielding varieties in tests, which in most breeding institutions last 2 or 3 years. The testing of new varieties in a large number of locations for 3 years, as practiced in many countries, gives good information on the behaviour of varieties in an array of environmental and acricultural conditions.

No matter how necessary these trials are, it is good to remember the words of Simmonds (1981) who said that genotype – environment interaction trials have little impact on plant breeding because experimental data "become available only when a variety is nearly finished or even already being marketed" (p. 196). The testing for yield performance, however, has a great importance for variety registration and subsequent production.

Two varieties may have equal genetic yield potentials but if they differ in winter-hardiness, disease resistance, vegetational period, etc., their interactions with locations and years will diverge and this is the source of a high portion of variance due to interaction variety/year/location (Tables 21.10 and 21.14).

Since one three-year period may be different from another, while yield performance trials longer than three years are inapplicable because they would hamper the introduction of superior varieties into commercial production, trials in a large number of locations (if a vast territory is considered) may in a way substitute for the year. This substitution does not mean that performance trials should last one or two years instead of three but that varieties whose average yields for all locations are significantly lower than that of the standard variety may be cancelled already in the first year without risk of making a mistake. This stands for varieties 5 and 7 in the 1967–1969 trials (Table 21.10) and varieties 4 and 7 in the 1976–1978 trials (Table 21.15), which was confirmed by the tests in the subsequent two years. Entries which are evidently inferior should be replaced by new entries in order to gain time in finding the best genotypes. Schutz and Bernard (1976) arrived at the same conclusion while testing soyabean varieties in different parts of the U.S.A. Similar conclusions have been drawn by many investigators.

Besides an early elimination of inferior varieties, another possibility of time saving is to release into commercial production any variety found to have significantly higher yields than the standard in all experimental locations in the first two years of trials. This was the case with varieties 6 and 4 in the 1967–1969 trials (Table 21.10) and varieties 10 and 6 in the 1976–1978 trials (Table 21.15).

As much as the yield performance trials of the Commission secure the choice of the best varieties, the real test of their values takes place when varieties come to the market. For example, while the variety Sava (NS-611) expanded rapidly to become the leading variety in the period 1970–1980, the variety Slavonka (OSK-6.9/11), although equal in yield to Sava, did not achieve prominence, primarily on account of poor winter-hardiness and grain quality. Some other varieties released at the same time have practically remained out of production, because their susceptibility to lodging or diseases made them unsuitable for intensive production.

As it is impossible to develop a cultivar that could counteract all limiting factors of the environment, a negative interaction cultivar/year or cultivar/location may be overcome by growing several varieties in an area or a region. Also, a positive interaction cultivar/year and cultivar/location should be turned to advantage, i.e., cultivars which respond better to high humidity or drought, superior in nitrogen uptake or tolerance to parasites should insistently be grown.

It is clear that the growing of several cultivars in a region allows for a better use of the diversities in the climate, soil and cultural practices; a more advantageous response to limiting factors of the environment, and, ultimately, to higher and more stable yields.

References

Allan, R. E., 1988. Performance of multiline versus pureline wheats in northwestern USA. Proc. 7th Intern. Wheat Genet. Sym., 13–19th July 1988, Cambridge (UK) p. 799–803.

Andjelković, M., 1964. Variranje prinosa pšenice u zavisnosti od sorte mesta i godine. Savremena poljoprivreda (Novi Sad), XVIII, (9): 693–706.

Allard, R. W., and Workman, P. L., 1963. Population studies in predominantly self-pollinated species. IV. Seasonal fluctuations in estimated values of genetic parameters in lima bean populations. Evolution 17: 470–480.

Allard, R. W., and Bradshaw, A. D., 1964. Implications of genotype-environmental interactions in applied plant breeding. Crop Sci. 4: 503–508.

Becker, H. C., 1981. Correlations among some statistical measures of phenotypic stability. Euphytica, 30: 835–840.

Borlaug, E. N., 1959. The use of multiline or composite varieties to control airborne epidemic diseases of self-pollinated crop plants. Proc. 1st. Inter. Wheat Genet. Sym., Winnipeg, Canada, p. 12–27.

Borojević, S., and Mišić, T., 1962. (On the value of growing wheat varieties in mixtures). Savremena Poljoprivreda, X, 1: 3–15.

Borojević, S., 1975. Ideotypes for high productivity, performance stability and adaptation. Proc. 2nd Inter. Winter Wheat Conf., Zagreb, p. 46–59.

Bradshaw, A. D., 1965. Evolutionary significance of phenotypic plasticity in plants. Adv. in Genetics 13: 115–155.

Browning, J. A., and Frey, K., 1969. Multiline cultivars as a means of disease control. Annu. Rev. Phytopathology, 7: 355–382.

Browning, J. A., Frey, K. J., McDaniel, M. E., Simons, M. D., and Wahl, I., 1979. The bio-logic

of using multilines to buffer pathogen populations and prevent loss. Indian J. Genet. & Pl. Breeding, 39: 3–10.
Duvick, D. N., and Brown, N. L., 1979. Current Breeding Methods: Maize. In: J. Sneep and A. J. T. Hendriksen (Editors), Plant Breeding Perspectives. Pudoc, Wageningen, p. 190–203.
Eberhart, S. A., and Russel, W. A., 1966. Stability parameters for comparing varieties. Crop Sci. 6: 36–40.
Eberhart, S. A., and Russel, W. A., 1969. Yield and stability of a 10-lines diallel of single-cross and double-cross maize hybrids. Crop Sci. 9: 357–361.
Eberhart, S. A., Debea, S., and Hallauer, P. B., 1973. Reciprocal selection in the BSSS and BSCBI maize populations and half-sib selection in BSSS. Crop Sci. 13: 451–456.
Falconer, D. S., 1964. Introduction to Quantitative Genetics. Ronald Press, New York, 365 pp.
Finlay, K. W., and Wilkinson, G. N., 1963. The analysis of adaptation in a plant breeding programme. Australian J. Agric. Res. 14: 743–754.
Frey, K. J., and Browning, J. A., 1976. Registration of Multiline E72, Multiline E73 and Multiline E74 oat cultivars. Crop Sci. 16: 311–312.
Frey, K. J., Browning, J. A., and Simons, M. D., 1979. Management systems for host genes to control disease losses. Ind. J. Genet. & Plant Breeding, 39: 10–21.
Frey, K. J., 1983. Plant population management and breeding. In: D. R. Wood (Editor), Crop Breeding, American Society of Agronomy, Madison, USA p. 55–88.
Gill, K. S., Nanda, G. S., and Singh, G., 1975. Studies on multilines in wheat. Stability of multilines of Kalyansona and PV18 cultivars. Crop Improv. 2: 1–16.
Gill, K. S., Nanda, G. S., Gurder, J., Singh, G., and Ajula, S. S., 1979. Multilines in wheat – a review. Ind. J. Genet. & Plant Breeding 39: 30–37.
Griffing, B., and Langridge, J., 1963. Phenotype stability of growth in the self-fertilized species. *Arabidopsis thaliana*. In: Statistical Genetics in Plant Breeding, National Academy of Sciences – Nat. Res. Council Publ. 982: 386–394.
Hallauer, A. R., 1972. Third phase in the yield evaluation of synthetic varieties of maize. Crop Sci. 12: 16–18.
International Code of Nomenclature for Cultivated Plants 1980. C. D. Brickell, E. G. Voss, A. F. Kelly, F. Schneider and R. H. Richens (Editors), The Intern. Bureau for Plant Taxonomy and Nomenclature, Utrecht, Netherlands. p. 12–13.
Jensen, N. F., 1952. Inter. varietal diversification in oat breeding. Agronomy J. 44: 30–34.
Kovács, G., and Abranyi, A., 1985. Study of the yielding ability of variety mixtures of winter wheat. Novénytermelés 34: 81–86.
Lin, C. S., Binns, M. R., and Lefkovitch, L. P., 1986. Stability analysis: Where do we stand? Crop Sci. 26: 894–899.
Lonnquist, J. H., 1949. The development and performance of synthetic varieties of corn. Agronomy Journal 41: 153–156.
Lonnquist, J. H., and Williams, N. W., 1967. Development of maize hybrids through selection among full-sib families. Crop Sci. 7: 369–370.
Mather, K., 1949. Biometrical Genetics. Methuen, London, 162 pp.
Mather, K., and Jinks, J. L., 1971. Biometrical Genetics. Second Edition. Chapman and Hall, London.
Matsuo, T. (Editor), 1975. Adaptability in plants. With special reference to crop yield. Japan. Comm. Inter. Biol. Program (Tokyo), Vol. 6, 217 pp.
Nasypayko, V. M., 1959. Sortosmesi ozimoi pšenici. Agrobiologia (Moskva), 3.
Oka, J. I., 1967. Adaptability for seasons and locations and yield stability in crop varieties; its mechanism and selection. Recent Adv. in Breeding (Tokyo), 8: 42–47.
Orlanskaya, P. P., i Polyakov, P. I., 1950. Vlianie smesi sortov na urožainost. Selekcia i semenovodstvo (Moskva), No. 1.
Russel, W. A., 1974. Comparative performance for maize hybrids representing different eras of maize breeding. Proc. 29th Ann. Corn Sorghum Res. Cont. Am. Seed Trade Assoc., Washington, D.C., p. 81–101.
Russell, W. A., Eberhart, S. A., and Vega, O., 1973. Recurrent selection for specific combining ability for yield in two maize populations. Crop Sci. 13: 257–261.

Simmonds, N. W., 1981. Principles of Crop Improvement. Longman, London, 408 pp.
Stroike, J. E., and Johnson, V. A., 1972. Winter wheat cultivar performance in an international array of environments. Univ. Nebraska Res. Bull. (Lincoln), No. 251.
Suneson, C. A., 1949. Survival of four barley varieties in a mixture. Agron. J. 41: 459–461.
Taniguchi, S., and Yokoyama, T., 1975. International rice adaptation experiment (IAEA). In: T. Matsuo (Editor), Adaptability in Plants. Jap. Comm. Inter. Biol. Program (Tokyo), Vol. 6: 7–17.
Van der Plank, J. E., 1963. Plant Diseases: Epidemics and Control. Academic Press, New York, 349 pp.
Wilhelmi, K. D., Kuhr, S. L., Johnson, V. A., Mattern, P. J., and Schmidt, J. W., 1979. Results of the International Winter Wheat Performance Nursery in 1974, 1976 and 1977. Research Bull. 279 (1976), 285 (1978), 288 (1979). Agr. Exp. Sta. Univ. Nebraska, Lincoln, USA.
Wricke, G., 1964. Zur Berechnung der Ökovalenz bei Sommerweizen und Hafer. Z. Pflanzenzüchtung 52: 127–138.

22. GENETIC BASIS OF SEED PRODUCTION AND INTRODUCTION OF CULTIVARS INTO PRODUCTION

To be able to rapidly introduce into commercial production a new cultivar that has just passed comparative trials against standard cultivars and has been registered, it is necessary to efficiently organize the multiplication of commercial seed. The purpose of seed production is as follows:

a) the introduction of the cultivar into commercial production,
b) the maintenance of the genetic identity of the cultivar, and
c) the maintenance of the productivity of the cultivar.

Introduction of the cultivar into commercial production

At the moment of registration of a new cultivar, the institute at which it was developed has on stock a limited amount of breeder's or elite seed of the cultivar. The seed must be multiplied for several years or generations in order to provide sufficient amounts for the beginning of commercial production. The number of reproductions or seed categories depends on the species, the genetic composition of the cultivar and market demand.

Categories of seed reproduction

In Sweden, which has a long tradition of producing commercial seed of field crops, there are the following seed categories: (1) elite seed – the basic stock produced by the breeder, (2) original seed – obtained by reproducing the elite seed, (3) controlled seed, i.e., seed resulting from the supervised reproduction of the original seed. For the last category, seed plots and seed lots are controlled because they must conform to certain standards. The controlled seed or the original seed are used to produce (4) quality seed and (5) commercial seed which need not be accompanied by a certificate of varietal purity but only a document on a laboratory analysis.

There are four seed categories in France: (1) the original selection with the requisite varietal purity of 999 per 1,000, (2) a selection with the purity of 998 per 1,000, (3) a reproduction with the purity of 990 per 1,000, and (4) commercial seed with the purity of 960 true plants per 1,000 tested plants.

In the Soviet Union, the source material is called elite seed and its multiplications are called reproductions. The last seed reproduction is the 2nd or the 3rd reproduction which is sold to seed processors for the production of commercial seed. The seed production is based on the controlled seed production at four-year intervals in the case of the cereals and other self-pollinated crops grown over large areas.

Four seed categories are recognized in the U.S.A. and Canada: (1) breeder seed produced and kept by the breeder or breeding company, (2) foundation seed obtained by multiplying the breeder seed, (3) registered seed multiplied from the foundation seed, and (4) certified seed which serves for the production of commercial seed.

In Yugoslavia, annual self-pollinated crops have the following seed categories:

(1) Elite seed – produced by the breeding institute or another institute authorized by the breeder. This is the source or basic material of each cultivar.
(2) Original seed obtained by reproducing the processed elite seed. Agricultural estates licenced for seed production produce the original seed under contract signed with the breeding institute.
(3) First (I) seed reproduction obtained by multiplying the processed original seed, produced by agricultural estates.
(4) Second (II) seed reproduction obtained by multiplying the processed seed of the first reproduction. This category is also produced by agricultural estates and it serves for the production of commercial seed.

In some cases, when a high level of organization of seed production has been reached, as in wheat and barley, the seed of the first reproduction is used for the production of commercial seed.

Each seed category must meet certain requirements with respect to varietal purity, presence of weed seeds, germination ability, health condition, etc. These requirements are stipulated in special laws and regulations on seed production and marketing.

Fig. 22.1 Breeder seed production of a cultivar

If seed production is organized well and the necessary amounts of seed are carefully calculated, a new cultivar may rapidly spread to a large acreage. Furthermore, farmers will be provided with a quality seed of the new cultivar. If the cultivar succeeds in achieving great initial success, its further seed production may be so organized that the cultivar becomes the production leader in 4 years after the release. If seed production is not adequately organized, there are risks of mixing the cultivar with other cultivars during harvest, transportation and storage, causing the loss of genetic identity and reductions in productivity of the new cultivar.

Maintenance of genetic identity of the cultivar

The purpose of seed production is not only to facilitate the promotion of new cultivars but also to ensure the maintenance of their genetic identity year after year and generation after generation. Farmers are looking for a certain cultivar not just any cultivar. Thus, it is necessary to maintain the genetic identity as it was when the cultivar was developed and registered. The maintenance of varietal identity depends on:

1) the genetic composition of the cultivar,
2) the system of pollination and reproduction of the cultivar, and
3) the methodology of seed production.

According to its genetic composition, the cultivar may vary from a pure line to a population or hybrid cultivar, as explained in Chapter 21.

Seed production invariably implies the selection of plants or seeds. The effect of the selection varies according to the genetic composition of the cultivar and the selection method.

In self-pollinated plants such as wheat and soyabean, the basic or elite seed of a cultivar is obtained by selecting several hundreds or thousands of plants each year. Each plant is threshed separately and all seeds are inspected. Seeds from the best plants are sown in separate rows. The plants are checked in the course of the growing season and rows which are atypical in respect to the basic type of the cultivar are removed before flowering. Deviant rows may occur on account of mechanical admixtures, natural hybridization with other cultivars, or mutation. The remaining rows are harvested and the seeds are bulked to make the elite seed. The elite crop is checked in the field for varietal purity and the seed produced is processed ready for sowing to produce Original seed. The Original seed undergoes the same procedure to produce the First and Second reproductions.

If the cultivar is a pure line, the selection of the best plants for the production of elite seed is the most effective means of maintaining the genetic identity and varietal purity of the cultivar.

When the cultivar is a pure cultivar and especially a population of a self-pollinated plant, composed of different genotypes (which may be fairly uniform morphologically), the selection of the best plants to obtain the elite seed may have various effects. For example, the selecting of early and short plants may cause a shift from the basic type of the cultivar recognized at registration. Changes in the genetic composition of the cultivar will be larger if the number of selected plants is smaller than if more plants are selected. A small number of plants represents only a portion of the genotypes present in the cultivar and a change in the frequency of the genotypes will occur the next year. This may introduce a considerable change in the genetic identity of the cultivar, especially if the cultivar is a population. The consequences may be positive, if more productive genotypes have been selected, or negative, if inferior genotypes have been selected. However, the cultivar as a whole retains some distinguishing heritable character.

As the purpose of seed production is not to change the cultivar but to maintain it and promote its expansion in commercial production, the selection of plants practiced in the production of elite seed should include genotypes which predominate in the cultivar. Some largely atypical plants may result from a natural hybridization or mutation and may be also selected but this subject is outside the scope of seed production.

Regarding the seed production of cross-pollinated plants, it is in fact a continual struggle to maintain the cultivar's identity and various interventions are required.

If the cultivar is a population comprised of different genotypes, as is the case in lucerne, red clover, rye, different grasses, etc., the identity of the cultivar will be successfully maintained only if a sufficient number of plants is selected for the production of elite seed to include the majority of the constituting genotypes of the cultivar. If a small number of plants is selected, a shift will result, its direction depending on whether early or late, tall or short genotypes have been selected. Inbreeding depression may also occur.

Therefore, seed production of cross-pollinated plants which are grown in conditions of random mating is a difficult business. Natural selection itself leads to a disequilibrium on account of the unequal fertility and vitality of the genotypes. The breeder causes changes in the process of reproduction by introducing an unconscious bias in the selection of genotypes. If this does not receive attention, the source cultivar may undergo considerable changes. The changes may lead in the direction of an increased adaptability to variable environmental conditions but also to a reduction in genetic yield potential.

The seed production of cross-pollinated plants should thus involve as few reproduction from the basic seed as possible because each subsequent generation may differ from the preceeding one. Wherever possible, the elite seed should be produced from selected families, mother clones, progenies, or lines.

If the cultivar is an F_1 hybrid, the seed production is fully controlled because inbred lines, which must be pure to ensure nearly 100% hybrid seeds, are produced separately. Hybrid seed for commercial production is produced each year.

Maintenance of the productivity of the cultivar

In the course of a large number of reproductions, under the effect of environmental factors, natural selection causes changes in the frequencies of genes and genotypes within a cultivar. More adaptable genotypes, which may not necessarily be the most productive ones, become prevalent after some time and the productivity of the cultivar decreases. Admixtures of less productive cultivars which are difficult to identify morphologically as well as negative mutations may also bring reductions in productivity. Nevertheless, the most frequent cause of reduced productivity is a change in the cultivar's reaction to pathogenic organisms. The genotypes resistant to the prevalent races of a pathogen increase the selection pressure against these races whose frequency drops considerably. New and virulent races take over and become prevalent. Consequently, the cultivar loses its resistance, especially if it is race-specific resistance, and reductions in the productivity of the cultivar are inevitable.

As there is an array of factors and forces exercising a negative effect on the productivity of the cultivar, it is an objective of seed production to maintain the genetic yield potential of the cultivar at the level it had at the time of its registration and introduction into commercial production.

Table 22.1 Grain yield in t/ha of different seed reproductions of wheat

Cultivar	Seed reproduction							Average	LSD	
	Year	*E**	*O*	I	II	III	*N*	$\bar{X}$	5%	1%
San Pastore	1961/62	*5.55*	5.12	5.63	5.41	–**	5.40	5.42	0.60	0.82
	1962/63	3.10	*3.06*	3.18	3.18	2.96	2.97	3.08	0.29	0.38
	1963/64	3.60	3.59	*3.60*	3.44	3.67	3.46	3.56	0.39	0.51
	Average	4.08	3.92	4.14	4.01	4.02	3.94	4.02		
				4.07	3.91	4.16				
Leonardo	1962/63	*3.96*	3.90	3.63	3.68	3.66	3.69	3.75	0.29	0.38
	1963/64	3.23	*3.25*	3.21	3.19	3.20	3.33	3.23	0.39	0.51
	Average	3.59	3.58	3.42	3.44	3.43	3.51	3.49		
			3.60	3.56	3.41	3.44				
Etoile de Choisy	1962/63	*3.64*	3.81	3.65	3.61	3.80	3.69	3.70	0.29	0.38
	1963/64	3.37	*3.29*	3.24	3.44	3.22	3.38	3.32	0.39	0.51
	Average	3.51	3.55	3.45	3.53	3.51	3.53	3.51		
			3.47	3.53	3.55	3.42				
Bezostaia-1	1963/64	3.68	3.96	3.70	3.60	3.74	3.69	3.73	0.39	0.51
Average for reproduction	1961/62	5.55	5.12	5.63	5.41	5.42	5.40	5.42	0.60	0.82
	1962/63	3.57	3.59	3.49	3.49	3.47	3.44	3.51	0.14	0.19
	1963/64	3.47	3.52	3.44	3.42	3.47	3.46	3.46	0.19	0.26
	Average	4.19	4.07	4.19	4.11	4.12	4.10	4.13		

* *E* – elite seed, *O* – original seed, I – first reproduction, II – second reproduction, III – third reproduction, *N* – unknown reproduction

** In 1961/62, the third reproduction was not grown. The average yield of all categories (5.42) was thus included in the calculation of the 3-year average.

Source: Borojević, 1965

Different views have been expressed with respect to the productivity of the various seed categories or reproductions. Yuriev (1950), Miladinović (1964) and others take the standpoint that the yield level decreases successively in further reproduction after the elite seed. Lukianenko (1948), Borojević (1965) and others found no differences in productivity amongst the close reproductions. If differences do occur, they should not be attributed to an alleged superiority of the elite seed to the second or the fifth reproduction but to other factors.

Our studies have shown that the elite seed is not superior in grain yield to the other reproductions of the same cultivar (Table 22.2).

In the case of the cultivar which is a pure line or which has a low genetic variability (as do the pure cultivars in our case), the selection of the best plants for the production of elite seed is in fact a mere selection of positive modifications. It cannot improve the genetic yield potential of the elite seed from one year in respect to the elite seed from the preceeding year or the other reproductions of the same cultivar (Table 22.1). The analysis of variance for the 3-year average yield of the variety San Pastore showed that the highest variability was due due to the year and the lowest due to the seed reproduction. As expected, there were no interactions between the seed reproduction and the year (Table 22.2).

Table 22.2 The analysis of variance for the yield over 3 years of different seed reproductions of the cultivar San Pastore from Table 22.1

Source of variation	Degrees of freedom	Mean squares	*F* test
Replications	4	0.0973	2.53
Years	2	11.1200	289.58**
Seed reproductions	4	0.0300	0.78
Seed reproductions × years	8	0.0168	0.43
Error	58	0.0384	

** Significant for 1%

In the case of cultivars with larger genetic variability (old cultivars, populations, etc.), the selection of the best plants may lead to the separation of higher as well as lower yielding genotypes from the most frequent genotypes of the cultivar.

The very existence of different genotypes in a cultivar of self-pollinated plants, resulting from its genetic composition, allows yield-superior elite seed to be obtained by means of individual selection and the establishment of inbred lines. But this elite seed will not be superior in yield to its first or third reproduction.

The possibility of selecting higher yielding genotypes depends also on the intensity of selection and the methodology of seed production.

If several thousand plants are selected from a crop with the purpose of obtaining elite seed, almost all genotypes constituting the cultivar may be represented in the sample. In such cases, it is not realistic to expect that the elite seed will be superior in yield to its reproductions or to the elite seed in the next year.

If the selection includes only several hundred plants, chances exists that not all genotypes will be represented in the sample. Consequently, certain lines may differ in productivity. The lines producing significantly higher yields than the elite, obtained by mass selection, should be put aside and they may be considered to be new varieties. However, this exceeds the framework of a regular seed production.

Production capacity of the seed

One of the main reasons for yield reductions occurring in a cultivar in the course of its commercial life is deterioration in seed quality. Poor seed-bed preparation, late sowing, insufficient nutrition, late harvest, or parasital attack make the plants poorly developed, with small seeds and a low yield. The consequences become evident in the next generation.

Table 22.3 Effect of different seed fractions (3-year average) on the quantitative traits of the cultivar San Pastore

Seed fractions	No. of plants per 1 m²		No. of spikes per 1 m²		No. of grains per spike		Mass of 1000 grains in grammes		Grain yield t/ha	
	$\bar{x}$	%	$\bar{X}$	%	$\bar{x}$	%	$\bar{x}$	%	$\bar{x}$	%
>2.8 mm	441	109	561*	110	32.1	101	41.5	103	6.28**	118
2.8–2.5 mm	441	109	567*	111	32.7	103	40.3	100	6.18**	116
2.5–2.2 mm	403	100	508	100	31.7	100	40.3	100	5.32	100
LSD for 5%	44		47		2.7		1.3		0.52	
1%	59		68		3.6		1.8		0.69	

* significantly higher for 5% and ** for 1% compared with small seeds (2.5–2.2mm)

Source: Borojević, 1964

To maintain the productivity of a cultivar, positive modifications of the seed should be exploited during seed production. Numerous investigations (Taylor, 1928; Sarić, 1960; Kaufmann and McFadden, 1963; Borojević, 1964; etc.) have indicated that large seeds are regularly superior (in the percent of germination, germination energy and the coefficient of tillering in cereals), when compared with small seeds of the same cultivar. These advantages bring about a larger number of plants or spikes per unit area and a higher yield (Table 22.3).

Differences in productivity exhibited by the various seed fractions of the same cultivar indicate that the seeds differ in their production capacity. This feature should be distinguished from the genetic yield potential which the cultivar possesses irrespectively of seed size.

The fact that seeds of various sizes, of the same cultivar and even of a single plant, have different production capacities is of great practical importance. The sowing of large seeds ensures the development of vigorous plants which efficiently utilize the available environmental conditions and respond positively to intensive cultural practices. Furthermore, large seeds of the cultivar ensure the development of uniform plants which is one of the main factors in reducing crop variability.

Therefore, special attention should be paid to seed quality during the process of seed production because well-developed and healthy seeds are the major prerequisite for the maintenance of the productivity of the cultivar.

References

Borojević, S., 1964. (The productive capacity of wheat seed and spikes of different sizes). Savremena poljoprivreda, (Novi Sad), XII: 331–350.

Borojević, S., 1965. (The productivity of various categories of seeds of wheat varieties). Savremena poljoprivreda, (Novi Sad), XIII: 107–120.

Borojević, S., 1969. (The genetics of seed production). Savremena poljoprivreda, (Novi Sad), XVII: 253–266.

Kaufman, M. L., and McFadden, A. D., 1963. The influence of seed size on results of barley yield trials. Canadian J. Pl. Sci. 43: 51–58.

Lukyanenko, P. P., 1948. Ob urožainosti semian elitoi ozimoi pšenici i metodah ego polučenia. Selekcia i semenovodstvo (Moskva), 5.

Miladinović, N., 1964. Produktivne i biološke osobine različitih semenskih reprodukcija. Savremena poljoprivreda, 3.

Sarić, M., 1960. Apsolutna težina zrna pojedinih klasaka u klasu pšenice kao pokazatelj njihove različite fiziološke vrednosti. Savremena poljoprivreda, 11: 873–882.

Taylor, J. W., 1928. Effect of continuous selection of small and large wheat seeds on yield, bushel weight, varietal purity and loose smut infection. J. American Soc. Agron. 20: 856–867.

Yuriev, V. A. i sotrud., 1950. Opščaia selekcia i semenovodstvo polevih kultur. Moskva.

AUTHOR INDEX

Page numbers printed in bold face refer to pages where references are listed

SUBJECT INDEX

R

S

T

V

Bei Fragen zur Produktsicherheit wenden Sie sich bitte an:
If you have any questions regarding product safety,
please contact:

Walter de Gruyter GmbH
Genthiner Straße 13
10785 Berlin
productsafety@degruyterbrill.com